Biocontrol of Oilseed Rape Pests

Biocontrol of Oilseed Rape Pests

Edited by

David V. Alford
BSc, PhD

Blackwell
Science

Editorial Offices:
9600 Garsington Road, Oxford OX4 2DQ, UK
Tel: +44(0)1865 776868
Blackwell Science, Inc., 350 Main Street,
Malden, MA 02148-5018, USA
Tel: +1 781 388 8250
Iowa State Press, a Blackwell Publishing
Company, 2121 State Avenue, Ames, Iowa
50014-8300, USA
Tel: +1 515 292 0140
Blackwell Publishing Asia Pty, 550 Swanston
Street, Carlton South, Victoria 3053, Australia
Tel: +61 (0)3 9347 0300
Blackwell Wissenschafts Verlag,
Kurfürstendamm 57, 10707 Berlin, Germany
Tel: +49 (0)30 32 79 060

First published 2003 by Blackwell Science Ltd

Library of Congress
Cataloging-in-Publication Data
Biocontrol of oilseed rape pests / edited by
David V Alford.
p. cm.
Includes bibliographical references (p.).
ISBN 0-632-05427-1 (hardback)
1. Rape (Plant)—Diseases and pests—
Biological control. 2. Insect pests—
Biological control. I. Alford, D. V.

SB608.R2 B56 2003
633.8′5396–dc21

2002013550

ISBN 0-632-05427-1

A catalogue record for this title is available from the British Library

Set in 10 on 12 pt Aldine 401 BT
by SNP Best-set Typesetter Ltd., Hong Kong
Printed and bound in Great Britain by
MPG Books Ltd, Bodmin, Cornwall

For further information on
Blackwell Publishing, visit our website:
www.blackwellpublishing.com

Contents

Preface

Oilseed rape is an important arable crop in many parts of the world, including much of central and northern Europe. The crop is attacked by a wide range of insect pests, many of which are of considerable economic importance. Traditionally, such pests have been targets for the application of insecticides. However, nowadays, there is an increasing demand to reduce chemical inputs on arable crops, and an increasing awareness of the potential benefits to be gained from the adoption of non-chemical methods for pest control within sustainable crop management strategies. Although thresholds for chemical control treatments are available for certain oilseed rape pests, these rarely if ever take account of the presence of naturally occurring biocontrol agents (BCAs), and oilseed rape crops are often sprayed unnecessarily. This is wasteful and can have unwarranted detrimental effects on parasitoids and other natural enemies; it can also be harmful to non-target organisms, including pollinating insects such as bees. Although knowledge and expertise of natural enemies of oilseed rape pests exist within Europe, information is fragmented, incomplete and largely parochial. It is also dispersed and not readily available to practitioners and farmers. This is a major barrier to the full exploitation of research on the natural enemies of oilseed rape pests that, otherwise, could enable a better-targeted approach to integrated crop management (ICM) on oilseed rape to be advocated and implemented.

In 1997, a three-year programme of work began in which participants from various European countries reviewed the natural enemies of oilseed rape insect pests. The work was done mainly with financial support from the Commission of the European Communities, Agriculture and Fisheries (FAIR) specific RTD programme, as Project CT96-1314 – *Minimizing pesticide use and environmental impact by the development and promotion of bio-control strategies for oilseed rape pests.* The various outputs from this project form the basis for the current monograph, but this publication does not necessarily reflect the Commission's views and in no way anticipates the Commission's future policy in this area.

The EC-funded project, under the acronym BORIS, reviewed the whole range of antagonists associated with insect pests on oilseed rape crops in western Europe, with emphasis being placed in three specific areas: parasitoids, predators and pathogens. The present monograph relies heavily on the output of the BORIS project but presents a broader geographical view, taking due regard of data from eastern Europe and elsewhere. The monograph also benefits from the wider experience and expertise of BORIS participants and from the inputs of several other specialists who were not part of the BORIS consortium.

Throughout the course of BORIS, and also during the preparation of this monograph, authors were frequently aware of gaps in current knowledge. In particular, taxonomic expertise on the parasitoids, predators and pathogens of oilseed rape insect pests clearly requires further development, with particular reference to entomopathogenic fungi and the larval stages of hymenopterous parasitoids, predatory diptera and staphylinid (rove) beetles.

Biocontrol of Oilseed Rape Pests, which is a synthesis of current knowledge within Europe, commences with reviews of the oilseed rape crop (Chapter 1), the pests (Chapter 2) and pest management strategies (Chapter 3). These introductory chapters are followed by specialist chapters on parasitoids of specific pests or groups of pests (Chapters 4–9). Rearing and trapping methods for parasitoids are also reviewed (Chapter 10). The reliable identification of parasitoids (most of which are hymenopterans) can be a formidable barrier to their further study. Therefore, a specific chapter (Chapter 11) on the identification of adult hymenopterous parasitoids has been included. It had been planned, also, to include a key to the larval stages of such parasitoids. However, unfortunately, information in this area is very limited and insufficient material was available for study. Authors of the various parasitoid chapters have been at pains to interpret published (and often questionable) names of organisms in line with current thinking; however, they are aware that future work may require amendments to be made. The situation is often confused, at least in the older literature, owing to some 'records' of supposed parasitoids of pests being invalid since, in reality, they refer to hyperparasitoids (i.e. to parasitoids of parasitoids).

Although there have been a vast number of studies on predators in arable cropping systems, relatively few of these relate to oilseed rape. Predators are often abundant in oilseed rape fields and, cumulatively, they are capable (at least potentially) of causing significant mortality amongst pest populations. However, unlike certain parasitoids, they do not specifically target oilseed rape pests. Also, unlike parasitoids (which often show a significant degree of host specificity), predators tend to be opportunist feeders that attack a wide range of prey. As an aid to their future study, information is presented in Chapters 12 and 13 on the nomenclature and taxonomy of the main groups of predators inhabiting oilseed rape fields. Methods for their study (Chapter 14) and the importance of on-farm landscape structure on predator populations (Chapter 15) are also reviewed, as is the impact (or potential impact) of predators on oilseed rape pests (Chapter 16).

Information on pathogens in arable cropping systems is generally scarce, with relatively little available of direct relevance to oilseed rape. However, for completeness, a brief review of current knowledge is given in Chapter 17.

Appropriate exploitation of naturally occurring BCAs has the potential to improve environmental quality on European farmland and to bring about socio-economic benefits to the industry as a whole. It is hoped, therefore, that this monograph will form a useful basis for further work on the biocontrol of oilseed rape pests, and will help to bring about a greater understanding of the beneficial role and potential of natural enemies of crop pests on arable farms. Hopefully, this will lead to the enhancement of populations of BCAs on arable farms and encourage their greater utilization in future pest management strategies.

It is a pleasure to thank my partners on the BORIS project, namely Dr. Yannick Ballanger (France), Dr. Rudulf Büchi (Switzerland), Dr. Wolfgang Büchs (Germany), Dr. Barbara Ekbom (Sweden), Lars Monrad Hansen (UK), Professor Heikki M. T. Hokkanen (Finland), Dr. Bernhard Kromp (Austria), Alan Lane (UK), Dr. Christer Nilsson (Sweden), Dr. Bernd Ulber (Germany), Professor Ingrid H. Williams (UK) and John E. B. Young (UK), and also guest participants Dr. Jörgen Eilenberg (Denmark), Dr. Anne Piirainen (Finland) and Professor Stefan Vidal (Germany), all of whom have advanced our knowledge on the biocontrol of oilseed rape pests and have contributed to a greater or lesser extent to this monograph. Thanks are also due to other named contributors and to Dr. Bernd Ulber (University of Göttingen), Mr. James E. Ashby and Dr. Andrew W. Ferguson (IACR Rothamsted) for the loan of specimens to supplement available material upon which drawings were based.

David V. Alford
Editor
April 2002

Publication of this monograph was supported financially by the European Commission. However, the content of the publication is the sole responsibility of the publisher, and in no way represents the views of the Commission or its services.

1

The Oilseed Rape Crop

David V. Alford
Orchard House, Oakington Road, Dry Drayton, Cambridge CB3 8DD, UK

Introduction

Oilseed rape is an important crop in European agriculture, having risen particularly to prominence during the latter part of the 20th century (Bunting, 1986). The European crop is dominated by *Brassica napus* (Brassicaceae), often cited as *Brassica napus* ssp. *oleifera*; to a lesser extent, turnip rape (*Brassica campestris*) is also grown, especially in Finland and Sweden. Turnip rape is sometimes afforded subspecific status, as follows: *B. campestris* ssp. *oleifera*. *Brassica napus* is grown extensively outside Europe, notably in Australia and Canada, where it is most often known as canola. In Asia (e.g. in China, India and Japan), the rape crop is dominated by *B. campestris*. *Brassica campestris* is a native plant throughout Europe, central Asia and the Middle East. *Brassica napus*, however, arose in cultivation in southern Europe and is not represented in the wild, except nowadays as feral 'weed' populations on, for example, fields margins and roadside verges; such populations have often originated from seeds spilt from trailers and combines following the harvesting of nearby oilseed rape crops. For further details see de Rougemont (1989).

On a world scale, the oilseed rape crop area approaches 30 million ha, about 5 million ha of which is grown in Europe (FAO, 2001). The crop is particularly important in central and northern Europe, notably in the Czech Republic, Denmark, France, Germany, Hungary, Poland, the UK and the Ukraine (Table 1.1).

Usually, seed from oilseed rape is crushed to extract the oil. This oil is used for culinary purposes, for fuel and as a lubricant, especially on machinery in the food-processing industry; it is also used in the production of soaps and synthetic rubber. The seed residue (seed-cake) is used for cattle fodder; the uncrushed seed also has a market as bird feed. Less significantly, the crop is sometimes grown as a green salad, as cattle feed and as on-farm standing food and shelter for game birds.

The Oilseed Rape Plant

Oilseed rape is an annual herbaceous plant, with erect branching stems, typically 1–2 m in height and with a long, thin tap root. The foliage is smooth, bluish-green, with deeply incised margins; in common with many brassicaceous plants, the leaves are coated with a waxen bloom. The four-petalled flowers are yellow and borne in clusters (rosettes), on both terminal and lateral shoots, in the form of racemes. The developing fruits are green,

Table 1.1 Area of oilseed rape grown in various European countries in 1999 (FAO, 2001).*

Country	Area grown ('000 ha)
Austria	66
Belarus	85
Czech Republic	350
Denmark	142
Finland	58
France	1,369
Germany	1,201
Hungary	182
Poland	466
Russia	160
Slovakia	113
Sweden	55
UK	542
Ukraine	222

* Countries with <50,000 ha excluded.

elongate pods (siliquas) that eventually ripen to brown. A siliqua is formed from an ovary of two united carpels (these form the pod walls), separated by an inner central wall (the septum) into two compartments (locules) within each of which a row of seeds develops; the ripe seeds, for which the crop is usually grown, are black and hard at maturity, similar in appearance to lead shot. A siliqua dehisces by an upward splitting and parting of the carpels from the base, leaving the septum (to which the seeds may, temporarily, remain attached) entire.

Crop growth stages

Keys to the growth stages of oilseed rape have been presented by various authors. Sylvester-Bradley & Makepeace (1984) introduced a decimal code, ranging from 0.0 to 9.9 that included gross classifications as follows:

- Stage 0 – germination and emergence;
- Stage 1 – leaf production;
- Stage 2 – stem extension;
- Stage 3 – flower bud development;
- Stage 4 – flowering;
- Stage 5 – pod development;
- Stage 6 – seed development;
- Stage 7 – leaf senescence;
- Stage 8 – stem senescence;
- Stage 9 – pod senescence.

This code was later revised (Sylvester-Bradley, 1985), but the essential elements remained the same. The Sylvester-Bradley codes are often cited in relevant UK-based pest control strategies, e.g. by Lane & Gladders (2000). Particularly on an international level, however, the system introduced by Lancashire *et al.* (1991) – commonly known as the BBCH (BASF/Bayer/Ciba/Hoechst) code – is more frequently adopted. The BBCH code, in common with that of Sylvester-Bradley, counts the number of internodes to measure stem elongation, but this is soon overtaken by assessments of bud development. Although bud development and flower development in both systems are similar, in the case of the former pod and seed development are combined. Gross categories in the BBCH scheme are as follows:

- Stage 0 – germination;
- Stage 1 – leaf development;
- Stage 2 – formation of side shoots;
- Stage 3 – stem elongation;
- Stage 4 – development of harvestable vegetative plant parts (not applicable to oilseed rape);
- Stage 5 – inflorescence emergence;
- Stage 6 – flowering;
- Stage 7 – development of fruit;
- Stage 8 – ripening;
- Stage 9 – senescence.

Each BBCH gross category is further subdivided into up to 10 detailed categories. The flowering stage, for example, runs from 60 (= 'first flowers open') to 69 (= 'end of flowering').

Crop Agronomy and Husbandry

It is not the intention here to provide an account of the agronomy and husbandry of oilseed rape but rather to review, briefly, current cultivation methods as a background to the accounts that follow on oilseed rape pests, their naturally occurring biocontrol agents (BCAs) and integrated crop management (ICM). For detailed reviews, readers are referred to, for example, Scarisbrick & Daniels (1986) and Kimber & McGregor (1995).

Crop rotation

In Europe, winter oilseed rape is normally grown as an annual break crop in three- to four-year arable rotations with cereals (winter wheat or winter barley) and break crops such as peas or beans. One-year set-aside also features in many rotations and often precedes oilseed rape.

Conventionally, crop rotation has been adopted partly to reduce the carry-over of pest and disease problems from one crop to another. Soil-borne diseases such as sclerotinia (*Sclerotinia sclerotiorum*), verticillium wilt (*Verticillium dahliae*) and club root (*Plasmodiophora brassicae*) often restrict the use of oilseed rape in close rotation with itself or with other host crops such as peas, potatoes, sunflowers and winter beans.

Soil cultivation, drilling and crop establishment

Primary cultivations depend, strongly, on soil type and weather conditions. Ploughing is often used in conventional systems. However, non-inversion methods of minimal tillage, including various combinations of discing, rotary cultivation and harrowing, may also be used. Other options include direct drilling or the broadcasting of seed into stubble. There are advantages and disadvantages attached to all of these techniques. Seed rates for conventional cultivars are in the region of 5–8 kg/ha, with a target plant population of at least 80 plants/m^2 (cf. hybrid cultivars, below).

Oilseed rape is either sown in late summer (= winter rape), for cropping in the following summer, or in spring (= spring rape). Winter rape is more vigorous than spring rape and has greater potential to compensate for damage caused by pests and other factors. However, a larger proportion of spring rape and spring turnip rape is grown in Finland and Sweden because of the more extreme winter conditions regularly experienced in these areas.

Occasionally, oilseed rape can be difficult to grow, and extensive losses may occur at establishment. In particular, slugs and wood pigeons can cause complete crop failure at this critical time; on spring-sown crops, cabbage flea beetles (*Phyllotreta* spp.) may also be important. The optimum sowing time for winter oilseed rape in Europe is from early August to early September. Yield penalties are incurred if sowing is delayed later than the first week of September. Early sowing can lead to problems from cabbage root fly (*Delia radicum*) or can create an excessively vigorous crop with increased risk from foliar diseases. Plant population density, row spacing and choice of cultivar, including hybrid and genetically modified (GM) cultivars (see below), may all exert indirect effects on the impact and incidence of pests. The architecture of the crop canopy is influenced by a number of factors, including plant population, growth habit and nitrogen use.

Nitrogen and other nutrients

Winter rape is responsive to nitrogen and most crops receive up to approximately 200 kg/ha, principally as spring applications.

The use of sulphur is becoming increasingly common on oilseed rape, and is generally targeted at crops growing in light, sandy soils in districts of low atmospheric sulphur deposition (e.g. non-industrial areas). However, the demand for supplementary use of sulphur is becoming more widespread. Phosphate and potash are applied to the seedbed to maintain levels within acceptable limits on a rotational basis.

Growth regulators

Growth regulators are not widely used, and in some countries (e.g. Sweden) they are not even registered for use on the crop. However, there is potential to manipulate crop physiology and canopy structure through the use of growth regulators and nitrogen. Conventional growth regulators, such as chlormequat, may be applied to shorten and thicken the stems, and help the canopy remain more erect during seed growth. The fungicide tebuconazole, applied for disease control in early spring, has a similar growth-regulating effect.

Cultivars

Agronomic changes during the mid- to late 1980s included the rapid move away from the growing of high erucic acid cultivars in favour of low erucic acid (single-low) cultivars and, finally, the adoption of double-low cultivars, i.e. to those with seeds low in both erucic acid and glucosinolates; these are also, alternatively, designated 'single-zero' or 'double-zero' (0 or 00) cultivars. Domination by double-zero cultivars in European agriculture reached virtually 100% by the autumn 1990 drilling. However, at present, some high erucic acid cultivars continue to be grown for industrial uses.

The top-yielding cultivars of oilseed rape are now hybrids. The commercially available hybrids are of two types: restored hybrids and cultivar associations. In the case of restored hybrids all plants are fully fertile, capable of producing pollen and setting seed without cross-pollination. Cultivar associations are composed of field mixtures of male-sterile hybrid plants which produce no pollen, and male-fertile plants of one or more cultivars whose principal function is to supply pollen and which contribute little to the overall yield. In the case of cultivar associations, recent work has shown that male-fertile plants are more heavily infested by both adult and larval pollen beetles (*Meligethes aeneus*), highlighting the need for control of this pest to ensure adequate pollination of the crop (Cook *et al.*, 1999).

Hybrid cultivars, offering higher yields, have a seed rate of 2–4 kg/ha, aiming at 40–50 plants/m^2 (cf. conventional cultivars, above). Nowadays, various disease-resistant cultivars are available (see under 'Disease control', below).

No insect pest preferences have been demonstrated for particular cultivars of oilseed rape, although characteristics of specific cultivars may make some more likely to be attacked than others. Büchi (1996), for example, showed that early-flowering (early-developing) oilseed rape cultivars, irrespective as to whether single-zeros or double-zeros, were more likely to be attacked by egg-laying females of rape stem weevil (*Ceutorhynchus napi*) than moderately-early-flowering to late-flowering ones. Büchi (1996) also related stem-splitting propensity of cultivars to the likelihood of their being attacked by rape stem weevil, especially in the face of high pest pressure, and in his experiments found that single-zero cultivars (cvs Bienvenu and Jet Neuf) had fewer split stems than double-zero ones (notably cvs Idol and Eurol).

None of the studies on the seed concentrations of glucosinolates in oilseed rape has demonstrated preferential pest preferences for ensuing crops: examples include Milford *et al.* (1989b) on pollen beetle; Rawlinson & Williams (1991) on cabbage stem flea beetle (*Psylliodes chrysocephala*), pollen beetle, cabbage seed weevil (*Ceutorhynchus assimilis*) and brassica pod midge (*Dasineura brassicae*), and Williams & Bartlet (1993) on cabbage stem flea beetle. However, this may be explained by differences in seed concentrations of glucosinolates not being reflected in concentrations of glucosinolates in the vegetative parts of the crops (Milford *et al.*, 1989a).

Glucosinolates and their catabolites, such as the isothiocyanates and nitriles, are important cues to host selection by cruciferous pests, aiding both orientation to and recognition of the host plant (Bartlet, 1996). Two strategies have been proposed for improving the glucosinolate content of the vegetative parts of oilseed rape for pest resistance (Bartlet *et al.*, 1999). The first involves rape lines with low constitutive, but high induced, glucosinolate levels; these would be unobtrusive to brassica-specialist pests in the absence of attack, but would be protected from generalist feeders and pathogens once damaged. The second involves rape lines with a high proportion of glucosinolate types that do not catabolize to isothiocyanates; the overall glucosinolate concentration of the plant would be maintained as protection from other herbivores and disease, but the plants would be less attractive to brassica-specialist pests. Recent advances in genetic engineering and plant breeding are now being used to alter the glucosinolate composition of oilseed rape tissues, to reduce susceptibility to both pests and diseases (Mithen & Murphy, 1994; Giamoustaris & Mithen, 1996).

Transgenic oilseed rape

The next decade is likely to see the widespread cultivation and use of genetically modified (GM) cultivars of oilseed rape in Europe. These are already widely cultivated in North America.

Herbicide-tolerant GM cultivars are likely to be among the first to be cultivated commercially in Europe. They have already been released extensively in field experiments throughout Europe and are currently being evaluated in farm-scale trials in, for example, the UK, for their effects on farmland biodiversity (Firbank *et al.*, 1999). At present, broad-leaved weeds in oilseed rape cannot be controlled effectively by farmers because the crop itself is susceptible to the required herbicide. The introduction of herbicide tolerance genes, therefore, promises more effective control of weeds. The genes code for plant enzymes capable of detoxifying specific herbicides, such as the broad-spectrum but short-lived glufosinate-ammonium and glyphosate (e.g. Sweet *et al.*, 1997). These crops offer more flexibility in weed management and improved control of the more difficult weeds and volunteers, which may at least benefit field populations of predators (see Chapter 15).

Insect-resistant GM cultivars of oilseed rape are also being developed. For example, plants are being modified to contain *Bt* toxins and proteinase inhibitors (PIs), and these have potential to reduce insecticide application.

Bt is a natural toxin produced by the soil-dwelling bacterium *Bacillus thuringiensis*. Different strains of *Bt* produce different toxins, usually specific to a particular order of insects. Although none of the major European pests of oilseed rape has yet been shown to be susceptible to *Bt* toxins, the minor pest cabbage root fly and certain Lepidoptera, e.g. diamond-back moth (*Plutella xylostella*), are known to be susceptible (Hokkanen & Wearing, 1995). Hokkanen & Wearing (1995) have also postulated that several major pests of European oilseed rape crops, such as brassica pod midge, cabbage seed weevil, cabbage stem flea beetle, cabbage flea beetles, pollen beetle and turnip sawfly (*Athalia rosae*), could be considered potentially susceptible to *Bt* toxins of a suitable strain.

PIs are proteins that inhibit the activity of proteinase enzymes that split proteins into peptides. Genes coding for PIs are being introduced into oilseed rape plants to target the digestive enzymes of insect pests, aiming to cause starvation and death following ingestion.

Harvesting

Harvesting may be accomplished through direct combining or by swathing prior to combining. Frequently, a desiccant herbicide (e.g. diquat) may be applied prior to direct combining.

Crop Protection

Pest control

In conventional rape crops, the main threat from pests arises from slugs, wood pigeons and various insects, notably cabbage seed weevil, cabbage stem flea beetle, cabbage stem weevil (*Ceutorhynchus pallidactylus*), pollen beetle and rape stem weevil. Slug control is obtained using molluscicide pellets applied before or after sowing during crop establishment. Chemical control of insect pests is now dominated by synthetic pyrethroid insecticides. In many countries there is a large element of prophylactic insecticide use against cabbage stem flea beetle in the autumn, ceutorhynchid stem weevils in the late winter or early spring, and pollen beetle and cabbage seed weevil in the spring. For further details see Chapter 2.

Disease control

The main diseases, often requiring treatment in the autumn, are canker (*Leptosphaeria maculans*, anamorph: *Phoma lingam*) and light leaf spot (*Pyrenopeziza brassicae*, anamorph: *Cylindrosporium concentricum*). Canker causes leaf spotting over the winter period and, later, cankers on the stem. During the autumn/winter period this disease is commonly referred to as 'phoma leaf spot'. Stem canker causes premature ripening and weakening of the stem, and this can cause lodging. In conventional crops, a conazole or a triazole fungicide is often applied as a spray in November, with a repeat application in spring (late February or early March) at early stem extension. During the spring/summer period, treatment is often required against alternaria (*Alternaria brassicae*) and sclerotinia stem rot (*Sclerotinia sclerotiorum*) (but not in Sweden or Finland). Sclerotinia control is of prime concern, and tends to dominate fungicide treatment decisions in the spring. Sclerotinia fungicides are often tank-mixed with insecticides applied for cabbage seed weevil control, although in some countries (e.g. Sweden) this is not allowed. Frequently used sclerotinia sprays include formulated mixtures containing an MBC compound. Alternatively, products such as a protectant dicarboximide fungicide may be applied. Crops are not often sprayed for alternaria alone, as the broad-spectrum fungicides applied against sclerotinia also provide control of this disease. For further details, see Lane & Gladders (2000).

Weed control

In conventional crops of winter oilseed rape, herbicides are applied during the autumn months to control broad-leaved weeds, grass weeds and volunteer cereals. The control of volunteer cereals is of primary concern. In some countries, including the UK, a residual anilide herbicide may be applied pre-emergence to control broad-leaved weeds. The additional use of a post-emergence graminicide may be required to control grass weeds and volunteer cereals. Alternatively, residual, pre-emergence herbicides (such as a residual amide herbicide) may be used in some countries for broad-spectrum weed control. For further details see, for example, Naylor (2002).

References

Bartlet, E. (1996). Chemical cues to host-plant selection by insect pests of oilseed rape. *Agricultural Zoology Reviews* **7**, 89–116.

Bartlet, E., Williams, I. H. & Pickett, J. A. (1999). The ideal glucosinolate profile for pest resistance in oilseed rape. *IOBC/wprs Bulletin* **22**, 13–17.

Büchi, R. (1996). Eiablage des Rapsstengelrüßlers *Ceutorhynchus napi* Gyll., in Abhängigkeit der Stengellänge bei verschiedenen Rapssorten. *Anzeiger für Schädlingskunde Pflanzenschutz Umweltschutz* **69**, 136–9.

Bunting, E. S. (1986). *Oilseed rape in perspective.* In: D. H. Scarisbrick & R. W. Daniels (eds). *Oilseed Rape.* Collins: London.

Cook, S., Murray, D. A. & Williams, I. H. (1999). Pollen beetle, *Meligethes aeneus* Fabricius, incidence in the composite hybrid winter oilseed rape, Synergy. *Proceedings of the 10th International Rapeseed Congress: 'New Horizons for an Old Crop', 26–29 September 1999, Canberra, Australia.*

FAO (2001). *Production Yearbook. Volume 53. FAO Statistics Series No. 156.* Food and Agriculture Organization of the United Nations: Rome.

Firbank, L. G., Dewar, A. M., Hill, M. O., May, M. J. Perry, J. N., Rothery, P., Squire, G. & Woiwod, I. P. (1999). Farm scale evaluation of GM crops explained. *Nature* **339**, 727–8.

Giamoustaris, A. & Mithen, R. (1996). Genetics of aliphatic glucosinolates. IV. Side-chain modification in *Brassica oleracea. Theoretical and Applied Genetics* **93**, 1006–10.

Hokkanen, H. M. T. & Wearing, C. H. (1995). Assessing the risk of pest resistance evolution to *Bacillus thuringiensis* engineered into crop plants: a case study of oilseed rape. *Field Crops Research* **45**, 171–9.

Kimber, D. S. & McGregor, D. J. (eds) (1995). *Brassica Oilseeds. Production and Utilization.* CAB International: Wallingford.

Lancashire, P. D., Bleiholder, H., Boom, T. van den, Langelüddeke, P., Strauss, R., Weber, E. & Witzenberger, A. (1991). A uniform decimal code for growth stages of crops and weeds. *Annals of Applied Biology* **119**, 561–601.

Lane, A. & Gladders, P. (2000). *Pests and diseases of oilseeds, brassica seed crops and field beans.* In: D. V. Alford (ed.). *Pest and Disease Management Handbook.* Blackwell Science: Oxford.

Milford, G. F. J., Fieldsend, J. K., Porter, A. J. R. & Rawlinson, C. J. (1989a). Changes in gluosinolate concentration during the vegetative growth of single and double low cultivars of winter oilseed rape. *Aspects of Applied Biology* **23**, 83–90.

Milford, G. F. J., Porter, A. J. R., Fieldsend, J. K., Miller, C. A., Leach, J. E. & Williams, I. H. (1989b). Glucosinolates in oilseed rape (*Brassica napus*) and the incidence of pollen beetles (*Meligethes aeneus*). *Annals of Applied Biology* **115**, 375–80.

Mithen, R. F. & Murphy, D. J. (1994). Application of biotechnology to the improvement of brassica oilseed crops. *Aspects of Applied Biology* **39**, 23–34.

Naylor, R. E. L. (ed.) (2002). *Weed Management Handbook.* Blackwell Science: Oxford.

Rawlinson, C. J. & Williams, I. H. (1991). Pests, diseases and effects of crop protection on single- and double-low winter rape. *Bulletin SROP/WPRS – Proceedings of the OILB Meeting, March 1990, Rothamsted, UK,* pp. 74–88.

Rougemont, G. M. de (1989). *A Field Guide to the Crops of Britain and Europe.* Collins: London.

Scarisbrick, D. H. & Daniels, R. W. (eds) (1986). *Oilseed Rape.* Collins: London.

Sweet, J. B., Shepperson, R., Thomas, J. E. & Simpson, E. (1997). The impact of releases of genetically modified herbicide tolerant oilseed rape in the UK. *The 1997 Brighton Crop Protection Conference – Weeds* **1**, 291–302.

Sylvester-Bradley, R. (1985). Revision of a code for stages of development in oilseed rape (*Brassica napus* L.). *Aspects of Applied Biology* **10**, 395–400.

Sylvester-Bradley, R. & Makepeace, R. J. (1984). A code for stages of development in oilseed rape (*Brassica napus* L.). *Aspects of Applied Biology* **6**, 399–419.

Williams, I. H. & Bartlet, E. (1993). Host plant factors affecting the feeding of cabbage stem flea beetles (*Psylliodes chrysocephala* L.) on oilseed rape. *IOBC/wprs Bulletin* **16** (9), 202–9.

2

Insect Pests of Oilseed Rape Crops

David V. Alford
Orchard House, Oakington Road, Dry Drayton, Cambridge CB3 8DD, UK

Christer Nilsson
Swedish University of Agricultural Sciences, Department of Crop Science, S-23053 Alnarp, Sweden

Bernd Ulber
Georg-August Universität, Institut für Pflanzenpathologie und Pflanzenschutz, Grisebachstraße 6, D-37077 Göttingen, Germany

Introduction

In northern Europe, oilseed rape is attacked by a wide range of insect pests. Some of these pests occur virtually wherever the crop is grown, whereas others have a more limited distribution. No pests are specific to oilseed rape, but several 'rape pests' are restricted to brassicaceous plants, including seed crops, forage crops, vegetable brassicas and wild hosts such as charlock (*Sinapis arvensis*).

In addition to insects, various nematodes are recorded as pests of oilseed rape. These include beet cyst nematode (*Heterodera schachtii*), brassica cyst nematode (*H. cruciferae*), needle nematodes (*Longidorus* spp.) and stubby-root nematodes (*Paratrichodorus* spp. and *Trichodorus* spp.). Slugs, especially field slug (*Deroceras reticulatum*), and wood pigeons (*Columba palumbus*) can also be of importance, especially in autumn and early winter. Apart from brief mention under integrated crop management (ICM) (Chapter 3) and passing mention in Chapter 16, non-insect pests are not discussed in this monograph.

Major pests

The following pests are considered of particular importance on oilseed rape crops in Europe, although not necessarily in all regions:

- brassica pod midge – *Dasineura brassicae*
- cabbage seed weevil – *Ceutorhynchus assimilis*
- cabbage stem flea beetle – *Psylliodes chrysocephala*
- cabbage stem weevil – *Ceutorhynchus pallidactylus*
- pollen beetles – *Meligethes* spp. (especially *M. aeneus*) [Note 1]
- rape stem weevil – *Ceutorhynchus napi*

Note 1. In this monograph, where the term 'pollen beetle' is used in the singular, reference to *M. aeneus* should be inferred.

The majority of observations on the biocontrol of oilseed rape pests relate to these species, and accounts of their natural enemies form the basis of the present monograph.

Minor pests

Several other insect pests occur in oilseed rape crops, but they are usually of only minor importance. These include:

- cabbage aphid – *Brevicoryne brassicae*
- cabbage flea beetles – *Phyllotreta* spp.
- cabbage root fly – *Delia radicum*
- peach/potato aphid – *Myzus persicae*
- rape winter stem weevil – *Ceutorhynchus picitarsis*
- turnip sawfly – *Athalia rosae*

In some situations, all of the above-listed species can be of some significance on oilseed rape. The two aphid species, for example, are potentially important vectors of virus diseases in autumn-sown (i.e. winter) rape crops. Further, larvae of cabbage root fly and adults of cabbage flea beetles can be a problem at establishment, especially on spring-sown oilseed rape. At least in France, weevils of the genus *Baris* are also associated with oilseed rape crops. Further details are given below.

Some insect pests, although very occasionally reported on oilseed rape crops, are of little or no importance and, in any case, are not primarily associated with them. Examples are given in Table 2.1. For further details of most of these species, which are not considered in any depth in the present monograph, see Alford (1999).

Incidental pests

A small number of non-specific, polyphagous pests may attack oilseed rape crops. These include: cutworms, e.g. the larvae of turnip moth, *Agrotis segetum* (Lepidoptera: Noctuidae); leatherjackets [the larvae of crane flies such as *Tipula oleracea* (Diptera:

Table 2.1 Examples of pests found on oilseed rape crops but associated mainly with other brassicaceous hosts.

Vernacular name	Scientific name	Order: Family
blue stem weevil	*Ceutorhynchus sulcicollis* (Paykull)	Coleoptera: Curculionidae
cabbage moth	*Mamestra brassicae* (Linnaeus)	Lepidoptera: Noctuidae
diamond-back moth	*Plutella xylostella* (Linnaeus)	Lepidoptera: Yponomeutidae
field thrips	*Thrips angusticeps* Uzel	Thysanoptera: Thripidae
green-veined white butterfly	*Pieris napi* (Linnaeus)	Lepidoptera: Pieridae
large white butterfly	*Pieris brassicae* (Linnaeus)	Lepidoptera: Pieridae
mustard beetle	*Phaedon armoraciae* (Linnaeus)	Coleoptera: Chrysomelidae
small white butterfly	*Pieris rapae* (Linnaeus)	Lepidoptera: Pieridae
swede midge	*Contarinia nasturtii* (Kieffer)	Diptera: Cecidomyiidae
turnip gall weevil	*Ceutorhynchus pleurostigma* (Marsham)	Coleoptera: Curculionidae
turnip root fly	*Delia floralis* (Fallén)	Diptera: Anthomyiidae
larva = a leaf miner	*Scaptomyza flava* (Fallén)	Diptera: Drosophilidae
larva = a leaf miner	*Scaptomyza pallida* (Zetterstedt)	Diptera: Drosophilidae
larva = cabbage leaf miner	*Phytomyza rufipes* Meigen	Diptera: Agromyzidae

Tipulidae)], mirid bugs (Heteroptera: Miridae, e.g. *Closterotomus norvegicus* and *Lygus rugulipennis*) (Hoßfeld, 1963) and wireworms [the larvae of click beetles such as *Agriotes lineatus* (Coleoptera: Elateridae)]. They are not considered further in this monograph.

The pest complex on winter and spring rape

Although some pests may occur on both winter rape and spring rape, owing to their phenology several pests occur only (or mainly) on one or the other. Also, as indicated in the previous chapter, the greater potential for a winter crop to compensate for pest damage can affect the relative importance of certain pests. The relative importance of the main pests of oilseed rape is summarized in Table 2.2.

Nomenclature of the Pests

Over the past many years, various scientific and vernacular names have been used in the literature for one and the same species of oilseed rape pest, and this has been at the very least a potential cause of confusion. To maintain an element of standardization, the names used in this monograph are those adopted and advocated by participants in the 1997–1999 EC-funded BORIS project (see Preface).

Scientific names

The following are the generally accepted scientific names for the main insect pests of oilseed rape mentioned in this monograph:

Table 2.2 Relative importance of the main pests of oilseed rape in Europe.

Pest	Winter rape	Spring rape
Baris weevils	– [Note 1]	×
brassica pod midge (*Dasineura brassicae*)	+	+
cabbage aphid (*Brevicoryne brassicae*)	(+)†	(+)
cabbage flea beetles (*Phyllotreta* spp.)	–	(+)
cabbage root fly (*Delia radicum*)	–	(+)
cabbage seed weevil (*Ceutorhynchus assimilis*)	+	+
cabbage stem flea beetle (*Psylliodes chrysocephala*)	+	×
cabbage stem weevil (*Ceutorhynchus pallidactylus*)	(+)	(+)
peach/potato aphid (*Myzus persicae*)	†	–
pollen beetles (*Meligethes* spp.)	+ [Note 2]	+
rape stem weevil (*Ceutorhynchus napi*)	+	×
rape winter stem weevil (*Ceutorhynchus picitarsis*)	(+)	×
turnip sawfly (*Athalia rosae*)	–	(+)

+ Often damaging in areas where it occurs.
(+) Occasionally or locally damaging.
– Present but of little or no importance.
† Potentially important as a virus vector.
× Not present.

Note 1. Recorded mainly in France.
Note 2. Rarely of significance on winter oilseed rape in the UK.

Order HEMIPTERA

Family APHIDIDAE
Brevicoryne brassicae (Linnaeus)
Myzus persicae (Sulzer)

Order COLEOPTERA

Family NITIDULIDAE
Meligethes spp., e.g. *M. aeneus* (Fabricius) and to a lesser extent *M. viridescens* (Fabricius)

Family CHRYSOMELIDAE
Phyllotreta spp., e.g. *P. nemorum* (Linnaeus)
Psylliodes chrysocephala (Linnaeus)

Family CURCULIONIDAE
Baris spp., e.g. *B. coerulescens* Scopoli
Ceutorhynchus assimilis (Paykull) [Notes 1 and 2] syn. *Ceutorhynchus obstrictus* (Marsham)
Ceutorhynchus napi Gyllenhal [Note 2]
Ceutorhynchus pallidactylus (Marsham) [Notes 1 and 2] syn. *Ceutorhynchus quadridens* (Panzer)
Ceutorhynchus picitarsis Gyllenhal [Note 2]

Order DIPTERA

Family CECIDOMYIIDAE
Dasineura brassicae (Winnertz)
syn. *Dasyneura brassicae* (Winnertz)

Family ANTHOMYIIDAE
Delia radicum (Linnaeus)
syn. *Chortophila brassicae* (Bouché)
syn. *Delia brassicae* (Bouché)
syn. *Delia brassicae* (Weidemann)
syn. *Erioischia brassicae* (Bouché)
syn. *Erioischia brassicae* (Weidemann)
syn. *Phorbia brassicae* (Bouché)
syn. & misspelling: *Hylemyia brassicae* (Bouché)

Order HYMENOPTERA

Family TENTHREDINIDAE
Athalia rosae (Linnaeus)
syn. *Athalia colibri* (Christ)

Note 1. See comments on homonyms, below.
Note 2. *Ceuthorrhynchus* or *Ceuthorhynchus* appear frequently in the literature, including many well-known reference works (e.g. Ritter, 1916; Balachowsky, 1963; Schmidt, 1970), but are misspellings.

Vernacular names

The English vernacular names adopted in this monograph for insect pests of oilseed rape are listed below; examples of vernacular names in several other European languages are also given:

Athalia rosae	English:	Turnip sawfly
	French:	Tenthrède de la rave
	German:	Rübsenblattwespe
	Danish:	Kålbladhveps

	Swedish:	Kålbladstekel
	Finnish:	Rapsipistiäinen
Baris spp.	English:	–
	French:	Baris des crucifères
	German:	Spitzmaulrüßler
	Danish:	–
	Swedish:	–
	Finnish:	–
Brevicoryne brassicae	English:	Cabbage aphid
	French:	Puceron cendré du chou
	German:	Mehlige Kohlblattlaus
	Danish:	Kålbladlus
	Swedish:	Kålbladlus
	Finnish:	Kaalikirva
Ceutorhynchus assimilis	English:	Cabbage seed weevil [Note 1]
	French:	Charancon des siliques
	German:	Kohlschotenrüßler
	Danish:	Skulpesnudebille
	Swedish:	Blygrå rapsvivel
	Finnish:	Rapsikärsäkäs
Ceutorhynchus napi	English:	Rape stem weevil
	French:	[Gros] Charancon de la tige du colza
	German:	Großer Rapsstängelrüßler
	Danish:	–
	Swedish:	–
	Finnish:	–
Ceutorhynchus pallidactylus	English:	Cabbage stem weevil [Note 2]
	French:	Charancon de la tige du chou
	German:	Gefleckter Kohltriebrüßler
	Danish:	Bladribbesnudebille
	Swedish:	Fyrtandad rapsvivel
	Finnish:	–
Ceutorhynchus picitarsis	English:	Rape winter stem weevil
	French:	Charancon du bourgeon terminal
	German:	Schwarzer Kohltriebrüßler
	Danish:	–
	Swedish:	–
	Finnish:	–
Dasineura brassicae	English:	Brassica pod midge [Note 3]
	French:	Cécidomyie des siliques
	German:	Kohlschotenmücke
	Danish:	Skulpegalmyg

	Swedish:	Skidgallmygga
	Finnish:	Litusääski
Delia radicum	English:	Cabbage root fly
	French:	Mouche du chou
	German:	Kleine Kohlfliege
	Danish:	Lille kålflue
	Swedish:	Liten kålfluga
	Finnish:	Pikkukaalikärpänen
Meligethes spp.	English:	Pollen beetles [Note 4]
	French:	Méligèthes des crucifères
	German:	Rapsglanzkäfer
	Danish:	Glimmerbøsse
	Swedish:	Rapsbaggar
	Finnish:	Rapsikuoriainen
Myzus persicae	English:	Peach/potato aphid [Note 5]
	French:	Puceron vert du pêcher
	German:	Grüne Pfirsichblattlaus
	Danish:	Ferskenbladlus
	Swedish:	Persikbladlus
	Finnish:	Persikkakiva
Phyllotreta spp.	English:	Cabbage flea beetles [Note 6]
	French:	Altises des crucifères/Petites altises
	German:	Kohlerdflöhe
	Danish:	Jordlopper
	Swedish:	Vanliga jordloppor
	Finnish:	Kirppa
Psylliodes chrysocephala	English:	Cabbage stem flea beetle
	French:	Altise d´hiver du colza/Grosse altise
	German:	Rapserdfloh
	Danish:	Rapsjordloppe
	Swedish:	Rapsjordloppa
	Finnish:	Rapsikirppa

Note 1. Acceptably abbreviated to 'seed weevil' in English advisory literature.

Note 2. Acceptably abbreviated to 'stem weevil' in English advisory literature, but care must be taken to avoid confusion with related species, notably 'rape stem weevil', and with the use of the term 'stem weevils' in a wider (i.e. non-specific) sense.

Note 3. Acceptably abbreviated to 'pod midge' in English advisory literature.

Note 4. *Meligethes aeneus* is often, acceptably, cited as 'pollen beetle' and *M. viridescens* as 'bronzed pollen beetle'.

Note 5. Use of a solidus (forward slash), as opposed to a hyphen or an en-dash, is preferred (see Alford, 1999).

Note 6. Acceptably abbreviated to 'flea beetles' where this does not cause confusion.

Homonyms

Recent proposed name changes for various species within the family Curculionidae include two well-known insect pests of oilseed rape, namely:

Ceutorhynchus assimilis (Paykull, 1792) **HOMONYM** [= cabbage seed weevil] syn. *obstrictus* (Marsham, 1802)
Ceutorhynchus quadridens (Panzer, 1795) **HOMONYM** [= cabbage stem weevil] syn. *pallidactylus* (Marsham, 1802)

This issue is of considerable significance with regards insect pests of oilseed rape because the names *obstrictus* Marsham (for *assimilis* Paykull) and *pallidactylus* Marsham (for *quadridens* Panzer) have been given priority by Colonnelli (1983, 1990); this usage has been followed in *Die Käfer Mitteleuropas* (Lohse & Lucht, 1994), the latter publication subsequently having resulted in their wider uptake, particularly in continental Europe.

Two observations are worthy of mention:

a) Until the publication of *Die Käfer Mitteleuropas*, the name *Ceutorhynchus obstrictus* (for cabbage seed weevil) appears virtually unused in entomological literature, and the name is certainly unknown to the vast majority of applied entomologists. On the other hand, a very large number of scientific and popular articles have been published on, or have made reference to, cabbage seed weevil – all under the name of *Ceutorhynchus assimilis*, without any mention of *obstrictus* as a synonym or otherwise.
b) The name *Ceutorhynchus pallidactylus* has been used occasionally in applied entomological literature, although until now it has usually been relegated to a synonym of *Ceutorhynchus quadridens* (e.g. by Seymour, 1989) – publications on this pest (usually as *C. quadridens* but sometimes as *C. pallidactylus*) are relatively infrequent.

In line with the comments on homonyms by Pope (1977), there would appear no good reason to adopt *obstrictus* at the expense of *assimilis*. Indeed, the confusion generated by so doing would be enormous. Further, UK taxonomists have in any case expressed doubt about the validity of using the name *obstrictus* for *assimilis*. A case to resist the change from *quadridens* to *pallidactylus* for cabbage stem weevil is less strong as, here, many of the objections to the adoption of the name *obstrictus* for cabbage seed weevil do not apply.

Therefore, within this monograph the name *Ceutorhynchus assimilis* is maintained for cabbage seed weevil; *C. obstrictus* is treated as a synonym. For cabbage stem weevil, however, the name *Ceutorhynchus pallidactylus* is adopted and *C. quadridens* treated as a synonym. To avoid unnecessary confusion in the crop protection literature, it is advocated strongly that this approach be taken by other applied entomologists when citing these pests.

Pest Status and Phenology

The following is a brief account of the status and lifecycles of the main pests associated with oilseed rape crops in Europe. For extensive information on the coleopterous pests of oilseed rape, which form the bulk of harmful insects associated with the crop, reference should be made to Balachowsky (1962, 1963). See also Paul (1992), Paul & Rawlinson (1992), Ekbom (1995), Garbe *et al.* (1996), Alford (1999) and Lane & Gladders (2000).

Inflorescence and bud pests

Brassica pod midge

This is a potentially important pest throughout Europe. Adults (Fig. 2.1) appear from mid-May onwards, emerging from cocoons in fields that grew oilseed rape in the previous four years (C. Nilsson, *unpublished*). Mating takes place at the emergence site and mated females migrate to rape crops. Adults live for only a few days. Eggs are deposited in batches of 20–30 in pods previously damaged by other pests, particularly cabbage seed weevils, utilizing in the first place weevil feeding punctures and, later in the season, larval exit holes. The midge eggs hatch a few days later. The larvae then feed gregariously on the inner wall of the pod for up to a month. Infested pods are discoloured and distorted, and may appear bloated (hence the name 'bladder pod midge', which has often been applied to this pest), although this symptom seems to be less evident in modern-day oilseed rape cultivars. Midge-infested pods split open or shatter prematurely to release the fully grown larvae, and this results in considerable loss of seed and, hence, crop yield. On falling to the ground, the larvae burrow beneath the soil to a depth of 5 cm or less, where they spin small silken cocoons in which to pupate. Some larvae pupate immediately, giving rise to adults about two weeks later, but an increasing proportion of the larvae enters winter diapause with each successive generation (Buhl, 1960; Czajkowska, 1978). The winter is spent as larvae within cocoons, pupation taking place in the spring just before adult emergence. The number of generations on winter oilseed rape is restricted to two, but, under favourable conditions and situations, the pest is capable of producing at least a partial third, especially when winter and spring rape are grown in the same area (Williams *et al*., 1987a, 1987b). Infestations are often most severe on the headlands of crops, and also especially important on spring-sown crops, as in Sweden. The adult midges are weak fliers and the mated females usually disperse no more than a few hundred metres from their emergence sites.

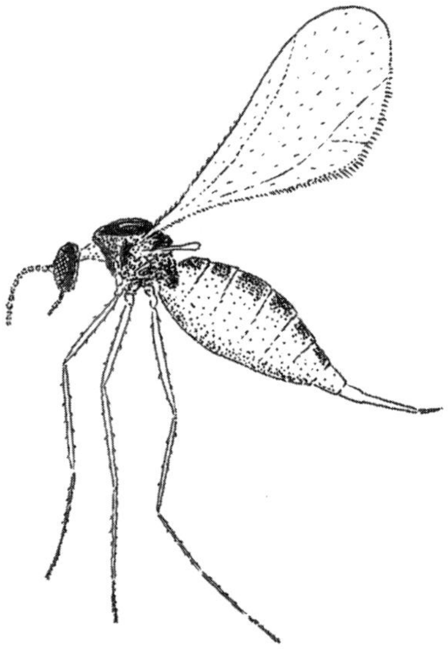

Fig. 2.1 Female brassica pod midge (*Dasineura brassicae*) (×20).

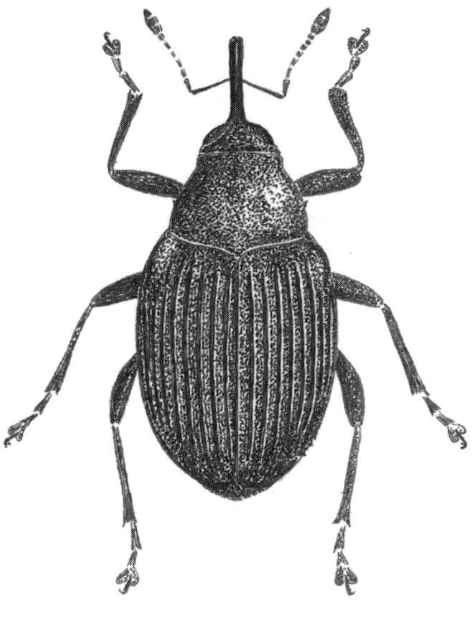

Fig. 2.2 Cabbage seed weevil (*Ceutorhynchus assimilis*) (×18).

Cabbage seed weevil

This widely distributed weevil (Fig. 2.2) is a major pest in both Europe and North America. It is of importance as a primary pest and also because its activities can predispose pods to subsequent invasion by brassica pod midge (see above). Adult weevils emerge from their overwintering sites during the flowering stage of winter rape. They feed on the developing pods, in which mated females eventually deposit eggs, typically one per pod. Following oviposition, the pod is 'marked' with a pheromone to deter egg-laying by rival females (Ferguson *et al.*, 1999). Eggs hatch one to two weeks later. Each larva then attacks the developing seeds, usually destroying about five seeds before becoming fully grown. It then bores through the wall of the pod, leaving a distinct emergence hole about the size of a pinhead. Individuals drop to the ground and pupate in the soil, new adults emerging from late July onwards. These new adults feed briefly on various brassicaceous plants, sometimes damaging nearby vegetable brassica crops such as broccoli and cabbage, and then hibernate in the shelter of perennial vegetation and leaf litter of field margins and woodlands. There is one generation annually.

Pollen beetles

Pollen beetles (notably *Meligethes aeneus*) (Fig. 2.3) are generally abundant throughout Europe on both winter and spring rape, but are usually more important on the latter crop. The adult beetles, which hibernate in woods and other sheltered sites, emerge in the spring and migrate to winter rape fields when air temperatures exceed 12–15°C. They feed on pollen from most plants, but eggs are laid only in buds of brassicaeous plants and when the buds are at least 3 mm long. The eggs hatch after a few days. Larvae, which pass through two instars, then feed on pollen before eventually dropping to the ground to pupate in the soil. The development from egg to new adult takes about a month and new adults appear from the middle of the summer, when they congregate on the flowers of various plants before eventually overwintering. There is just one generation annually.

Fig. 2.3 Pollen beetle (*Meligethes aeneus*) (×25).

Adult and larval damage can result in the presence of blind stalks (in place of pods) on the developing racemes. Damage caused by the beetles triggers a compensation reaction from the plant, which leads to an increased production of new racemes and buds. There is then often a normal, final pod number even after extensive damage. In many cases this results in pod filling becoming unsynchronized with the photosynthetic area of the plant; this leads to a lower seed number per pod and, later, to compensate for this, a slightly higher 'thousand kernel' weight. Backward crops, and spring crops with a faster development than winter crops, suffer most. Crops are particularly susceptible at the early bud stages and become less and less sensitive as the plants develop (Nilsson, 1988, 1994). At flowering, control measures are rarely implemented (Nilsson, 1987). Winter rape is often past the susceptible bud stages before the main invasion of beetles to the crop occurs. Pollen beetles are of particular importance in countries, such as Finland and Sweden, where spring rape dominates. Although *M. aeneus* is the dominant species on oilseed rape, *Meligethes viridescens* also occurs, but, compared with *M. aeneus*, requires higher temperatures for both oviposition and development (Fritzsche, 1957). In Scandinavia and central Europe, other species of *Meligethes* on oilseed rape are usually present in only small numbers (Nolte & Fritzsche, 1952; Jurek, 1972; Karltorp & Nilsson, 1981).

Stem borers

Cabbage stem flea beetle

Cabbage stem flea beetle (Fig. 2.4) is an important pest of winter rape; spring-sown crops are not attacked. The pest is widely distributed in northern parts of Europe with a maritime climate. It also occurs in Asia, Canada and North Africa (Balachowsky, 1963; Bonnemaison, 1965). At least in central parts of northern Europe it has a cyclic appearance with peaks in populations at intervals of about seven years. Its importance as a pest in the British Isles, however, has declined following the extensive use of autumn-applied pyrethroid insecticides.

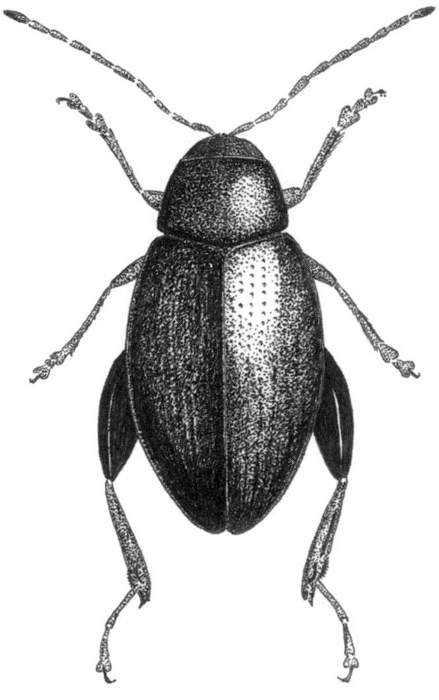

Fig. 2.4 Cabbage stem flea beetle (*Psylliodes chrysocephala*) (×12).

Young adults appear in the summer at about harvest time. Shortly afterwards they enter a period of aestivation, reappearing and then invading newly germinating autumn-sown rape crops from late August or early September onwards. Adults on the young winter crop are usually most abundant during September and October, but if conditions are favourable individuals may survive in decreasing numbers throughout the winter period. Adults graze the tissue of host plants and can cause considerable damage to the leaves and cotyledons of slowly growing plants. Eggs are laid in the soil, close to host plants, mainly during September and October. Under favourable weather conditions, oviposition continues throughout the winter up to the spring (Schulz, 1985). Eggs hatch from late September onwards, the duration of incubation varying considerably according to temperature (Bonnemaison & Jourdheuil, 1954; Alford, 1979). The young larvae bore into the petioles, where they form extensive galleries. Infestations may also spread to the stems and growing points. The larvae feed gregariously and heavy infestations can result in considerable crop damage (Graham & Alford, 1981; Ballanger, 1984; Lane & Cooper, 1989; Nilsson, 1990; Winfield, 1992). Feeding continues throughout the winter period and is completed in the spring, in some years as late as the end of May or in early June. Fully fed larvae (then in their third and final instar) pupate in the soil, a few centimetres from the surface. There is just one generation annually. For further details see, for example, Günthardt (1949), Dosse (1951) and Ballanger (1984).

Cabbage stem weevil

Adults of this generally common and widely distributed weevil (Fig. 2.5) appear from the bud stages of winter rape onwards. They then invade brassicaceous plants, including oilseed rape. Eggs are laid in small groups, preferably in the underside of petioles and,

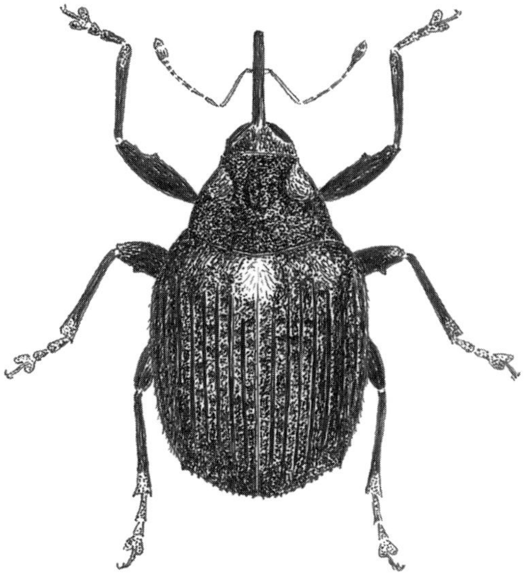

Fig. 2.5 Cabbage stem weevil (*Ceutorhynchus pallidactylus*) (×15).

rarely, in the young shoots, usually in April or May. Initially, larvae feed gregariously in the petioles; they then invade the stems and lateral shoots, often forming extensive galleries. Larvae are fully grown in three to five weeks. They then vacate the host plant and pupate in the soil. There is a single generation annually. Larval infestations cause distortion of tissue and loss of vigour, such damage being especially severe on spring rape. On winter rape, larvae are most often found in the main and lateral shoots and direct damage caused is rarely of any significance. However, the incidence of stem canker (*Leptosphaeria maculans* – anamorph: *Phoma lingam*) can be enhanced substantially by feeding wounds produced by adults and larvae (Broschewitz & Daebeler, 1987). At least in Sweden, larvae feed only in the lower part of both winter and spring oilseed rape, and economic damage is rare (C. Nilsson, *unpublished*). For further details see Graham & Gould (1980) and Kelm & Klukowski (2000). Adults emerge from pupae in the summer and, prior to entering hibernation, can cause extensive damage to vegetable brassicas, such as cabbages, growing in the vicinity of oilseed rape crops by grazing on the underside of the leaf veins and petioles. Cabbage stem weevil is often considered to be of minor importance in winter rape (Broschewitz & Daebeler, 1987; Winfield, 1992); however, in spring-sown rape and in backward, poorly developed winter rape the plant growth, leaf area and seed yield can be reduced significantly (Graham & Gould, 1980; Palosz, 1980; Kelm & Walczak, 1998).

Rape stem weevil

This continental species is one of the most damaging pests of winter rape in continental Europe, especially in Austria, France, Germany, Poland and Switzerland. Adults occur in February and March. They then invade host plants to deposit eggs singly in the developing stems, close to a terminal bud during the period of stem elongation (March to April). The process of egg deposition (as a consequence of the deposition of secretions by the egg-laying female) later results in the twisting and splitting of tissue, followed by distortion and considerable disruption to growth. Infested tissue is often invaded by secondary organisms, especially the fungal pathogen *Leptosphaeria maculans* (canker). Eggs hatch one to two weeks after deposition and the larvae then feed within the pith for three to five weeks before eventually dropping to the ground and pupating in the soil. The adult stage is reached about a month later, but, unlike those of related species, they remain within their earthen chambers and do not emerge until the following spring. For further details see Ballanger (1983), Büchi (1988, 1996), Debouzie & Ballanger (1993) and Lerin (1993).

Rape winter stem weevil

This weevil is a locally and sporadically important pest of winter rape; infestations are most frequently reported in England, France, Germany and Switzerland. Aestivated adults are active in the autumn, and may then invade emerging oilseed rape crops. Eggs are laid mainly from September to mid-November. Following egg hatch, larvae bore into the young plants where they feed gregariously in the petioles; later, they invade the stems and developing crowns. Heavy infestations cause considerable damage, resulting in distortion, stunting and even the death of plants. Larval development continues throughout the winter, and pupation takes place in the spring. New adults appear in mid-May or early June and then aestivate. There is just one generation annually. For further details see Büchi (1986) and Lane & Cooper (1989).

Miscellaneous pests

Aphids

In general, cabbage aphid is of little or no direct importance on established winter rape, although, in spring, it has recently become an important pest of winter rape in parts of France (Y. Ballanger, *pers. comm.*), and it is often abundant on racemes after flowering. Autumn infestations can have a deleterious effect on the establishment of recently drilled crops, especially when the aphids arrive shortly after crop emergence. However, the aphid is more often troublesome on spring-sown crops. The heads of infested plants are discoloured and distorted, and liberally coated with mealy wax; also, infested seed pods (which often assume a purplish colour) fail to develop properly. Although such damage is particularly associated with spring rape, mild winters may enable colonies to survive the winter and cause such symptoms on winter crops – this has occurred frequently in recent years in England. Cabbage aphid overwinters in the egg stage or, under favourable conditions, as wingless aphids (apterae). Colonies build up rapidly on infested plants during the spring and summer, winged aphids (alatae) being produced from about May onwards. As a primary pest, cabbage aphid is of particular significance on vegetable brassicas, such as broccoli, Brussels sprout and cabbage rather than on oilseed rape.

On brassica crops such as oilseed rape, peach/potato aphid is of significance only as a potential vector of virus diseases, such as turnip yellows luteovirus (TuYV). The aphid itself can cause the same damage to winter rape seedlings as cabbage aphid, and plants infested with TuYV may lack vigour and produce lower yields (Hardwick *et al.*, 1994). Cabbage aphid also transmits TuYV, but is more important as a vector of cauliflower mosaic virus (CaMV) and turnip mosaic virus (TuMV). These two viruses, although less common than the largely symptomless TuYV, produce characteristic symptoms of infection (including leaf distortion and stunted growth), usually located in small patches throughout the crop. Yield losses from heavily infected plants can be considerable. For further details see Graichen & Rabenstein (1996); Ellis *et al.* (1999) and Schliephake *et al.* (2000).

In France, turnip aphid [*Lipaphis pseudobrassicae* (Kaltenbach) = *L. erysimi* (Kaltenbach)] has recently become an autumn pest of oilseed rape (Y. Ballanger, *pers. comm.*). This species is very similar to cabbage aphid. Colonies develop on the underside of leaves, infested foliage curling and turning yellow; in spring and summer, infestations may also occur on the inflorescences.

Baris *weevils*

Species of *Baris* appear to be of only minor significance on oilseed rape crops (e.g. in Belgium and, particularly, France). In France, *B. coerulescens* is the main species of *Baris* associated with oilseed rape; there are also frequent references to *B. laticollis* Marsham, particularly in the older literature. Nowadays, the former species predominates and is widely distributed; the latter species is less common, and also less damaging (Y. Ballanger, *pers. comm.*). Other species cited as pests include *B. chorizans* Germar (in France, Germany, Poland and the former Yugoslavia). Several species of *Baris* (but not *B. coerulescens*) are present in the UK; however, none has been recorded attacking oilseed rape crops.

In favourable areas, adults of *Baris* appear from mid-March or early April onwards. Eggs are laid in basal parts of stems and they hatch after about two weeks. Larvae then feed within the stems of host plants for up to two months before eventually pupating *in situ*. Second-generation adults appear in the summer and these eventually deposit eggs; these eggs give rise to larvae that feed within the stems of brassicaceous plants from late summer or early autumn onwards. The lifecycle is completed in the following spring. In less favourable areas, there is just one generation annually. The biology of *B. coerulescens* is different from other *Baris* species, as the larvae feed within the roots and only one generation occurs per year, irrespective of conditions (Koubaiti, 1992).

Cabbage flea beetles

Flea beetles in the genus *Phyllotreta* are well-known pests of brassicaceous plants, especially vegetable brassicas, and various species have also been reported to attack oilseed rape crops. The most frequent on oilseed rape are *Phyllotreta atra* (Fabricius), *P. nemorum* (Fig. 2.6), *P. nigripes* (Fabricius) and *P. undulata* (Kutschera) in Denmark and Germany (Nielsen, 1977; Sommer, 1981); *P. atra*, *P. consobrina* (Curtis), *P. cruciferae* (Goeze), *P. nemorum*, *P. nigripes* and *P. undulata* in France and the UK (Jourdheuil, 1960; Gratwick, 1992); *P. undulata* in Sweden (Ekbom, 1990); *P. atra* and *P. undulata* in Switzerland (Günthardt, 1949; Sommer, 1981); *P. cruciferae* and *P. striolata* (Fabricius) in North America (Lamb, 1989).

Details of the lifecycle and habits vary slightly according to species but are exemplified by that of *P. cruciferae*, as follows. Adults appear in the spring from April onwards, often then migrating to spring-sown oilseed rape crops. They feed on the young green tissue of host plants, making numerous pit-like punctures that subsequently develop into holes as the tissue enlarges. The feeding of adults on cotyledons, young leaves and stems

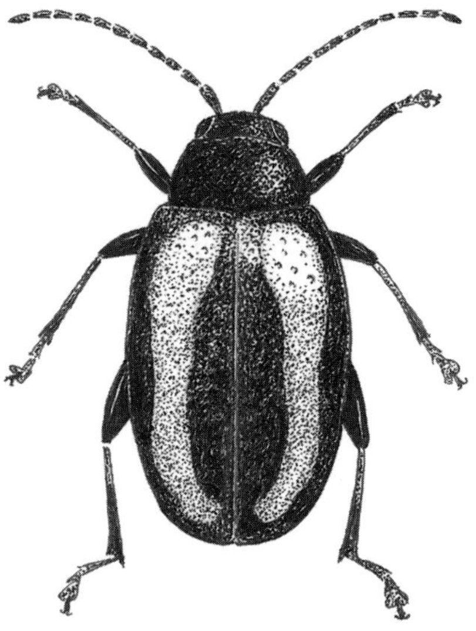

Fig. 2.6 Large striped flea beetle (*Phyllotreta nemorum*) (×18).

is greatly enhanced by warm and dry weather in spring, and substantial plant losses or complete failure of the crop has been observed under such conditions (Lamb, 1989). Eggs are eventually laid in batches in the soil close to host plants. After egg hatch the larvae attack the roots to feed externally. Pupation eventually takes place in the soil, and new adults emerge from late June onwards. There is a single generation annually. Larvae of some species of *Phyllotreta*, e.g. those of *P. nemorum*, mine within the leaves, petioles and stems of host plants (Sommer, 1981), but this feeding is not of economic importance on oilseed rape. Damage to spring oilseed rape seedlings in dry areas can be of great economic importance. Leaf damage by adults, up to the 4- to 5-leaf stage (even prior to plant emergence) decreases plant stand considerably. Larval feeding, however, is of no importance on oilseed rape.

Cabbage root fly

Cabbage root fly is a notorious and widely distributed pest of vegetable brassicas. It can also attack oilseed rape, but its presence is often overlooked. In recent years, early sowings of winter rape (with seedlings emerging in late August and early September) have been increasingly subject to damage. The adult flies deposit eggs close to emerging seedlings and larvae then invade the roots, causing stunting and wilting. Infestations on winter rape are especially serious in dry conditions when growth of seedlings is slow. Although the pest has several generations in a year, winter rape is rarely attacked.

Turnip sawfly

This is a sporadic pest of brassicaceous crops such as swede and turnip. Attacks may also occur on oilseed rape. Adults are present from May to October, there being up to three generations annually. Eggs are laid singly in the leaves of host plants, usually several per leaf. The eggs hatch about a week later. At first the larvae graze on the underside of leaves, leaving the upper epidermis undamaged. Later, they feed externally and will rapidly reduce the foliage to a skeleton of major veins. Fully grown larvae pupate in the soil, each in a silken cocoon, and adults are produced two to three weeks later. Larvae of the final brood overwinter in their cocoons and pupate in the spring. Numbers of this widely distributed pest declined noticeably in north-western Europe during the late nineteenth century. However, populations recovered from the 1940s onwards, and problems have since occurred on oilseed rape, especially in central Europe.

Oilseed Rape Pests – A Simplifiled Key to Adults [Note 1]

1. Wingless or with two pairs of membranous wings 2
– Fore wings membranous or modified as hardened covers (elytra); if fore wings membranous, then hind wings absent 4
2. Body delicate, less than 3 mm long; hind end of abdomen with a pair of siphunculi and with a distinct cauda; wingless or fully winged. 3
– Body robust, 6–8 mm long; mainly yellow to reddish-yellow; fore wings and hind wings fully developed; females with a saw-like ovipositor . . . . **turnip sawfly (*Athalia rosae*)**
3. Body greyish-green, more-or-less coated in mealy wax; siphunculi barrel-shaped; typically inhabiting large, dense colonies; cauda broadly triangular . . . . **cabbage aphid (*Brevicoryne brassicae*)** [Note 2]
– Body mainly green to olive-green, sometimes mottled with pink or red; siphunculi elongate; typically present in only small colonies; cauda elongate . . . . **peach/potato aphid (*Myzus persicae*)**
4. Fore wings membranous; hind wings absent; body delicate, greyish to reddish, 1.2–1.5 mm long; antennae many-segmented, moniliform and with distinct whorls of hairs; wing venation reduced and indistinct; female with a long, protrusible ovipositor . . . . **brassica pod midge (*Dasineura brassicae*)** [Note 3]
– Fore wings hardened and forming elytra 5
5. Front of head with a more-or-less conspicuous snout (rostrum) which bears the antennae and mouthparts; antennae geniculate, the terminal segments forming a club 6
– Head without a rostrum; antennae not geniculate, either clubbed or filiform 7
6. Body more-or-less barrel-shaped and devoid of scales, most species 3–4 mm long; elytra metallic blue-black . . . . ***Baris spp.***
– Body more-or-less quadrate and often at least partly clothed in hair-like scales, 2–4 mm long; rostrum relatively long and trunk-like . . . . ***Ceutorhynchus*** [Note 4]
7. Hind legs greatly enlarged and modified for jumping; antennae filiform 8
– Hind legs normal; terminal segments of antenna forming a distinct club; body 1.5–2.5 mm long. . . . . ***Meligethes spp.***
8. Antennae 10-segmented; body 4–5 mm long; elytra usually metallic greenish-black or bluish-black but sometimes bronzy . . . . **cabbage stem flea beetle (*Psylliodes chrysocephala*)**

– Antennae 11-segmented; body 1.5–3.0 mm long; elytra black or metallic greenish-black (sometimes with two conspicuous yellow longitudinal stripes) . . . . ***Phyllotreta spp.***

Note 1. *Delia radicum*, *Phytomyza rufipes* and *Scaptomyza* are not included, as adults of these pests are rarely seen in the crop; further, they would in any case need to be distinguished from the many other species of Diptera that occur on the crop, most of which are of no significance as either pests or predators. Taxonomic details of predatory Diptera are included in Chapter 13 of this monograph.

Note 2. Turnip aphid (*Lipaphis pseudobrassicae*) may be distinguished from cabbage aphid by its tongue-like cauda.

Note 3. This description does not allow separation from other midge species, several of which may occur on the crop (although not as pests).

Note 4. Species of *Ceutorhynchus* may be separated as follows:

1. Legs greyish to black. 2
– Legs at least partly reddish 3
2. Body 2.2–3.0 mm long; mainly lead-grey and relatively narrow-bodied, with two rows of whitish hairs between longitudinal furrows on the elytra. . . . **cabbage seed weevil (*C. assimilis*)** [Note 5]
– Body 3.2–4.0 mm long; greyish, with three rows of whitish hairs between the longitudinal furrows on the elytra . . . . **rape stem weevil (*C. napi*)**
3. Body 2.5–3.5 mm long; greyish-brown, with a whitish patch of hairs just behind the thorax; legs reddish. . . . . **cabbage stem weevil (*C. pallidactylus*)**
– Body 2.2–3.5 mm long; mainly shiny black, with a pale yellowish mark on the shoulder of each elytron; legs partly reddish. . . . . **rape winter stem weevil (*C. picitarsis*)**

Note 5. Adults of cabbage seed weevil may be distinguished from those of the smaller *C. floralis* (Paykull) by their 7-segmented antennae; in *C. floralis* the antennae are 6-segmented. The latter species is associated with wild brassicaceous hosts (such as *Capsella bursa-pastoris*) and has no pest status; however, it is often found on flowering oilseed rape crops and can be mistaken for *C. assimilis*, especially when present in water-trap samples.

Oilseed Rape Pests – A Simplified Key to Larvae

1. Three pairs of true legs present on thorax 2
– True legs absent 4
2. Eight pairs of abdominal prolegs (unlike those of lepidopterous larvae: without crochets); head distinct and orientated more-or-less vertically; body greenish-grey and later velvet-black, up to 18 mm long; feeding externally on foliage. . . . . **turnip sawfly (*Athalia rosae*)**
– Abdomen without prolegs; head orientated more-or-less horizontally and partly sunken into the prothorax; body whitish to creamish-white; head, thoracic legs, pinacula and prothoracic and anal plates black or brownish; not feeding externally on foliage. 3

3. Up to 5 mm long; each segment with two to three brownish plates; yellowish gut contents often visible; feeding in buds and flowers. . . . . ***Meligethes* spp.**
– Up to 8 mm long; anal plate with two upwardly curved hooks; feeding within shoots, petioles, etc. in autumn, winter and spring . . . . **cabbage stem flea beetle (*Psylliodes chrysocephala*)**
4. Whitish to creamish-white, with a distinct, yellowish or brownish head capsule; body arched 5
– Body whitish and more-or-less maggot-like; head indistinct. 9
5. Feeding, usually singly, within developing pods. . . . . **cabbage seed weevil (*Ceutorhynchus assimilis*)**
– Feeding, often in groups, in roots, stems, shoots, petioles or growing points . . . 6
6. Up to 6 mm long; feeding during late autumn, winter and spring. . . . . **rape winter stem weevil (*Ceutorhynchus picitarsis*)** [Note 1]
– Up to 6 or 8 mm long; feeding during spring and early summer 7
7. Feeding inside roots and stem bases during the summer; head capsule with upper half of clypeus with a median line and clypeal margin irregular . . . . ***Baris* spp.**
– Feeding inside stems and shoots in spring and early summer; head capsule with outline of clypeus more or less triangular and the margin straight 8
8. Up to 6 mm long; body whitish . . . . **cabbage stem weevil (*Ceutorhynchus pallidactylus*)** [Note 1]
– Up to 8 mm long; body creamish-white . . . . **rape stem weevil (*Ceutorhynchus napi*)** [Note 1]
9. Feeding gregariously within developing pods; body hyaline and translucent to whitish, up to 2 mm long [Note 2] . . . .**brassica pod midge (*Dasineura brassicae*)**
– Not feeding within pods 10
10. Mining within leaves or petioles; posterior spiracles borne on more-or-less distinct processes 11
– Tunnelling within roots and lowermost parts of plants; up to 10 mm long; creamish-white; head deeply retracted into prothorax, with two dark-pigmented mouth-hooks; hind end of body with numerous papillae; posterior spiracles large and 3-pored, not borne on processes . . . . **cabbage root fly (*Delia radicum*)**
11. Forming elongated mines within leaves and petioles; body up to 6 mm long, whitish and more or less opaque; posterior spiracles many-pored and borne on a pair of short, but distinct, posteriorly directed stalks. . . . . **cabbage leaf miner (*Phytomyza rufipes*)**
– Forming large, whitish blotch mines in leaves; body up to 5 mm long, translucent, with white tracheae clearly visible, and with several broad, posterior tubercles; posterior spiracles located at the tips of a pair of close-set, pointed, cone-like processes . . . . ***Scaptomyza* spp.**

Note 1. See Dosse (1952) for further details.
Note 2. Pinkish or reddish midge larvae may also occur inside developing pods, but these are fungal feeders and of no economic importance. Parasitoid larvae (e.g. those of the pteromalid wasp *Trichomalus perfectus*) may also occur inside pods – see Chapter 6.

Pest Management

It is not the purpose here to provide advice on how to 'control' pests of oilseed rape but to discuss current pest management strategies in general terms. Recent reviews of pest management on oilseeed rape have been provided by Winfield (1992) and by Lane & Gladders (2000), and a decision-support system promulgated by Johnen & Meier (2000); see also Anon. (1996). In addition, Alford (2000) recently summarized the prospects for biological control of oilseed rape pests in Europe. Integrated crop management (ICM) and integrated pest management (IPM) in oilseed rape are considered in Chapter 3.

In general, chemical control measures on winter oilseed rape are most frequently required (a) in the autumn against cabbage stem flea beetle and against aphid virus vectors; and (b) in the spring against pollen beetle, ceutorhynchid weevils and brassica pod midge. In both winter and spring rape, application of an insecticide against one pest may have activity against another. However, oilseed rape crops rarely, if ever, require protection against all of the above-mentioned pests; control thresholds (based on estimates of infestation levels which, if not controlled, will result in yield losses equal to the cost of treatment) have been developed for all oilseed rape pests of major importance (see Tables 2.3–2.7).

Aphids

Until relatively recently, insecticide sprays were rarely used to combat aphids on oilseed rape crops, although a specific aphicide e.g. (pirimicarb) was sometimes applied as a spray against cabbage aphid. Such treatments were usually required (or thought to be required) in the spring when many aphid-infested plants were found in crops. Milder winters have certainly resulted in the increased significance of cabbage aphid in the spring on winter rape in England; however, in general, spraying is more often justified on spring rape than on winter rape (Ellis *et al.*, 1999).

On oilseed rape, aphids (in this instance cabbage aphid, peach/potato aphid and turnip aphid) are of greater significance as potential vectors of viruses, and nowadays in many winter rape crops an autumn treatment with a pyrethroid insecticide has become a routine. Coincidentally, such applications have probably had an impact on cabbage stem flea beetle, populations of which appear in some regions to have declined in recent years.

Brassica pod midge

Appropriate crop management is often advocated to overcome this weak-flying pest: namely, the growing of oilseed rape in a wide rotation, and the avoidance of growing winter and spring rape in the same locality; however, such crop management practices are by no means universally adopted by farmers. With regards chemical control, few or no pesticides are recommended for the direct use against this pest. However, on midge-vulnerable crops chemical control of cabbage seed weevil (a major precursor for subsequent midge attacks) is often advocated and treatment thresholds for the weevil lowered where brassica pod midge is perceived as a problem (see below).

Cabbage seed weevil

The importance of this pest can be overrated and sprays are often applied unnecessarily, especially on winter rape. As exemplified in the British Isles, for many years, sprays of the organophosphate (OP) triazophos were applied immediately after flowering, virtually as a routine; in the late 1980s the OP phosalone was also available as an end-of-flowering treatment (Gladders *et al.*, 1989). In England and Wales over 72,000 ha of oilseed rape were treated with triazophos in 1991, most of which was targeted against cabbage seed weevil. From 1992, however, the recommendation for the use of triazophos against this pest was withdrawn. Today, recommended treatments include either a pyrethroid during flowering or phosalone at the late-flowering stage.

For many years, crop extension workers have advocated that adult numbers on the crop should exceed a specific threshold before a spray becomes justified (see Table 2.3); this threshold is halved on farms where brassica pod midge (see above) is a regular problem (e.g. Lane & Walters, 1993; Garbe *et al.*, 1996; Lane & Gladders, 2000). However, relatively few commercial crops in the British Isles ever exceed these thresholds; the same is true elsewhere in Europe – in Germany, for example, fields exceeding the threshold are estimated at no more than from 10% to 30% in various years and regions (B. Ulber, *unpublished*).

Flea beetles

Seed treatment has been the most used control measure against flea beetles. The former use of the persistent organochlorine gamma-HCH has now been abandoned and in parts of Europe replaced with carbosulfan and other insecticides, often applied in dust-free coatings. This is still the most used method against cabbage flea beetles (i.e. *Phyllotreta* spp.), as attacks can come very suddenly and before the plants have emerged. More and more, seed treatments against cabbage stem flea beetle are being replaced by pyrethroid sprays.

Infestations of cabbage stem flea beetle may warrant early treatment against adults or a later treatment against the young larvae, and sometimes both. To keep treatments to the minimum, recommendations to spray against larvae are usually made only if larval numbers exceed specific thresholds (typically five or more larvae per plant in the autumn/early winter period, but three larvae per plant in the case of backward, poorly

Table 2.3 Examples of pest control thresholds advocated for cabbage seed weevil (*Ceutorhynchus assimilis*) on winter rape.

Country (source)	Threshold [Note 1]
Austria (BFL, 2000)	0.5 to 1 weevils/plant
Denmark (Nielsen, 2000)	6 weevils/plant
France (Pilorgé *et al.*, 1997)	0.5 weevils/plant
Germany (Garbe *et al.*, 1996)	0.5 to 1 weevils/plant
Poland (Anon., 2000)	4 weevils/25 plants
Sweden (Twengström, 2000)	0.5 to 1 weevils/plant
Switzerland (Büchi, 1988)	0.5 to 1 weevils/plant
UK (Lane & Gladders, 2000)	2 weevils/plant

Note 1. Lower values are advocated where brassica pod midge (*Dasineura brassicae*) is a regular problem.

Table 2.4 Examples of pest control thresholds advocated for cabbage stem flea beetle (*Psylliodes chrysocephala*) on winter rape.

Country (source)	Threshold (in autumn/winter)
Austria (BFL, 2000)	2 to 3 adults/m^2
Denmark	No threshold established
Finland	Pest not present
France (Pilorgé *et al.*, 1997)	2 to 3 larvae/plant [Note 1]
Germany (Garbe *et al.*, 1996)	3 to 5 larvae/plant
Poland (Z. Klukowski, *pers. comm.*)	1 adult/m of plant row [Note 2]
Sweden (Twengström, 2000)	1 to 2 adults/m of row
Switzerland (Anon., 1997)	>2 larvae/plant
UK (Lane & Gladders, 2000)	5 larvae/plant [Note 3]

Note 1. A threshold of 3 plants in 10 with adult feeding punctures is sometimes advocated, to avoid excessive adult damage, to reduce egg-laying potential and, hence, to lower subsequent larval infestations.
Note 2. Threshold for *Psylliodes* and *Phyllotreta* spp. combined.
Note 3. Equivalent to 60% of leaf petioles with feeding scars.

established crops) (see Table 2.4); in areas at particular risk from cabbage stem flea beetle, sowing crops before mid-August should be avoided (Lane & Gladders, 2000). In the Scandinavian area the threshold could be as low as one larva per plant (Nilsson, 1990), but this varies significantly between years. There is apparently no relationship between adult numbers and subsequent larval infestations (Hoßfeld, 1993), but autumn temperatures can be used to help assess the need for and timing of treatments (Johnen & Meier, 2000); see also Alford (1979).

Pollen beetle

In continental Europe, pollen beetle is generally considered an important pest on both winter and spring rape. In the British Isles, however, its impact as a pest on winter rape is considered limited and the use of an insecticide is rarely advocated (Lane & Walters, 1993; Lane & Gladders, 2000). Control thresholds used in several European countries are summarized in Table 2.5. Depending on the registration of insecticides in various countries, a few OPs and several pyrethroids are available for chemical control of the pest. Recently, the intensive use of insecticides has provoked resistance to pyrethroids in parts of Denmark, France and Sweden.

Stem weevils

Advocated treatment thresholds for cabbage stem weevil and rape stem weevil are summarized in Tables 2.6 and 2.7, respectively.

In continental Europe (e.g. Germany and Poland), cabbage stem weevil can be damaging to both winter and spring rape. Under UK conditions, however, it is rarely a problem on the winter crop; nor is cabbage stem weevil considered of economic importance on winter rape in France or Switzerland. Various pyrethroids are available as treatments against adults, applied at the green-bud to yellow-bud stage.

In Germany, with regards cabbage stem weevil and rape stem weevil, Büchs (1998) has advocated changes to conventional control practices as follows:

Table 2.5 Examples of pest control thresholds advocated for pollen beetle (*Meligethes aeneus*).

	Winter rape		Spring rape	
Country (source)	GS 51	GS 52–53	GS 51	GS 52–53
Austria (BFL, 2000)	–	4 to 5	No thresholds established	
Denmark (Nielsen, 2000)	3	6	1	3
Finland (Hokkanen *et al.*, 1988)	–	–	1	2 to 3
France (Pilorgé *et al.*, 1997)	1	2 to 3	Crop rarely present	
Germany (Garbe *et al.*, 1996)	3 to 4	7 to 8	1	2 to 3
Poland (Anon., 2000)	1	3 to 5	1	3 to 5
Sweden (Twengström, 2000)	2 to 3	3 to 4	0.5 to 1	1 to 2
Switzerland (Anon., 1997)	5	–	Crop not present	
UK (Lane & Gladders, 2000)	–	15	–	3

GS = growth stage (codes according to Lancashire *et al.*, 1991).

Table 2.6 Examples of pest control thresholds advocated for cabbage stem weevil (*Ceutorhynchus pallidactylus*) on winter rape.

Country (source)	Threshold
Austria (BFL, 2000)	10 weevils/yellow water trap in three days [Note 1]
Denmark	No threshold established
Finland	No threshold established
France (Pilorgé *et al.*, 1997)	No threshold established
Germany (Garbe *et al.*, 1996)	10 weevils/yellow water trap in three days [Note 1]
Poland (Z. Klukowski, *pers. comm.*)	20 weevils/yellow water trap in three days or 6 weevils/25 plants
Sweden	No threshold established
Switzerland (Büchi, 1988)	No threshold established
UK	No threshold established

Note 1. *Ceutorhynchus napi* (Table 2.7) and *C. pallidactylus* not monitored separately.

Table 2.7 Examples of pest control thresholds advocated for rape stem weevil (*Ceutorhynchus napi*) on winter rape.

Country (source)	Threshold
Austria (BFL, 2000)	10 weevils/yellow water trap in 3 days [Note 1]
Denmark	Pest not present
Finland	Pest not present
France (Pilorgé *et al.*, 1997)	No threshold established [Note 2]
Germany (Garbe *et al.*, 1996)	10 weevils/yellow water trap in 3 days [Note 1]
Poland (Anon., 2000)	10 weevils/yellow water trap in 3 days, or 2 to 4 weevils/25 plants
Sweden	Pest not present
Switzerland (Büchi, 1988)	1 puncture/plant, or 40% to 60% of plants with punctures
UK	Pest not present

Note 1. *Ceutorhynchus napi* and *C. pallidactylus* (Table 2.6) not monitored separately.
Note 2. Treatment (using a pyrethroid) advised within 8 days following first captures of adults.

- owing to differences in the pest phenology, the practice amongst farmers to combine the two weevil species into a single threshold of 10 weevils/trap in a three-day period should be discontinued;
- in regions where rape stem weevil occurs commonly, an insecticide application immediately after the damage threshold of 10 weevils/yellow water-trap in a three-day period is exceeded is still required;
- in regions where cabbage stem weevil predominates (mainly north-western Germany), a single application of a pyrethroid insecticide is necessary two or three weeks after the first major flight activity of cabbage stem weevil is recorded by yellow traps, even if the action threshold is exceeded several times – i.e. peaks of flight activity in early spring should be ignored.

This scenario, however, requires further evaluation before it can be regarded as commercially acceptable.

Owing to its local and sporadic nature, treatments against rape winter stem weevil are rarely required, but application of a pyrethroid spray is usually advocated if larvae infest 10% or more of plants in the late autumn or early winter (Lane & Gladders, 2000).

Natural Enemies of Oilseed Rape Pests

Oilseed rape pests are subject to attack by a wide range of beneficial organisms or natural enemies (parasitoids, predators and pathogens), many of which help to limit or reduce pest populations. Most parasitoids associated with oilseed rape pests are members of the order Hymenoptera, notably members of the families Braconidae, Ichneumonidae and Pteromalidae. Their effects on specific pests or specific pest groups are considered in detail in Chapters 4–8. Non-hymenopterous parasitoids, such as tachinid flies (order Diptera: family Tachinidae), either are of little significance in oilseed rape crops or attack only minor pests (see Chapter 9). Parasitoids often show some degree of host specificity. Predators, however, tend to be general feeders. The more important groups of predators in oilseed rape fields include ground beetles (order Coleoptera: family Carabidae), rove beetles (order Coleoptera: family Staphylinidae), various dipterans, e.g. dance flies (order Diptera: family Hybotidae) and spiders (see Chapter 12). Their impact on oilseed rape pests is considered in detail in Chapter 16. Pathogens (namely entomopathogenic fungi and nematodes, bacteria and viruses) are often collectively termed BCAs (biological control agents). Their significance in oilseed rape crops is largely unknown (see Chapter 17).

Pesticide Side-Effects on Beneficial Organisms

Side-effects of pesticides used in oilseed rape crops depend on two important factors: (a) type of pesticide, and (b) how the pesticide is used. Most pesticides have effects on beneficial organisms when tested *in vitro*, but this is not always the case when the pesticides are used on the farm (i.e. under field conditions).

In vitro tests

It is evident that most insecticides have effects on parasitoids and predators (see Chapter 15) and that fungicides have effects on pathogens (see Chapter 17). Many herbicides are also known to have biological effects on organisms other than higher plants. Members of the European IOBC/wprs Working Group on pesticide side-effects, and other scientists (Aukema *et al.*, 1990; de Clercq *et al.*, 1991; Hassan, 1992; Heimbach, 1992; Lewis *et al.*, 1992; Wiles & Jepson, 1992; Samsoe-Petersen, 1993; Hassan *et al.*, 1994; Oomen *et al.*, 1994; Thacker *et al.*, 1995; Çilgi *et al.*, 1996; Haskell & McEwen, 1998; Hassan *et al.*, 1998; Sterk *et al.,* 1999) have developed and standardized methods for the *in vitro* testing of pesticide effects on different organisms, mostly parasitoids and predators. Most pesticides tested are insecticides, but some fungicides (Heimbach, 1988) and herbicides (Samsoe-Petersen, 1995) have also been evaluated. Groups of pesticides with the same mode of action, e.g. pyrethroids, are also known to differ only slightly in their side-effects. Many of the test organisms used are important natural enemies in oilseed rape ecosystems. These tests are relevant for most crops, as the difference between the chosen indicator organisms, and other parasitoids and predators, can be expected in most cases to be small (Birnie *et al.*, 1998). There is hardly any testing procedure for pathogens and almost no papers dealing with this topic (but see Hassan *et al.*, 1994; Sterk *et al.*, 1999).

Field tests

The evaluation of pesticide side-effects under field conditions is much more complicated, and must take into account factors such as application technique (Dewar *et al.*, 1990), dose and formulation (Heimbach & Abel, 1991; Wiles & Jepson 1995; Longley & Jepson, 1997), spatial and temporal factors (Jepson & Thacker, 1990; Duffield & Jepson, 1994; Longley & Izquierdo, 1994; Duffield *et al.*, 1996; Longley *et al.*, 1997; Berraondo *et al.*, 1998), ecological and behavioural effects and population dynamics of the parasitoids and predators (Jepson, 1989). The effects of different farming systems, soil cultivation and other agronomic factors should also be considered (Everts *et al.*, 1989; Basedow, 1995; Volkmar, 1996). Ploughing, as an example, could be as detrimental to the spider population within an arable field as an insecticide spray. Taking the long-term effects (Inglesfield, 1989; Krooß & Schaefer, 1998) into account is even harder, and involves parameters such as on-farm landscape structure and biodiversity (Körner, 1990; Jepson & Sherratt, 1991; Sunderland, 1992).

Considerable work in this direction has been done in cereals (Everts, 1990; Al Hussein & Wetzel, 1993a, 1993b, 1993c), initially in winter wheat (Casteels & de Clercq, 1990; Al Hussein *et al.*, 1990; Cocquempot *et al.*, 1991; Al Hussein *et al.*, 1991; Giller *et al.*, 1995), and only very few papers deal with oilseed rape crops (Winfield, 1963; Lehmann, 1965b; Coutin & Coulon, 1966; Kaczmarek, 1987; Raw, 1989; Ulber *et al.*, 1990; Goltermann & Pelz, 1995; Palosz, 1995a, 1995b; Seta *et al.*, 1995). In spite of these studies, there is still a need to investigate (in field trials) the selectivity of different pesticides to the key beneficial organisms in oilseed rape crops.

In oilseed rape, pesticides are applied in several ways:

- as seed treatments (Lehmann, 1964; Schmuck, 1992);
- as granules (Finlayson, 1979; Büchs *et al.*, 1989; Purvis, 1996);
- as a spray, either as a whole-field or a part-field application.

Seed treatments and granules could have side-effects. However, owing to the limited volume of space affected by the application, these effects are limited to affecting mostly adult predators or their larvae and, generally only indirectly, parasitoids.

Side-effects of pesticide spraying depend on the density of the crop stand (i.e. the number of plants/m^2) and crop-plant architecture, the crop growth stage and the time of application. The filter effect of the crop (Raw, 1989; Çilgi & Jepson, 1992; Wiles & Jepson 1995) at later growth stages is important to understand the effects on the epigeal fauna. Also, the coincidence between pest, predator, pathogen, parasitoid and spraying will determine most of the side-effects. The use of pesticides on areas outside the oilseed rape field, or in the crop following the previous year's oilseed rape crop, could be of importance for those parasitoids and predators that hibernate therein. By way of an example: in Sweden, barley is often sown after a crop of spring rape, and an aphid spray in barley can coincide with the emergence of parasitoids of pollen beetle (Nilsson, 1994). Similar effects can be found when insecticides are applied to control aphids in winter wheat grown after winter rape (Nitzsche, 1998).

If the migration of parasitoids into an oilseed rape field takes place at the time of spraying or shortly before or after such treatment, most could be killed (Lehmann, 1965a; Nitzsche, 1998). Many parasitoid species attacking the larvae of stem weevils, pollen beetle, cabbage seed weevil and brassica pod midge are particularly active in the crop during flowering; these can be heavily affected by insecticide sprays applied one to two weeks before or during flowering (Nitzsche, 1998). A spray residue on the crop could also have a lethal effect; however, for some of the most frequently used insecticides on oilseed rape (namely, pyrethroids) there seems to be a repellent effect (Umoru *et al.*, 1996), at least to some of the parasitoids. Predators can often escape such effects, either because they migrate to the field after the application of an insecticide, or because they are in hiding in the soil during the time of application or are on the soil surface (which the pesticide might not reach) – often as another life stage, e.g. as soil-dwelling larvae. For some predators, re-population of a field after application of a pesticide spray can take a considerable time and is a function of the population dynamics (Rzehak & Basedow, 1982).

In the autumn, the crop canopy has no filtering effect on the pesticide penetration, when the crop usually is sprayed with a herbicide and sometimes, also, with an insecticide to combat cabbage stem flea beetle and aphid virus-vectors. This could, of course, have a strongly negative effect on beneficial organisms on the soil surface, e.g. linyphiid spiders (Çilgi & Frampton, 1994).

When parasitoids arrive in an oilseed rape field well after an insecticidal spray has been applied the only effect that can be seen is the functional response of the parasitoid, owing to the lower pest population density created by the spray (Nilsson & Andreasson, 1987). If the parasitoid population is low this could be beneficial to the parasitization level, as the number of eggs a female can produce per unit of time is limited.

Methods to reduce side-effects

Most pesticides have adverse side-effects on natural enemies, which means that there are only two ways to have viable natural control and still use pesticides (when they are needed) and that is to pay attention to the phenology of the most important natural enemies (Lehmann, 1965a; Alford *et al.*, 1994, 1995, 1996; Murchie *et al.*, 1997; Nitzsche, 1998) and to use an application technique that is as much on target and selective as possible.

Conclusions on the side-effects of pesticides

- Most pesticides have adverse side-effects on natural enemies.
- Selectivity tests done *in vitro* and in cereals are in most cases probably also valid for oilseed rape.
- More knowledge is needed on the differences in penetration of a pesticide spray that variations in plant stand and plant architecture produce.
- Application methods for pesticides with higher precision should be developed. Experiments should be done to investigate the phenology of natural enemies and to discover if spray windows can be constructed that might allow the use of certain pesticides without harm to natural enemies.

Acknowledgements

This work was part-funded by the European Commission Concerted Action: *Minimizing pesticide use and environmental impact by the development and promotion of bio-control for oilseed rape pests, FAIR 96-1314.*

References

Alford, D. V. (1979). Observations on the cabbage stem flea beetle, *Psylliodes chrysocephala*, on winter oil-seed rape in Cambridgeshire. *Annals of Applied Biology* **93**, 117–23.

Alford, D. V. (1999). *A Textbook of Agricultural Entomology.* Blackwell Science: Oxford.

Alford, D. V. (2000). Biological control of insect pests on oilseed rape in Europe. *Pesticide Outlook* **11**, 200–2.

Alford, D. V., Murchie, A. & Williams, I. H. (1994). Observations on the impact of standard insecticide treatments on *Trichomalus perfectus*, a parasitoid of seed weevil on winter oilseed rape in the UK. *IOBC/wprs Bulletin* **18**, 122–6.

Alford, D. V., Williams I. H., Murchie A. K. & Walters K. F. A. (1995). *The Status and Potential of Parasitoids of Seed Weevil and Pod Midge on Winter Ooilseed Rape.* HGCA Oilseeds Project Report No. OS14. Home-Grown Cereals Authority: London.

Alford, D. V., Walters, K. F. A., Williams, I. H. & Murchie, A. K. (1996). A commercially viable low-cost strategy for the management of seed weevil population on winter oilseed rape in the UK. *Proceedings of the Brighton Crop Protection Conference – Pests and Diseases* **2**, 609–14.

Anon. (1996). Guideline on good plant protection practice. Rape. *Bulletin OEPP/EPPO Bulletin* **26**, 349–67.

Anon. (1997). Krankheiten und Schädlinge im Raps. *UFA-Revue*, pp. 7–8.

Anon. (2000). *Zalecenia Ochrony Roslin na Lata 2000/01, dotyczace zwalczania chorób, szkodników oraz chwastów róslin uprawnych.* Rosliny Rolnicze, Czesc II, Instytut Ochrony Roslin: Poznan, Poland.

Aukema, B., Berg, J. H. J. van den, Leopold, A., Jagers, G. A. J. M. & Everts, J. W. (1990). A method for testing toxicity of residues of pesticides on a standardized substrate to erigonid and linyphiid spiders. *Journal of Applied Entomology* **109**, 76–80.

Balachowsky, A. S. (ed.) (1962). *Entomologie Appliquée à l'Agriculture. Tome I. Coléoptères. Premier Volume.* Masson: Paris.

Balachowsky, A. S. (ed.) (1963). *Entomologie Appliquée à l'Agriculture. Tome I. Coléoptères. Second Volume.* Masson: Paris.

Ballanger, Y. (1983). Nuisibilité du charancon de la tige du colza. *Phytoma* **384**, 35–7.

Ballanger, Y. (1984). *L´altise d´hiver du colza:* Psylliodes chrysocephala *L.* INRA PG – thése. INRA: Paris.

Basedow, T. (1995). Ackerbauschädlinge, ihre Antagonisten und die Arthropodenvielfalt auf Feldern langfristig unterschiedlich intensiv bewirtschafteter Betriebe – eine vergleichende Übersicht. *Mitteilungen der Deutschen Gesellschaft für Allgemeine und Angewandte Entomologie* **10**, 565–72.

Berraondo, P. E. Z., Morgan, D., Walters, K. F. A., Young, J. E. B. & Lane, A. (1998). Spatial recovery of two species of Carabidae following cumulative pesticide applications in winter wheat. *Proceedings of the Brighton Crop Protection Conference – Pests and Diseases* **2**, 613–8.

BFL (Bundesamt und Forschungszentrum für Landwirtschaft) (2000). *Richtlinien für die Pflanzenschutzarbeit.* BMLF: Wien.

Birnie, L., Shaw, K., Pye, B. J. & Denholm, I. (1998). Considerations with the use of multiple dose bioassays for assessing pesticide effects on non-target arthropods. *Proceedings of the Brighton Crop Protection Conference – Pests and Diseases* **1**, 291–6.

Bonnemaison, L. (1965). Insect pests of crucifers and their control. *Annual Review of Entomology* **10**, 233–50.

Bonnemaison, L. & Jourdheuil, P. (1954). L'altise d'hiver du colza (*Psylliodes chrysocephala* L.). *Annales des Épiphyties* **5**, 345–524.

Broschewitz, B. & Daebeler, F. (1987). Beitrag zur Biologie und Schadwirkung des Gefleckten Kohltriebrüsslers (*C. quadridens* Panz.) an Winterraps. *Nachrichtenblatt Pflanzenschutz der DDR* **41**, 34–7.

Büchi, R. (1986). Biologie und Bekämpfung des Schwarzen Triebrüsslers *Ceutorhynchus picitarsis* Gyll. (Col., Curculionidae). *Anzeiger für Schädlingskunde, Pflanzenschutz und Umweltschutz* **59**, 51–6.

Büchi, R. (1988). Neue Bekämpfungsschwelle für den Rapsstengelrüßler *Ceutorhynchus napi* Gyll. *Mitteilungen der Schweizerischen Landwirtschaft* **35**, 110–7.

Büchi, R. (1996). Eiablage des Rapsstengelrüßlers *Ceutorhynchus napi* Gyll. In: Abhängigkeit von der Stengellänge bei verschiedenen Rapssorten. *Anzeiger für Schädlingskunde, Pfanzenschutz und Umweltschutz* **69**, 136–9.

Büchs, W. (1998). Strategies to control the cabbage stem weevil (*Ceutorhynchus pallidactylus* [Mrsh.]) and the oilseed rape stem weevil (*Ceutorhynchus napi*) by a reduced input of insecticides. *IOBC/wprs Bulletin* **21**, 205–20.

Büchs, W., Heimbach, U. & Czarnecki, E. (1989). Effects of snail baits on non-target carabid beetles. *British Crop Protection Council Monograph* **41**, 245–52.

Buhl, C. (1960). Beobachtungen über vermehrtes Schadauftreten der Kohlschotenmücke (*Dasyneura brassicae* Winn.) an Raps und Rübsen in Schleswig-Holstein. *Nachrichtenblatt des Deutschen Pflanzenschutzdienstes (Braunschweig)* **12**, 1–6.

Casteels, H. & Clercq, R. de (1990). The impact of some commonly used pesticides on the epigeal arthropod fauna in winter wheat. *Mededelingen van de Faculteit Landbouwwetenschappen, Rijksuniversiteit Gent* **55**, 477–82.

Çilgi, T. & Frampton, G. K. (1994). Arthropod populations under current and reduced-input pesticide regimes: results from the first four treatment years of the MAFF "SCARAB" project. *Proceedings of the Brighton Crop Protection Conference – Pests and Diseases* **2**, 653–60.

Çilgi, T. & Jepson, P. C. (1992). The use of tracers to estimate the exposure of beneficial insects to direct pesticide spraying in cereals. *Annals of Applied Biology* **121**, 239–47.

Çilgi, T., Wratten, S. D., Robertsson, J. L., Turner, D. E., Holland, J. M. & Frampton, G. K. (1996). Residual toxicities of three insecticides to four species (Coleoptera: Carabidae) of arthropod predator. *Canadian Entomologist* **128**, 115–24.

Clercq, R. de, Casteels, H. & Janssens, J. (1991). Influence of pesticides on the epigeal arthropod fauna in laboratory tests. *IOBC/wprs Bulletin* **14**, 110–4.

Cocquempot, C., Chambon, J. P., Reynaud, P. & Fischer, L. (1991). Incidence des traitements insecticides sur la faune aranéologique d'un champ de blé à l'épiaison. *Agronomie* **11**, 423–34.

Colonnelli, E. (1983). Alcune note de sistematica generale dei Ceutorhynchinae, con descrizione di un nuovo genere e di tre nuove specie (Coleoptera, Curculionidae). *Fragmenta Entomologica, Roma* **17**, 159–79.

Colonnelli, E. (1990). Curculionidae Ceutorhynchinae from the Canaries and Macronesia (Coleoptera). *Vieraea* **18**, 317–37.

Coutin, R. & Coulon, J. (1966). Toxicités comparées de plusiers insecticides vis-à-vis de *Dasyneura brassicae* Winn. et de son parasite principal *Prosactogaster oebalus* Walk. *Phytiatrie – Phytopharmacie* **15**, 105–12.

Czajkowska, M. (1978). Badania nad prszczarkiem kapustnikiem (*Dasyneura brassicae* Winn.) IV. Sklad gatunkowy I znaczenie pasozytów. *Roczniki nauk Rolniczych, Seria E* **8**, 159–75.

Debouzie, D. & Ballanger, Y. (1993). Dynamics of a *Ceutorhynchus napi* population in winter rape fields. *Acta Oecologica* **14**, 603–18.

Dewar, A. M., Thornhill, W. A. & Read, L. A. (1990). The effects of tefluthrin on beneficial insects in sugar beet. *Proceedings of the Brighton Crop Protection Conference – Pests and Diseases* **3**, 987–92.

Dosse, G. (1951). Der Große Kohltriebrüßler (*Ceuthorrhynchus napi* Gyll.). Biologie, Schadauftreten und Bekämpfung unter besonderer Berücksichtigung der Gallbildung an Kohlpflanzen. *Zeitschrift für Angewandte Entomologie* **32**, 489–566.

Dosse, G. (1952). Zur Biologie und Morphologie des schwarzen Triebrüßlers *Ceuthorrhynchus picitarsis* Gyll., mit differentialdiagnostischen Angaben zur Unterscheidung der Larven von *Ceuthorrhynchus napi* Gyll., *C. quadridens* Panz und *C. picitarsis* Gyll. *Zeitschrift für Angewandte Entomologie* **34**, 303–12.

Duffield, S. J. & Jepson, P. C. (1994). The effect of spatial scale of treatment with dimethoate on invertebrate population recovery in winter wheat. *Journal of Applied Ecology* **31**, 263–81.

Duffield, S. J., Jepson, P. C., Wratten, S. D. & Sotherton, N. W. (1996). Spatial changes in invertebrate predation rate in winter wheat following treatment with dimethoate. *Entomologia Experimentalis et Applicata* **78**, 9–17.

Ekbom, B. (1990). Flea beetles (*Phyllotreta* spp.) in spring oilseed rape in Sweden. *IOBC/wprs Bulletin* **13** (4), 57–61.

Ekbom, B. (1995). *Insect pests of brassica oilseeds*. In: D. Kimber & J. McGregor (eds). *Brassica Oilseeds – Production and Utilization.* CAB International: Wallingford.

Ellis, S. A., Oakley, J. N., Parker, W. E. & Raw, K. (1999). The development of an action threshold for cabbage aphid (*Brevicoryne brassicae*) in oilseed rape in the UK. *Annals of Applied Biology* **134**, 153–62.

Everts, J. W. (1990). *Sensitive Indicators of Side-Effects of Pesticides on the Epigeal Fauna of Arable Land.* Landbouuniversiteit Wageningen: The Netherlands.

Everts, J. W., Aukema, B., Hengeveld, K. & Koeman, J. H. (1989). Side-effects of pesticides on ground-dwelling predatory arthropods in arable ecosystems. *Environmental Pollution* **159**, 203–25.

Ferguson, A. W., Solinas, M., Ziesmann, J., Isidoro, N., Williams, I. H., Scubla, P., Mudd, A., Clark, S. J. & Wadhams, L. J. (1999). Identification of the gland secreting oviposition-deterring pheromone in the cabbage seed weevil, *Ceutorhynchus assimilis* Payk. (Coleoptera: Curculionidae) and the mechanism of pheromone deposition. *Journal of Insect Physiology* **45**, 687–99.

Finlayson, D. G. (1979). Combined effects of soil-incorporated and foliar-applied insecticides in bed-system production of brassica crops. *Canadian Journal of Plant Science* **59**, 399–410.

Fritzsche, R. (1957). Zur Biologie und Ökologie der Rapsschädlinge aus der Gattung *Meligethes*. *Zeitschrift für Angewandte Entomologie* **40**, 222–80.

Garbe, V., Broschewitz, B., Erichsen, E., Hoßfeld, R., Lauenstein, G., Steinbach, P., Ulber, B. & Zellner, M. (1996). Schadensschwellen bei Rapsschädlingen. Instrumente einer wirtschaftlichen Winterrapsproduktion. *Raps* **14**, 58–63.

Giller, P. S., Ryan, B., Kennedy, T. & Connery, J. (1995). Aphid–parasitoid interactions in a winter cereal crop: field trials involving insecticide application. *Journal of Applied Entomology* **119**, 233–9.

Gladders, P., Lane, A. & French, N. (1989). *Pests and diseases of oilseed rape, brassica seed crops and field beans.* In: N. Scopes & L. Stables (eds). *Pest and Disease Control Handbook.* BCPC: Thornton Heath.

Goltermann, S. & Pelz, A. (1995). Nützlingsschonender Insektizideinsatz im Winterraps? *Raps* **13**, 153–4.

Graham, C. W. & Alford, D. V. (1981). The distribution and importance of cabbage stem flea beetle (*Psylliodes chrysocephala* L.) on winter oilseed rape in England. *Plant Pathology* **30**, 141–5.

Graham, C. W. & Gould, H. J. (1980). Cabbage stem weevil (*Ceutorhynchus quadridens*) on spring oilseed rape in southern England and its control. *Annals of Applied Biology* **95**, 1–10.

Graichen, K. & Rabenstein, F. (1996). Europäische Isolate des Westlichen Rubenvergilbungsvirus (BWYV) vom Raps (*Brassica napus* L. ssp. *napus*) sind nicht pathogen für Zuckerrübe (*Beta vulgaris* L. var. *altissima*) und repräsentieren Isolate des Wasserrübenvergilbungsvirus (TuYV). *Zeitschrift für Pflanzenkrankheiten und Pflanzenschutz* **103**, 233–45.

Gratwick, M. (ed.) (1992). *Crop Pests in the UK.* Chapman & Hall: London.

Günthardt, E. (1949). Beiträge zur Lebensweise und Bekämpfung von *Ceuthorrhynchus quadridens* Panz. und *Ceuthorrhynchus napi* Gyll., mit Beobachtungen an weiteren Kohl- und Rapsschädlingen. *Mitteilungen der Schweizerischen Entomologischen Gesellschaft* **22**, 441–591.

Hardwick, N. V., Davies, J. M. L. & Wright, D. M. (1994). The incidence of three virus diseases of winter oilseed rape in England and Wales in the 1991/92 and 1992/93 growing seasons. *Plant Pathology* **43**, 1045–9.

Haskell, P. T. & McEwen, P. (1998). *Ecotoxicology. Pesticides and Beneficial Organisms.* Chapman & Hall: London.

Hassan, S. A. (1992). Guideline for the evaluation of side-effects of plant protection product on *Trichogramma cacoeciae*. *IOBC/wprs Bulletin* **15**, 18–39.

Hassan, S. A., Bigler, F., Bogenschütz, H., Boller, E., Brun, J., Calis, J. N. M., Coremans-Pelseneer, J., Duso, C., Grove, A., Heimbach, U., Helyer, N., Hokkanen, H., Samsoe-Petersen, L., Sauphanor, B., Stäubli, A., Sterk, G., Vainio, A., Veire, M. van de, Viggiani, G. & Vogt, H. (1994). Results of the sixth joint pesticide testing programme of the IOBC/WPRS working group 'Pesticides and beneficial organisms'. *Entomophaga* **39**, 107–19.

Hassan, S. A., Hafes, B., Degrande, P. E. & Herai, K. (1998). The side-effects of pesticides on the egg parasitoid *Trichogramma cacoeciae* Marchal (Hym., Trichogrammatidae), acute dose-response and persistence tests. *Journal of Applied Entomology* **122**, 569–73.

Heimbach, U. (1988). Nebenwirkungen einiger Fungizide auf Insekten. *Nachrichtenblatt des Deutschen Pflanzenschutzdienstes* **40**, 180–2.

Heimbach, U. (1992). Laboratory method to test effects of pesticides on *Poecilus cupreus* (Coleoptera: Carabidae). *IOBC/wprs Bulletin* **15**, 103–9.

Heimbach, U. & Abel, C. (1991). Nebenwirkungen von Bodeninsektiziden in verschiedenen Applikationsformen auf einige Nutzarthropoden. *Verhandlung der Gesellschaft für Ökologie* **19**, 163–70.

Hokkanen, H., Husberg, G.-B. & Söderblom, M. (1988). Natural enemy conservation for the integrated control of the rape blossom beetle *Meligethes aeneus* F. *Annales Agriculturae Fenniae* **27**, 281–94.

Hoßfeld, R. (1963). Synökologischer Vergleich der Fauna von Winter- und Sommerrapsfeldern. *Zeitschrift für Angewante Entomologie* **52**, 209–54.

Hoßfeld, R. (1993). The use of yellow dishes as a decision aid for the control of *Psylliodes chrysocephala* L. *Gesunde Pflanzen* **45**, 291–5.

Hussein, M. L. Al & Wetzel, T. (1993a). Nebenwirkungen der Fungizide Afugan, Desmel, Desgan und des Insektizids Decis auf räuberische Käfer (Carabidae; Staphylinidae) in Wintergerste. *Archiv für Phytopathologie und Pflanzenschutz* **28**, 155–67.

Hussein, M. L. Al & Wetzel, T. (1993b). Nebenwirkungen des Fungizids Desgan und einiger Insektizide auf räuberische Käfer (Carabidae; Staphylinidae) in Winterweizen. *Archiv für Phytopathologie und Pflanzenschutz* **28**, 249–62.

Hussein, M. L. Al & Wetzel, T. (1993c). Untersuchungen zu Nebenwirkungen des Fungizids Furesan (Flusilazol + Pyrazophos) auf räuberische Käfer (Carabidae; Staphylinidae) in Wintergerste. *Archiv für Phytopathologie und Pflanzenschutz* **28**, 447–52.

Hussein, I. A. Al, Lübke, M. von & Wetzel, T. (1990). Nebenwirkungen von Insektiziden auf Kurzflügelkäfer (Col., Staphylinidae) in Winterweizenfeldern. *Journal of Applied Entomology* **109**, 226–32.

Hussein, I. A. Al, Heyer, W. & Wetzel, T. (1991). Untersuchungen zum Auftreten der Webespinnen (Araneae) in Winterweizen und ihre Beeinflussung durch ausgewählte Insektizide. *Archiv für Phytopathologie und Pflanzenschutz* **27**, 219–28.

Inglesfield, C. (1989). A long-term field study to investigate effects of alpha-cypermethrin on predatory and parasitic arthropods. *Mededelingen van de Faculteit Landbouwwetenschappen, Rijksuniversiteit Gent* **54**, 895–904.

Jepson, P. C. (ed.) (1989). *Pesticides and Non-Target Invertebrates.* Intercept: Andover.

Jepson, P. C. & Sherratt, T. N. (1991). Predicting the long-term impact of pesticides on predatory invertebrates. *Proceedings of the Brighton Crop Protection Conference – Weeds* **2**, 911–9.

Jepson, P. C. & Thacker, J. R. M. (1990). Analysis of the spatial component of pesticide side effects on non-target invertetrate populations and its relevance to hazard analysis. *Functional Ecology* **4**, 349–55.

Johnen, A. & Meier, H. (2000). A weather-based decision support system for managing oilseed rape pests. *Proceedings of the BCPC Conference – Pests and Diseases* **2**, 793–800.

Jourdheuil, P. (1960). Influence de quelques facteurs écologiques sur les fluctuations de population d'une biocénose parasitaire: étude relative à quelques Hyménoptères (Ophioninae, Diospilinae, Euphorinae) parasites de divers Coléoptères inféodés aux Crucifères. *Annales des Épiphyties* **11**, 445–658.

Jurek, M. (1972). [*Meligethes* spp. on winter rape – *Brassica napus* L. var. *oleifera* Metz. f. *biennis* Thel.] *Polskie Pismo Entomologiczne* **42**, 483–90. [In Polish.]

Kaczmarek, S. (1987). [Effect of the preparation 'Enolofos 50' on Carabidae and Staphylinidae in winter rape.] *Polskie Pismo Entomologiczne* **57**, 377–81. [In Polish.]

Karltorp, M. & Nilsson, C. (1981). Rapsbagger I mellansvenska våroljeväxtodlingar. *Växtskyddnotiser* **45**, 146–54.

Kelm, M. & Klukowski, Z. (2000). The effect of stem weevil (*Ceutorhynchus pallidactylus* Marsh.) infestation on oilseed rape. *IOBC/wprs Bulletin* **23**, 125–30.

Kelm, M. & Walczak, B. (1998). The relationship between the stem weevil (*Ceutorhynchus pallidactylus* Marsh.) injury and losses of the flower buds. *IOBC/wprs Bulletin* **21**, 147–51.

Körner, H. (1990). Der Einfluß der Pflanzenschutzmittel auf die Faunenvielfalt der Agrarlandschaft (unter besonderer Berücksichtigung der Gliederfüßler auf der Oberfläche der Felder). *Bayerisches Landwirtschaftliches Jahrbuch* **67**, 375–496.

Koubaiti, K. (1992). *Reproduction, Développement et Nuisibilité de* Baris coerulescens *Scop. (Coléoptère, Curculionidae) au Contact du Colza* (Brassica napus *var.* oleifera). Thesis, Tours University, France.

Krooß, S. & Schaefer, M. (1998). The effect of different farm systems on epigeic arthropods: a five-year study on the rove beetle fauna (Coleoptera: Staphylinidae) of winter wheat. *Agriculture, Ecosystems & Environment* **69**, 121–33.

Lamb, R. J. (1989). Entomology of oilseed Brassica crops. *Annual Review of Entomology* **34**, 211–29.

Lancashire, P. D., Bleiholder, H., Boom, T. van den, Langelüddeke, P., Strauss, R., Weber, E. & Witzenberger, A. (1991). A uniform decimal code for growth stages of crops and weeds. *Annals of Applied Biology* **119**, 561–601.

Lane, A. & Cooper, D. (1989). Importance and control of insect pests of oilseed rape. *Aspects of Applied Biology* **23**, 269–76.

Lane, A. & Gladders, P. (2000). *Pests and diseases of oilseeds, brassica seed crops and field beans.* In: D. V. Alford (ed.). *Pest and Disease Management Handbook.* Blackwell Science: Oxford.

Lane, A. & Walters, K. F. A. (1993). Recent incidence and cost effective control of pests of oilseed rape in England and Wales. *IOBC/wprs Bulletin* **16** (9), 185–92.

Lehmann, W. (1964). Einfluß chemischer Bekämpfungsmaßnahmen auf einige Rapsschädlinge und ihre Parasiten. I. Saatgutinkrustierung. *Nachrichtenblatt Deutscher Pflanzenschutzdienst (Berlin) N.F.* **18**, 53–6.

Lehmann, W. (1965a). Einfluß chemischer Bekämpfungsmaßnahmen auf einige Rapsschädlinge und ihre Parasiten. II. Knospen- und Stengelschädlinge. *Archiv für Pflanzenschutz* **1**, 209–19.

Lehmann, W. (1965b). Der Einfluß chemischer Bekämpfungsmaßnahmen auf einige Parasiten von Rapsschädlingen. *Mitteilungen aus der Biologischen Bundesanstalt für Land- und Forstwirtschaft* **115**, 35–42.

Lerin, J. (1993). Influence of the growth rate on the splitting of the stem after an attack of *Ceutorhynchus napi* Gyll. *IOBC/wprs Bulletin* **16** (9), 160–3.

Lewis, G. B., Brown, R. A., Edwards, P. J. & Canning, L. C. (1992). A laboratory test method for assessing the toxicity of pesticides to the ground beetle *Pterostichus melanarius* (L.) (Carabidae, Coleoptera). *IOBC/wprs Bulletin* **15**, 110–6.

Lohse, G. A. & Lucht, W. H. (1994). *Die Käfer Mitteleuropas.* Goecke & Evers: Krefeld.

Longley, M. & Izquierdo, J. I. (1994). Spatial and temporal changes in aphid/parasitoid distributions following an insecticide application to winter wheat. *Proceedings of the Brighton Crop Protection Conference – Pests and Diseases* **2**, 951–2.

Longley, M. & Jepson, P. C. (1997). Cereal aphid and parasitoid survival in a logarithmically diluted deltamethrin spray transect in winter wheat: field-based risk assessment. *Environmental Toxicology and Chemistry* **16**, 1761–7.

Longley, M., Jepson, P. C., Izquierdo, J. & Sotherton, N. (1997). Temporal and spatial changes in aphid and parasitoid populations following applications of deltamethrin in winter wheat. *Entomologia Experimentalis et Appticata* **83**, 41–52.

Murchie, A. K., Williams, I. H. & Alford, D. V. (1997). Effects of commercial insecticide treatments to winter oilseed rape on parasitism of *Ceutorhynchus assimilis* Paykull (Col., Curc.) by *Tricomalus perfectus* (Walker) (Hym., Pteromalidae). *Crop Protection* **16**, 199–202.

Nielsen, G. C. (2000). *Vejledende bekaempelsestaerskler for skadedyr 2001. Landskontoret for Planteavl: Arhus, Denmark.*

Nielsen, J. K. (1977). Host plant relationships of *Phyllotreta nemorum* L. (Coleoptera; Chrysomelidae). I. Field studies. *Journal of Applied Entomology* **84**, 396–406.

Nilsson, C. (1987). Yield losses in summer rape caused by pollen beetles (*Meligethes* spp.). *Swedish Journal of Agricultural Research* **17**, 105–11.

Nilsson, C. (1988). Pollen beetles (*Meligethes aeneus* F.) and flowering in rape. *Swedish Journal of Agricultural Research* **18**, 113–8.

Nilsson, C. (1990). Yield losses in winter rape caused by cabbage stem flea beetle larvae (*Psylliodes chrysocephala* (L.)). *OILB/wprs Bulletin* **13**, 53–6.

Nilsson, C. (1994). Pollen beetles (*Meligethes* spp.) in oil seed rape crop (*Brassica napus* L.): biological interactions and crop losses. *Department of Plant Protection Sciences, SLU. Dissertations* **1**, 39 pp.

Nilsson, C. & Andreasson, B. (1987). Parasitoids and predators attacking pollen beetles (*Meligethes aeneus* F.) in spring and winter rape in southern Sweden. *IOBC/wprs Bulletin* **10**, 64–73.

Nitzsche, O. (1998). *Auftreten und Effizienz von Parasitoiden als natürliche Gegenspieler von Schadinsekten im Winterraps unter besonderer Berücksichtigung unterschiedlicher Bodenbearbeitungsmaßnahmen nach Winterraps.* PhD thesis, Georg-August-University of Göttingen, Germany.

Nolte, H.-W. & Fritzsche, R. (1952). Untersuchungen über das Vorkommen verschiedener *Meligethes*-Arten auf Raps. *Beiträge Entomologie* **2**, 434–48.

Oomen, P. A., Jobsen, J. A., Romeijn, G. & Wiegers, G. L. (1994). Side-effects of 107 pesticides on the whitefly parasitoid *Encarsia formosa*, studied and evaluated according to EPPO guideline no. 142. *Bulletin OEPP/EPPO Bulletin* **24**, 89–107.

Palosz, T. (1980). Einschätzung der Schadenskoeffizienten einiger Rapsschaedlinge. *Tagungsberichte der Akademie für Landwirtschafts-Wissenschaften der DDR, Berlin* **181** 135–43.

Palosz, T. (1995a). [Intensive technologies in agriculture and the carabid fauna.] *Ochrony Roslin* **39**, 8. [In Polish.]

Palosz, T. (1995b). [The species composition of ground beetles (Col., Carabidae) on winter rape crops with different technology and intensity of cultivation.] *Ochrony Roslin* **35**, 108–15. [In Polish.]

Paul, V. H. (1992). *Krankheiten und Schädlinge des Rapses.* Th. Mann: Gelsenkirchen-Buer.

Paul, V. H. & Rawlinson, C. J. (1992). *Diseases and Pests of Rape.* Th. Mann: Gelsenkirchen-Buer.

Pilorgé, E., Maisonneuve, C. & Ballanger, Y. (1997). *Les Ravageurs du Colza d´Hiver.* CETIOM, France.

Pope, R. D. (1977). A check list of British insects. Coleoptera and Strepsiptera. *Handbooks for the Identification of British Insects* **11** (3), 1–105.

Purvis, G. (1996). The hazard posed by methiocarb slug pellets to carabid beetles: understanding population effects in the field. *BCPC Symposium Proceedings No. 66, Slug and Snail Pests in Agriculture,* pp. 189–96.

Raw, K. A. (1989). The effects of summer applications of insecticides on the non-target invertebrate fauna of oilseed rape. *Aspects of Applied Biology* **23**, 295–302.

Ritter, E. (1916). *Fauna Germanica. Die Käfer des Deutschen Reiches. V. Band.* Lutz': Stuttgart.

Rzehak, H. & Basedow, T. (1982). Die Auswirkungen verschiedener Insektizide auf die epigäischen Raubarthropoden in Winterrapsfeldern. *Anzeiger für Schädlingskunde, Pflantzenschutz und Umweltschutz* **55**, 71–5.

Samsoe-Petersen, L. (1993). Effects of 45 insecticides, acaricides and molluscicides on the rove beetle *Aleochara bilineata* (Col.: Staphylinidae) in the laboratory. *Entomophaga* **38**, 371–82.

Samsoe-Petersen, L. (1995). Effects of 67 herbicides and plant growth regulators on the rove beetle *Aleochara bilineata* (Col.: Staphylinidae) in the laboratory. *Entomophaga* **40**, 95–104.

Schliephake, E., Graichen, K. & Rabenstein, F. (2000). Investigations on the vector transmission of the beet mild yellowing virus (BMYV) and the turnip yellows virus (TuYV). *Zeitschrift für Pflanzenkrankheiten und Pflanzenschutz* **107**, 81–7.

Schmidt, G. (1970). Die deutschen Namen wichtiger Arthropoden. *Mitteilungen aus der Biologischen Bundesanstalt für Land- und Forstwirtschaft* **137**, 1–222.

Schmuck, R. (1992). Auswirkungen einer Rapssaatbehandlung mit dem Saatgutpuder Oftanol T auf Laufkäfer. *Gesunde Pflanzen* **44**, 264.

Schulz, R.-R. (1985). Untersuchungen zur Vermehrungsrate des Rapserdflohs (*Psylliodes chrysocephala* L.). *Archiv für Phytopathologie und Pflanzenschutz* **21**, 305–11.

Seta, G., Mrowcynski, M. & Widerski, K. (1995). [Combined application of the insecticide Fury 100 EC and foliar fertilizers on winter rape.] *Rosliny Oleiste* **16**, 217–20. [In Polish.]

Seymour, P. R. (1989). *Invertebrates of Economic Importance in Britain. Common and Scientific Names.* 4th edn. HMSO: London.

Sommer, G. (1981). *Biologie und Parasitenkomplex der Halticinen Gattung* Phyllotreta. Ph.D. thesis, Albert-Ludwigs-Universität zu Freiburg, Freiburg, Germany.

Sterk, G., Hassan, S. A., Baillod, M., Bakker, F., Bigler, S., Blümel, S., Bogenschütz, H., Boller, E., Bromand, B., Brun, J., Calis, J. N. M., Coremans-Pelseneer, J., Duso, C., Garrido, A., Grove, A., Heimbach, U., Hokkanen, H., Jacas, J., Lewis, G., Moreth, L., Polgar, L., Roversti, L., Samsoe-Petersen, L., Sauphanor, B., Schaub, L., Stäubli, A., Tuset, J. J.,

Vainio, A., van de Veire, M., Viggiani, G., Vinuela, E. & Vogt, H. (1999). Results of the seventh joint pesticide testing programme carried out by the IOBC/WPRS-Working group 'Pesticides and beneficial organisms'. *BioControl* **44**, 99–117.

Sunderland, K. D. (1992). Effects of pesticides on the population ecology of polyphagous predators. *Aspects of Applied Biology* **31**, 19–28.

Thacker, J. R. M., Young, R. D. F., Stevensson, S. & Curtis, D. J. (1995). Effects of a change in pesticide droplet size on topical toxicity of chlorpyrifos and deltamethrin to *Myzus persicae* (Homoptera: Aphididae) and *Nebria brevicollis* (Coleoptera: Carabidae). *Journal of Economic Entomology* **88**, 1560–5.

Twengström, E. (ed.) (2000). *Behovsanpassad bekämpning av skadegörare i jordbruksgrödor.* Faktablad om växtskydd, 1J. Swedish University of Agricultural Sciences, 61 pp.

Ulber, B., Stippich, G. & Wahmhoff, W. (1990). Möglichkeiten, Grenzen und Auswirkungen des gezielten Pflanzenschutzes im Ackerbau: III Auswirkungen unterschiedlicher Intensität des chemsichen Pfianzenschutzes auf epigäische Raubarthropoden in Winterweizen, Zuckerrüben und Winterraps. *Zeitschrift für Pflanzenkrankheiten und Pflanzenschutz* **97**, 263–83.

Umoru, P. A., Powell, W. & Clark, S. J. (1996). Effect of pirimicarb on the foraging behaviour of *Diaeretiella rapae* (Hym. Brac.) on host-free and infested oilseed rape plants. *Bulletin of Entomological Research* **86**, 193–201.

Volkmar, C. (1996). Spinnengemeinschaften auf einem typischen Ackerbaustandort Mitteldeutschlands und deren Beeinflußung durch unterschiedliche Pflanzenschutz-intensität im Verlauf einer Fruchtfolgerotation (Spinnen in AgroÖkosystemen). *Revue Suisse de Zoologie Hors Serie* **2**, 683–90.

Wiles, J. A. & Jepson, P. C. (1992). The susceptibility of a cereal aphid pest and its natural enemies to deltamethrin pesticide. *Science* **36**, 263–72.

Wiles, J. A. & Jepson, P. C. (1995). Dosage reduction to improve the selectivity of deltamethrin between aphids and coccinellids in cereals. *Entomologia Experimentalis et Applicata* **76**, 83–96.

Williams, I. H., Martin, A. P. & Kelm, M. (1987a). The phenology of the emergence of brassica pod midge (*Dasineura brassicae* Winn.) and its infestation of winter oil-seed rape (*Brassica napus* L.). *Journal of Agricultural Science, Cambridge* **108**, 579–89.

Williams, I. H., Martin, A. P. & Kelm, M. (1987b). The phenology of the emergence of brassica pod midge (*Dasineura brassicae* Winn.) and its infestation of spring oil-seed rape (*Brassica napus* L.). *Journal of Agricultural Science, Cambridge* **109**, 309–14.

Winfield, A. L. (1963). A study on the effects of insecticides on parasites of larvae of blossom beetles (*Meligethes aeneus* F., Coleoptera: Nitidulidae). *Entomologia Experimentalis et Applicata* **6**, 309–18.

Winfield, A. L. (1992). Management of oilseed rape pests in Europe. *Agricultural Zoological Review* **5**, 51–95.

3

Management of Oilseed Rape Pests

Keith F. A. Walters
Central Science Laboratory, Sand Hutton, York YO41 1LZ, UK
John E. B. Young
ADAS Boxworth, Boxworth, Cambridge CB3 8NN, UK
Bernhard Kromp
Ludwig Boltzmann-Institute for Biological Agriculture and Applied Ecology, Rinnböckstraße 15, A-1110 Wien, Austria
Patrick D. Cox
Central Science Laboratory, Sand Hutton, York YO41 1LZ, UK

Introduction

During recent years there has been heightened public concern about the effects of modern agricultural practices on the environment. This has coincided with the farming industry being put under immense financial pressures as farm incomes have dropped in conjunction with the move towards global trading and pricing (Meikle & Scarisbrick, 1995). Arable farmers have long recognized the need to make efficient use of variable inputs, such as insecticides, fungicides, fertilizers, seeds and energy, and many acknowledge the need to minimize the environmental impact of the key inputs that are required to improve the economic viability of major crops such as oilseed rape. Their aim is to create a long-term sustainable farming system, which will also remain profitable and bring environmental benefits. These aims are embodied in the concept of integrated crop management (ICM), which is increasingly being promoted as the best means of combining efficient, profitable production with greater environmental responsibility and safety at the farm level.

Historically, pesticides have provided a reliable and cost-effective approach to pest and disease control in arable crops. The development of resistance (to pesticides) within pest populations has often been addressed by improving management techniques or through the development of new classes of pesticides. Improvements in management techniques have included strategic selection and integration of pesticide products, establishment of effective pest assessment methods, adoption of optimal timing for pesticide applications, introduction of improved farm machinery and the development of computer models that offer improved integration of the range of information upon which decision making is based (Green *et al.*, 1991; Hardwick, 1998; Morgan *et al.*, 2000). However, despite these advances, prophylactic use of relatively inexpensive pesticides remains a popular option (Walters & Hardwick, 2000).

As outlined in Chapter 2, oilseed rape crops are subject to attack by a range of insect pests. Other pests, such as slugs and pigeons, can also be important. In the UK, for example, autumn-sown crops are liable to be attacked by cabbage stem flea beetle (*Psylliodes chrysocephala*) and slugs at or soon after crop emergence, by virus diseases spread by peach/potato aphid (*Myzus persicae*) throughout the autumn period, by rape winter stem weevil (*Ceutorhynchus picitarsis*) during the winter, and by pollen beetle (*Meligethes aeneus*), cabbage seed weevil (*Ceutorhynchus assimilis*), brassica pod midge (*Dasineura brassicae*) and cabbage aphid (*Brevicoryne brassicae*) from early spring onwards, all of which are potentially damaging (Walters & Lane, 1994a). In addition, there is considerable variability in the size of pest populations and the incidence of damaging outbreaks (both between and within years) between different geographical areas that is only partly explained by different soil type, macroclimate, topography and hydrogeology. For example, oilseed rape is grown in all of the main crop production areas of the UK and, although the major pests are now widespread, surveys confirm that infestation levels of pests such as pollen beetle and cabbage seed weevil can vary widely from season to season (Lane & Walters 1994; Walters & Lane, 1994b). Similarly, in most years, variability has been noted in pest incidence and damage recorded between even geographically close regions (Table 3.1), and the difficulties of accurately forecasting such variable incidence (and, thus, risk) on either a local or a national scale have resulted in insurance spraying becoming an attractive option. Despite there being well-established (Lane & Walters, 1993, 1994; Walters *et al.*, 2001) and practical thresholds available for pesticide application, which are based on extensive research (Walters & Lane 1992), this has led to widespread, and extreme, over-use of insecticides (Table 3.2), which is wasteful, environmentally damaging and unsustainable.

The need to reduce reliance on chemical options for pest management has lead to intensive investigation of the potential for developing and using commercially viable biological control methods. Classical biological control techniques and their integration within crop production systems were originally developed in Europe for use on high-value greenhouse-grown (protected) crops, such as tomatoes. There is now widespread use of biocontrol in many horticultural commodities, particularly on protected crops, and a range of biocontrol agents (BCAs) (e.g. the parasitoid wasp *Encarsia formosa* for control of whiteflies, the predatory beetle *Cryptolaemus montrouzieri* for control of mealybugs, and entomopathogenic nematodes for control of soil pests) are commercially available (Copping, 2001). Use of some of these agents can be integrated with applications of

Table 3.1 Infestation levels of cabbage seed weevil (*Ceutorhynchus assimilis*) and pollen beetles (*Meligethes* spp.) on winter oilseed rape crops in various parts of England during spring/summer 2000.

	Mean number of adults/plant	
Region	Cabbage seed weevil	Pollen beetles
Eastern	0.26	2.02
Midlands (West)	0.52	3.91
Northern	0.99	1.47
South eastern	0.12	6.11
South western	0.43	1.76

Table 3.2 Area of winter oilseed rape treated with insecticides in England during the spring/summer (1988–1998) (after Garthwaite & Thomas, 1998) and the proportion of commercial crops found from pest monitoring to exceed established spray thresholds.

Year	1988	1990	1992	1994	1996	1998
Area treated ('000 ha)	260	546	463	370	552	710
Area grown ('000 ha)	346	390	420	403	356	505
Overall extent of treatment	75%	140%	110%	92%	155%	141%
Proportion of crops exceeding threshold	<10%	<10%	<10%	*c*. 25%	<10%	<10%

chemical pesticides (Head *et al*., 2000), and their performance compares well with that of conventional approaches (Williams & Walters, 2000).

There is intense interest in developing biocontrol approaches in arable crops such as oilseed rape (Alford, 1998). However, owing to the costs involved, it is currently uneconomic to rear and artificially introduce biocontrol agents to field-grown arable crops. Consequently, biocontrol strategies in arable crops focus on ways of preserving and enhancing the activity of naturally occurring BCAs (Kromp, 1999). A wide range of these (parasitoids, predators and, to a lesser extent, pathogens) have considerable potential for limiting or reducing pest populations. Therefore, ICM provides an ideal opportunity for the development and promotion of biocontrol in oilseed rape and other arable crops.

Within Europe, the International Organisation for Biological Control (IOBC) has published guidelines for integrated production (El Titi *et al*., 1993). The IOBC guidelines define integrated production (i.e. ICM) as a farming system that produces high-quality food and other products by using natural resources and regulating mechanisms to replace polluting inputs and to secure sustainable farming. Emphasis is placed on a 'holistic systems' approach that involves the whole farm as a basic unit, on the central role of ecosystems, on balanced nutrient cycles and on the welfare of all species in animal husbandry. The preservation and improvement of soil fertility and of a diversified environment are essential components. Biological, technical and chemical methods are balanced carefully, taking into account the protection of the environment, profitability and social requirements.

Defining single 'blueprints' for conventional and ICM systems is not straightforward, as there is a high degree of site-specificity involved in each approach. Both systems can vary widely at the farm level, according to local factors such as soil type, crop rotation and the incidence of weeds, pests and diseases. Furthermore, a considerable overlap can occur between conventional and ICM systems. The final design of an ICM system cannot be determined at farm level unless the basic rules and principles of ICM are applied to match local needs and circumstances.

This chapter identifies the main differences between the conventional and ICM systems currently in use for growing winter oilseed rape in various EU countries, and discusses the role and benefits of utilizing biocontrol methods against insect pests of oilseed rape within ICM.

On-Farm Landscape Structure and Habitat Management

Arthropod diversity in agro-ecosystems is strongly influenced by site-specific factors such as soil type, macroclimate, topography and hydrogeology, and factors influenced by husbandry methods such as intensity of cultivation, tillage method, variable inputs, crop diversity and crop type (Kromp, 1999). These are superimposed on the effects of factors relating to the on-farm landscape structure, such as uncultivated areas between and around fields.

Integrated pest management (IPM) (itself a key component of ICM) involves planning a programme for the whole farm, including the non-farmed land, to enhance biodiversity and landscape features. This may include developing and maintaining a network of hedges, ditches, field margins, beetle banks and conservation headlands, which enable wild species to establish and migrate, and to provide recreational areas for people. A greater diversity of broad-leaved weed species may be left within crops to provide food sources for birds and insects, so long as the numbers of aggressive, crop-damaging weeds are contained. The conservation, creation and improvement of habitats for parasitoids and predators of crop pests may regulate pest populations by increasing the natural level of biocontrol, thereby reducing the need for insecticidal intervention.

Much recent work has been directed at defining the effect of landscape structure and the potential for manipulating it to favour the development of large populations of natural enemies of pests and, thus, to enhance the potential of reducing the level of dependence on insecticides. Field margins in particular have excited much interest as possible sources of natural control agents. However, the different methods of utilizing such resources has lead to some confusion in the use of terminology, and in an attempt to establish clear definitions it has been proposed that at least three distinct features should be clearly distinguished (Greaves & Marshall, 1987; Thomas, 1996):

- 'boundaries' – which separate one field from the next or from its adjacent land use (e.g. hedges, windbreaks, ditches, streams, grass baulks, verges, forest edges);
- 'boundary strips' – which are situated between the boundary structure and the cropped area of the field (e.g. grass margins, herbaceous strips, fallow strips, farm tracks);
- 'crop margins' – which are the outer edges of the cropped area of the field but can be cultivated separately as conservation, managed or unfertilized headlands.

These definitions are followed throughout this chapter.

Field margins and pest populations

There is an increasingly large volume of literature on the potential contribution that field margins make to biological control of pests in agro-ecosystems. The contribution of field margins towards natural pest suppression is usually assumed to be due to the enhancement of beneficial populations and, thereby, to their increased efficacy on pest population densities.

Farmers and advisors frequently raise concerns about uncultivated field margins adjacent to arable fields also offering a habitat for pest populations. A recent detailed study addressed this issue (Hausamman, 1996) and did not find any negative impacts of sown and managed weed strips established in the middle of fields of oilseed rape crops; for example, there was no increase in pest incidence. By contrast, a crop adjacent to an old (unmanaged) grassy boundary strip showed increased densities of several oilseed rape pests, particularly at the edges of the cropped area. Similarly, in a different study, populations of cabbage stem flea beetle and various species of ceutorhynchid weevils in oilseed rape crops were not found to be enhanced by such weed strips (Lethmayer *et al.*, 1997). Recently emerged (i.e. young) pollen beetle adults were abundant on yellow-flowering plants such as wild white mustard (*Sinapis alba*) growing in crop margins adjacent to a rape fields. However, adult pollen beetles, which damage oilseed rape crops only in the late spring or early summer, were attracted to weed strips only transiently where they then fed on the available pollen – the habitat offered by these strips was neither suitable for, nor utilized for, overwintering. Thus, weed strips adjacent to oilseed rape fields probably do not contribute towards the population build-up of pollen beetles during critical periods of the crop growth cycle.

With the exception of slugs, detrimental effects on populations of oilseed rape pests resulting from managed crop margins, set-aside or other similar features, have not been identified. Therefore, habitat manipulation to encourage the development of larger populations of natural enemies of pests remains a viable option. However, to achieve this objective reliably an improved understanding of the population dynamics (Wratten & Thomas, 1990) and dispersal behaviour (Berraondo *et al.*, 1998) of antagonists is essential.

Trap crops in field margins for agro-technical pest control

Certain crop cultivars or wild plants are particularly attractive to insect pests and can be used to concentrate them in non-crop areas, thus reducing pest incidence in adjacent fields (Hokkanen, 1991). When pests are diverted successfully to (and concentrated in) such areas the latter can be treated with pesticides or other agents, once again reducing the amount of active ingredient that might otherwise be applied to commercial crops (Walters & Hardwick, 2000). For example, an agro-technical system for the control of pollen beetles attacking cauliflower crops has been developed, which involves surrounding the field with strips of trap crops (including broccoli, chinese cabbage, turnip and oilseed rape) and establishing pest population thresholds at which the trap crops are treated with an insecticide (Hokkanen, 1989).

Similar systems for pest control in oilseed rape have also been investigated. Strips of turnip and winter rape established in crop margins resulted in from 28.4% to 80.3% of the total population of pollen beetles in the field being concentrated in the trap crop during the early immigration phase (Büchi, 1995). Subsequently, they also contributed to the enhancement of parasitoid numbers in and around the field. However, the strips did not make a significant contribution to the control of rape stem weevil (*Ceutorhynchus napi*).

To develop an approach for reducing the very early attack of pollen beetle on spring rape, Nilsson (1996) mixed seeds of turnip rape (2%–4%) with spring rape. The turnip rape would then act as catch plants, to reduce the effect of very early attack of pollen beetles on the spring rape. Using this strategy, twice as many pollen beetles could be tolerated in crops before insecticide applications were required; importantly, yields were not affected adversely by the presence of the turnip rape plants.

To optimize the effect of trap crops, careful selection of plants is essential. For example, turnip sawfly (*Athalia rosae*) responds to brassicacious plants with a defined range of glucosinolate content (R. Büchi, *pers. comm.*). Further work will be required to determine if this response can be used to develop a non-chemical method for the control of this pest.

On-farm landscape structures and beneficial arthropods

Although many studies have confirmed the general value of uncultivated field margins for beneficial arthropod populations (Kromp, 1999), few investigations have centred on natural enemies of oilseed rape pests.

Work on the use of flowering strips to enhance natural enemy incidence in agro-ecosystems has indicated that the age and composition of the structures seem to be important for the development and efficacy of parasitoid and predator populations. For example, parasitization rates of pollen beetles have been shown to be significantly increased from 24% to 58% by the presence of six-year-old boundary strips containing naturally established vegetation, whereas sown one-year-old strips showed no such effect (Thies *et al.*, 1997). Further, the flowers of various species of plants are known to be differentially attractive to adult hover flies (Diptera: Syrphidae) (MacLeod, 1999). Larvae of hover flies (e.g. the generally abundant species *Episyrphus balteatus*) are important predators of aphids (see Chapter 12), and the adult females require pollen and nectar to develop their eggs. Recent work has shown that the number of eggs produced by each female is related to the species of plant upon which they have fed (Fig. 3.1).

Unless manipulated to increase the rate of parasitoid or predator emigration, the effect of boundary strips and crop margins on the incidence of natural enemies in the crop tends to decline rapidly with distance from the field edge. For example, the mortality of pollen beetle larvae in oilseed rape fields at distances of 3 and 30 m from two categories of ecological compensation areas (wild flower strips and extensive meadows) has been established (Büchi, 2002). In fields with wild flower strips, over a three-year period parasitism of pollen beetle larvae by *Phradis* spp. was increased at distances of 3 m from the field edge, whereas a similar increase was not recorded at 30 m.

A recent study has addressed the incidence of predation by aphidophagous syrphid larvae in a cereal/oilseed rape rotation (K. F. A. Walters & P. N. Northing, *unpublished*). In this work, numbers of adult aphidophagous syrphids caught in yellow water traps did not differ significantly between trapping stations in the margin and those located at distances of 0, 10 and 30 m from a wild flower border, indicating significant penetration of the predator into the field (Fig. 3.2). Numbers of aphids in fields with flower margins

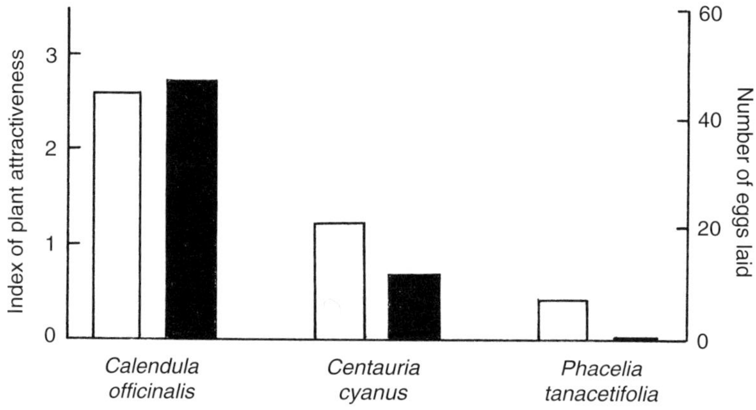

Fig. 3.1 Relationship between the number of eggs laid by adult hover flies (■) (*Episyrphus balteatus*) and plant attractiveness (□), i.e. the plant upon which they have fed.

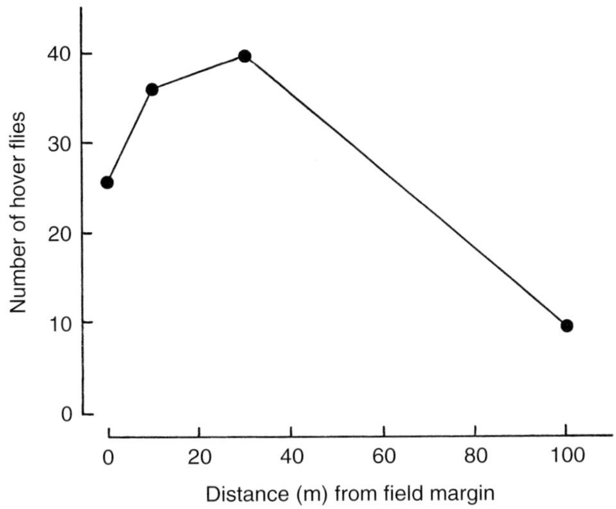

Fig. 3.2 Mean number of adult aphidophagus hover flies caught in yellow water traps during July and August. Data from replicated trapping stations set in a wild flower margin (i.e. at 0 m) and within an adjacent winter rape crop at 10, 30 and 100 m from the margin.

were less than half those found in nearby (control) fields that did not have manipulated margins. A second study in cereal fields indicated that the apparent depression of aphid populations might have been the result of efficient utilization of the prey source through more effective targeting of oviposition (Fig. 3.3).

At a landscape level, Thies & Tscharntke (1999) found that a higher incidence of parasitization of pollen beetle, and a lower incidence of crop damage, occurred in oilseed rape crops located in structurally complex landscapes compared with those growing in simple (monotonous) landscapes. These authors considered that prairie farming areas (i.e. farms with very large fields and a general absence of hedgerows and other environmentally advantageous boundary structures) were less likely to support beneficial invertebrates than those that were less-intensively managed. Some progress towards the development of practical on-farm techniques, based on landscape approaches, has been made; (Bommarco, 1997), for example, found that the beneficial effects of the abundant predatory ground beetle *Poecilus cupreus* correlated with specific landscape factors, such as the ratio of field area to the length of the field margin. However, there is a general lack of understanding of the factors that cause natural enemies to aggregate in uncultivated linear habitats such as field boundaries and, subsequently, to migrate into fields. Some of these factors are likely to be species-specific, as has been found for several parasitoids of pollen beetle (R. Büchi, *pers. comm.*), and improved knowledge of the ecology of key natural enemies may offer the potential for manipulating them to improve natural control and reduce reliance on pesticides. Similarly, a detailed understanding of the biology can enable appropriately timed push-pull strategies (e.g. the introduction of mowing or the use of pheromones – see below) to be developed to move populations of beneficial arthropods from field margins into the centre of the field where they can exert an effect on pest populations. Single-species studies elucidating autecology and habitat requirements of key antagonists of oilseed rape pests, therefore, are urgently required to provide a basis for further development of integrated pest management appro-

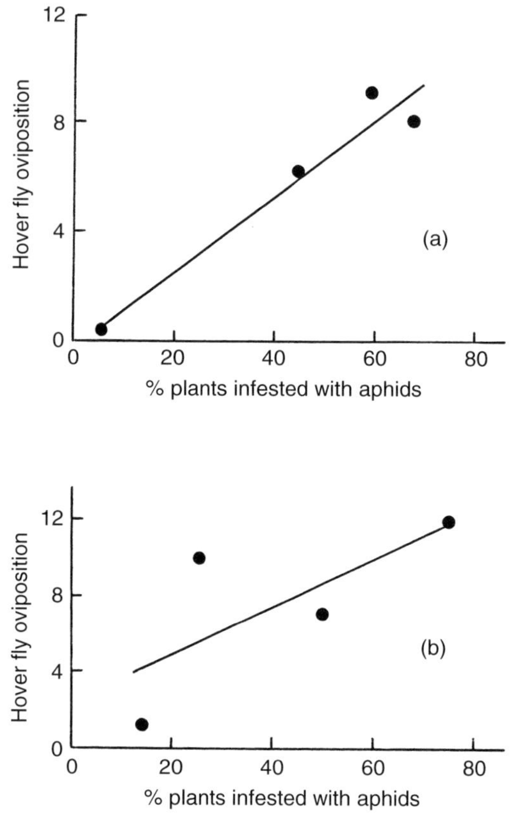

Fig. 3.3 Effect of flowering strips around the edge of commercial cereal fields on the relationship between the number of eggs laid by hover flies and aphid infestation levels on cereals. (a) Fields with flowering strips: $r^2 = 0.95$, $P < 0.05$; (b) fields without flowering strips: $r^2 = 0.53$, $P > 0.05$.

aches (Bommarco, 1997). However, when such techniques have been developed, agro-political issues (e.g. subsidies and future policies on set-aside) will have crucial effects on the future development of cultural landscapes and the role that they will play in commercial farming practice.

Crop Husbandry

Crop rotation

Diverse crop rotations are considered desirable within ICM systems for the purposes of minimizing pests and diseases, and to provide good opportunities to control problem weeds. For example, grass weeds (Poaceae) that cause problems in cereals, e.g. black-grass (*Alopecurus myosuroides*) and wild oats (*Avena fatua*), can be targeted more effectively in a broad-leaved crop, such as oilseed rape, than in a cereal crop.

The content and location of crop rotations at farm level may influence the spatial distribution and incidence of pest and beneficial species, particularly species such as cabbage stem flea beetle that have only limited dispersion potential. Additionally, rotations can promote natural enemies in arable fields. For example, in cereal/oilseed rape rotations, first- and second-wheat crops are often subject to a series of insecticide applications. Populations of several species of predatory epigaeic (ground-dwelling) beetles (e.g. families Carabidae and Staphylinidae) have been shown to recover more slowly after each successive insecticide application (Berraondo *et al.*, 1998). However, these species are also found commonly in the succeeding oilseed rape fields (Büchs & Nuss, 2000), where the more complex structure of the crop during the summer period will reduce penetration of insecticide sprays to the lower horizons more effectively than in cereals (Kjaer & Jepson, 1995). During the period in which oilseed rape is grown in fields, the recovery time of the beetles quickly returns to normal.

There may be scope, annually, to manipulate the placement and distribution of oilseed rape crops on large farms to optimize the spatial distribution and impact of mobile beneficial species, especially hymenopterous parasitoids (Powell, 2000). In order to benefit BCAs, therefore, placement of rape crops in a dispersed 'patchwork' of fields may be a more desirable strategy than growing crops in uniform 'blocks'. The benefits of this strategy may well outweigh the sometimes proposed strategy of blocking oilseed rape crops in order to reduce the perceived risk of carry-over from one year to the next of relatively immobile pests such as brassica pod midge. These topics clearly merit further investigation, particularly in the context of ICM.

Soil cultivation

In ICM systems, minimal cultivation techniques tend to be favoured for a variety of reasons, which include:

- reduction of energy inputs;
- reduction in mineralization and leaching of nitrogen;
- improvement of the physical properties of the top soil;
- introduction of alternative strategies for weed control;
- reduction of soil erosion;
- increase in populations of beneficial organisms.

There is evidence that minimal (non-inversion) tillage throughout the rotation may favour populations of epigaeic beneficial insects, particularly polyphagous predators (Thiele, 1977). Furthermore, various hymenopterous parasitoids of pests such as pollen beetles, ceutorhynchid stem weevils and flea beetles have been shown to be favoured by minimal cultivations, owing to a reduction in mechanical damage to their larvae in the soil. Increased activity of epigaeic predators, such as ground beetles (family Carabidae), may exert a useful regulatory effect on populations of oilseed rape pests that enter the soil before pupation, e.g. larvae of brassica pod midge, cabbage stem flea beetle, ceutorhynchid weevils and pollen beetle (see Chapter 16). Long-term suppression of these pests through the establishment of large and stable populations of predatory and parasitioid species could well result in environmental benefits by reducing the need to apply insecticides.

Crop establishment

The manipulation of the crop canopy and microclimate in ICM systems may offer a means of improving the impact of BCAs, especially that of parasitoids and pathogens.

There is likely to be scope for (and advantage to be gained from) manipulating sowing dates, seed rates and choice of crop cultivar in ICM systems, but detailed investigations yielding more information are required before any potential benefits can be realized in this area.

Nutrients

Nitrogen usage in oilseed rape can be reduced in ICM systems through improved targeting and precision. This can be achieved through the field-specific measurement of residual nitrogen levels in the soil. Techniques for canopy management are also being developed, in which nitrogen use is regulated in order to create a crop canopy that will give an optimum, high-yielding structure. Nitrogen usage is unlikely to have any direct effect on the biocontrol of oilseed rape pests. However, indirectly, the effect of nitrogen on the structure and microclimate of the crop canopy could influence factors such as the searching efficiency and behaviour of hymenopterous parasitoids and the development of entomopathogenic fungi.

The use of other nutrients (such as phosphate, potash and sulphur) in ICM systems does not differ greatly from that in conventionally managed oilseed rape crops and they probably have no effect on BCAs.

Growth regulators

Changes in the duration of the flowering period (normally about six weeks) and canopy architecture, which may be influenced by the use of growth regulators, might interact with the incidence of insect pests and with the activity of hymenopterous parasitoids and other BCAs. However, at present, the impact of growth regulators on biocontrol and their relevance to ICM is unclear.

Harvesting

With regards harvesting, the use of a desiccant is generally avoided, but not precluded, in ICM systems. However, there are no data on the specific effects on BCAs of using desiccants in oilseed rape crops.

Crop Protection

Pest control

In ICM systems, application (or otherwise) of insecticides is governed by crop monitoring, combined with the use of action thresholds. Specific action thresholds have been developed in some countries for a range of oilseed rape pests (for details see Chapter 2). Well-defined guidelines are also used as a basis of treatment decisions for several other pests, including brassica pod midge, slugs and aphid vectors of virus diseases.

ICM systems aim to minimize insecticide and molluscicide usage through the use of action thresholds and through the enhancement of naturally occurring BCAs in the form of parasitoids, predators and pathogens. Decision support systems (DSSs), combined with crop monitoring information, pest forecasting and population modelling techniques, also have a vital role in ICM systems. Although, currently, no fully selective insecticides are available, the use of broad-spectrum products is avoided, whenever possible, and the use of semi-selective products (e.g. the carbamate pirimicarb for aphid control) is encouraged, to minimize harmful effects and impact on non-target organisms, including BCAs. In addition, behavioural responses of some parasitoids and predators may reduce the negative effects of insecticides on natural enemies (Umoru *et al.*, 1996; Singh *et al.*, 2001) and current work is investigating methods of applying pesticides that take advantage of such behavioural traits.

Disease control

In ICM systems, the use of fungicides can be minimized through the growing of rape cultivars with resistance to stem canker (*Leptosphaeria maculans* – anamorph: *Phoma lingam*), downy mildew (*Peronospora parasitica*) and light leaf spot (*Pyrenopeziza brassicae*). Further improvements in disease forecasting and risk assessment are also likely to play an important role in oilseed rape ICM systems by reducing the extent of unnecessary fungicide use. Further, as the application of sprays to control stem rot (*Sclerotinia sclerotiorum*) on oilseed rape tends to overlap with sprays applied against cabbage seed weevil, there is a tendency in some parts of Europe (e.g. in the UK) to include a prophylactic insecticide with this operation. This practice may be reduced in ICM systems through increased knowledge of disease risk and by improved forecasting. Fungicides are generally thought unlikely to have a direct effect on the biocontrol potential of most parasitoids and predators, although responses of beneficial insects to some products have been recorded by Sterk *et al.* (1999); however, most fungicides investigated under their testing programme were found to be toxic to entomopathogenic fungi and, therefore, should be used with care in ICM systems.

Weed control

The use of herbicides remains a vital part of ICM systems in oilseed rape, since row spacing and the unsuitability of the crop prevent the use of mechanical weeding techniques. If a wider row spacing is used (e.g. to save on seed or pesticide costs) then band spraying and mechanical weeding can be an alternative for ICM. Pre-sowing cultural control methods may be employed to control volunteer cereals, as these can increase in conjunction with minimal cultivation. In this instance, applying a surface cultivation immediately after the previous crop is harvested creates a 'stale' seedbed. Provided some moisture is available in the soil, this early cultivation stimulates a flush of germinating weeds, which may then be destroyed by pre-sowing cultivations or following application of a contact herbicide such as glyphosate. In favourable circumstances, this cultural practice may remove the need to apply a post-emergence graminicide.

Choice of pesticide product

Within ICM systems, there is an obvious need to choose pesticides that have maximum selectivity and minimum environment impact. Residual products with broad-spectrum activity against non-target flora or fauna, and those that may be a threat to groundwater purity, are avoided in ICM systems. In oilseed rape, insecticide choice is predominantly limited to the synthetic pyrethroid insecticides. Consequently, there is only limited scope to select products with superior safety to non-target species. However, the synthetic pyrethroids are deemed to have better operator safety than many of the older organochlorine (OC) and organophosphorus (OP) insecticides. Initiatives are currently being developed within the UK agrochemical industry to provide the user with more detailed information on the environmental and ecological safety profiles of certain pesticides, and this will assist users in selecting the most appropriate and safest pesticide for their needs. The implementation of holistic ICM ultimately depends on the efficient interpretation and delivery of information to the end user: namely, the farmer.

The safe integration of pesticide use with BCAs is a vital factor in the development of ICM systems. The IOBC gathers and collates information and data on the safety of all types of pesticides to biocontrol organisms commercially available for release in biological control systems on horticultural crops. The same information and service could be extended to improve the safe integration of pesticide use with the main naturally occurring BCAs found in oilseed rape crops.

Reduced rates of pesticides

In recent years, it has become an increasingly common practice to apply many pesticides at rates of use below those recommended by manufacturers on product labels (i.e. at less than published 'label rates') (Poehling, 1987). This practice is adopted in conventional and ICM regimes, and is often referred to as 'appropriate dose' technology. The main advantage of 'reduced rates' is in the reduction of variable costs whilst maintaining acceptable levels of weed, pest or disease control. Reduced rates are most frequently used for fungicides and herbicides, as the financial benefits are more rewarding with these materials than with lower-costing insecticides.

Nevertheless, reduced-rate insecticide use may be of value in reducing the toxicity of certain products to beneficial parasitoid and predatory arthropods involved in the biocontrol of oilseed rape pests. There is evidence from work with cereal aphids in the UK, for example, that reduced-rate applications of certain synthetic pyrethroids are safer to hymenopterous parasitoids than full-rate applications (Oakley *et al.*, 1996). Reduced-rate applications may, therefore, allow greater survival of certain beneficial species, which may then (through exercising an element of biocontrol) be able to prevent resurgence of the pest population. This technique may be applicable to synthetic pyrethroids applied to control oilseed rape pests such as cabbage seed weevil, where the activity of hymenopterous parasitoids is known to be an important factor in regulating the pest population (see below).

Novel techniques of pest control

There is great scope to introduce novel, non-insecticidal, pest control techniques into ICM systems. For example, trap or sacrificial cropping and the use of insect semiochemicals or anti-feedants may be possible methods to divert pests (such as adult pollen beetles and cabbage seed weevils) away from commercial crops. Semiochemicals may also be employed to attract insect parasitoids and predators into a crop. Push-pull techniques (stimulo-deterrent diversionary strategies) (Hardie *et al.*, 1991; Pickett *et al.*, 1992; Vet & Dicke, 1992; Glinwood *et al.*, 1998; Powell *et al.*, 1998; Powell, 1999, 2000) have a potentially useful role in ICM systems and, in the future, could well be applied to insect pests of oilseed rape.

Technology Transfer: Making Complex Integrated Pest Management Strategies Accessible to Farmers

Throughout Europe, increasing economic and environmental pressures on the agricultural industry have resulted in the need to take full advantage of the recent developments in non-chemical control and forecasting of oilseed rape pests, in order to rationalize pesticide usage in the crop. Although the specific results of such work have been transferred rapidly into advisory practice, they have highlighted complications in the development of an overall management strategy for oilseed rape; this is due to the incidental effects of treatments applied to control target pests on populations of other pests and on naturally occurring antagonists (Lane & Walters, 1995). In addition, the number and nature of these advances can be difficult to apply in practice. For example, if too much information or if labour-intensive inputs (e.g. field walking) are required, their implementation can become uneconomic; in combination with existing approaches, they may also be too time consuming to be of practical use. Accordingly, to utilize the current knowledge of the biology and control of pest species, a method of integrating the results of new studies and presenting them in a suitable form for farmers and their consultants is required.

Computerized models and decision support systems (DSSs) represent a means of assimilating and synthesizing existing and new data, and these can provide a mechanism for transferring appropriate information to decision-makers (Morgan & Walters, 1999). Many models and DSSs have been developed for use in IPM schemes on other crops, and they range in their complexity from a single statistical equation to a comprehensive and complex software package that incorporates expert systems, simulation models, databases and text/graphics encyclopaedias (Knight, 1997). To enable holistic management decisions relating to the pest complex damaging oilseed rape to be made effectively, DSS technology has been used to facilitate the application of recent work in the field. This has resulted in interactive systems that automate the selection and running of constituent models, reducing confusion when applying recent advances. These interactive systems also share inputs common to different models, in order to minimize time spent on expensive field walking and gathering of other input data (such as weather records). Critically, they can take account of the incidental effects of treatments, enabling more complete integration of decision-making for different pests and thus supporting the option of integrating chemical and biological control.

A major example of such models is the weather-based DSS for managing oilseed rape pests (PRO_PLANT) which has been produced in Germany (Johnen & Meier, 2000). The system was first developed for managing cereal diseases and later extended to cover growth regulators in cereals, diseases of sugar beet, weeds in maize crops, potato late blight and, most recently, pests of oilseed rape (Epke *et al.*, 1996; Frahm *et al.*, 1996). The system considers cabbage stem flea beetle in autumn, and pollen beetle, rape stem weevil (*Ceutorhynchus napi*), rape winter stem weevil, cabbage seed weevil and brassica pod midge in spring. Prior to the development of the oilseed rape pest module, control in Germany was based on pest catches in yellow traps and assessment of associated damage thresholds. Problems occurred when thresholds were related to non-critical stages (for example, the number of adults of cabbage stem flea beetle) since, as a consequence, variable reproduction rates, and subsequent differential survival or rate of development of eggs and of the damaging larval stages (Bonnemaison & Jourdheuil 1954; Hoßfeld, 1993) were ignored. There can also be over-reliance on adult assessments when larvae are the critical damaging stage in the lifecycle (Johnen & Meier, 2000). Spring pests were also problematic, since these can have long invasion periods (two to three months), resulting in the application of several insecticidal sprays.

PRO_PLANT utilizes temperature-based models to predict the need for, and optimal timing of, chemical treatments. The system requires information on the crop (growth stage, growing conditions, previous insecticide sprays), field assessments of pest densities, and weather data as inputs. The main component of the application is a phenology model, which is used to determine the optimal timing of field inspection and, subsequently, the justification for, and timing of, any treatment application. The first calculations determine the flight phase (start and end of adult migration into the crop) and provide an estimation of the percentage of adults that have already migrated at any point during this phase. After arrival in the crop, the period required for ovary maturation is determined. Where appropriate, as in the case of cabbage stem flea beetle during autumn, the beginning of egg hatch and the rate of larval development are also calculated, to predict the dates of occurrence of the second- and third-instar larvae (owing to differential effectiveness of insecticides between larvae of different ages). Finally, the need for treatment is determined by a field-specific risk assessment, based on observed pest index, crop parameters, weather-based phenological data and established thresholds. If treatment is deemed necessary, selection of an appropriate insecticide is also addressed. Use of PRO_PLANT has been shown to reduce substantially farmers' reliance on insecticides. Current 'normal' practice in Germany results in two to three applications of insecticides per season to control oilseed rape pests, whereas a strategy using PRO_PLANT recommended only one to two applications (Johnen & Meier, 2000). This represents a considerable improvement and offers environmental benefits, including the encouragement of naturally occurring BCAs, which will further reduce the need for pesticide use.

Case Study: User-Guided Development of a Decision Support System for the Management of Pests of Oilseed Rape

An analogous DSS, known as DORIS (*D*ecisions for *O*ilseed *R*ape *I*nsect *P*est*S*), has been developed in the UK. It addresses the main damaging pests of oilseed rape in the UK: namely, aphids acting as vectors of plant viruses, cabbage stem flea beetle, rape winter stem weevil and slugs in autumn/winter; and brassica pod midge, cabbage aphid, cabbage seed weevil and pollen beetle in spring/summer. Two versions are available: one an internet application and the other a free-standing PC-based model (Morgan *et al.*, 1998). In common with PRO_PLANT, DORIS is designed to utilize available information to optimize insecticide usage on oilseed rape, whilst maintaining economic control of damaging insects and thereby reducing overall pesticide usage on the crop. It is also intended to promote, and improve accessibility to, non-chemical methods of pest control on oilseed rape.

An important feature of the development of DORIS was a close interaction with industry representatives, including potential end-users such as farmers, consultants, the agrochemical sector and supermarkets. Two-way consultation commenced at the initial concept phase and was repeated at intervals throughout the developmental stages. This ensured that advice given was correctly interpreted and that the system was updated when requirements were altered, owing to the rapidly changing economics and constraints faced by the industry. The result has been a system that is sufficiently flexible to accommodate readily any future, as yet undefined, additions or changes, and which is widely viewed as addressing the current key concerns of the industry. Initial consultation indicated that the current approach of treating each specific pest as an individual problem was not sustainable. Also, implementation of recent advances in pest management was constrained by the time required for pest incidence assessment/decision-making for each of the eight key invertebrate pests; further, the relatively low cost of insecticides, coupled with the potential for sharing application costs by tank-mixing with fungicides, made prophylactic control an attractive option. Therefore, to optimize insecticide use, pests needed to be treated as a single problem (spanning the whole growth cycle of the crop) rather than as a sequence of different problems. In addition, all time-consuming calculations needed to be automated, the provision of information should be timely and concise, and the system capable of relying on only minimal data input by the user. It was essential, ultimately, that decision-making still rested with the practitioner, the system thereby playing a supporting role. Finally, commercial considerations made security of field/farm-specific information important for some users, and all users required an approach that did not rely on their having to run the application on a daily or a weekly basis.

To avoid regular abortive access, the internet application uses a rule-based model to alert e-mail users when there is a need to run DORIS. System access is via a standard username/password approach, providing commercial security. An individual farmer may have more than one field of oilseed rape, and consultants will have many farms (each of which may have multiple fields). Therefore, upon entering the application at the beginning of a cropping season, the user is able to establish a hierarchy of farms and fields, which enables field-specific information to be calculated, stored and used (Morgan *et al.*, 1998). This obviates the need for users to maintain multiple copies of the program to accommodate multiple farms or fields. At subsequent access, the user selects the farm/field for assessment and the system loads appropriate information into the model. An unexpected advantage of this approach was identified by supermarket representatives who indicated that this could automate provision of some of the traceability data that they now require.

DORIS has been built as a dynamic structure, incorporating computer models, databases, pest action thresholds, cost benefit/economic calculations and encyclopaedic information. It can be viewed as a series of interacting models, linked via a central core program (Morgan *et al.*, 2000) that incorporates intelligent automation and is designed to prevent inappropriate use of components.

Having selected a field, the user is offered a choice of pest models to run. These are constrained by the system to include only those for pests that may currently represent an economic risk to the crop, thereby avoiding time wasting. Inevitably, the amount of information available for each pest is variable, making integration of decision-making into a coherent holistic application difficult. DORIS addresses this problem by dealing with each pest individually, but by taking account of activities of, and actions against, other pests which may influence the decision. For example, brassica pod midge adults lay their eggs through feeding damage to pod walls caused by cabbage seed weevil. Thus, in the presence of brassica pod midge, the weevil represents a greater risk than on its own, and the potential for economic damage and, therefore, the need for control are adjusted accordingly. Complexity was avoided, wherever possible, to ensure that model run-time is minimized (another requirement identified by the consultation process). This approach presented the added advantage of simplifying the validation of DORIS. Models vary from simple rule-based approaches to more complicated population or behavioural ones. This approach to the development of DORIS has met user requests that pests should be treated as a single problem, that all time-consuming calculations should be automated, that provision of information should be timely and concise, and that the need for the application to be run on a daily or weekly basis should be avoided.

Another important requirement highlighted by user groups was the need for the system to integrate readily with other on-farm practices and decision-making. At a basic level, it was essential that the system did not generate extra labour requirements (e.g. increased time spent assessing pest incidence) during periods when other priority work needed to be done. More complicated was the requirement that it should be capable of being combined with crop disease control. Currently, this is addressed by requiring the user to input his/her disease control intentions. For example, DORIS calculates the need for control of cabbage seed weevil on the basis of standard inputs, identifying three decision categories. If pest incidence alone justifies the use of a control measure, then the recommendation and supporting information are delivered to the user. Alternatively, if control is not justified by pest populations alone, a further calculation is made to determine whether shared application costs resulting from tank-mixing with, for example, a fungicide would make treatment economically viable; if not, then a 'no-treatment' recommendation is delivered, but if treatment becomes a viable option the user is asked whether or not tank-mixing is intended and an appropriate response is then delivered by the system. This has received such a positive response from the industry that further work is now underway to develop appropriate disease models that will enable the construction of a combined pest and disease DSS for oilseed rape.

A major difficulty when developing a holistic DSS is the effective and economic provision of sufficient input data to enable robust conclusions to be reached. An early requirement of the industry was that input by users should be minimized and time spent collecting data and running the model should be less than for current approaches. To meet this constraint, automated remote data acquisition is used wherever possible. This currently reduces user-input requirements to a maximum of five data items for a single model. Data are also stored by the system and, where appropriate, re-used. To achieve this, the pest models are supported by a series of databases that record previous decisions, weather data and other information. In addition, databases have been developed to support the encyclopaedic components of DORIS; these contain information on oilseed rape pests and their natural enemies (Walters *et al.*, 2000), details of pesticides (including registration constraints, compatibility with natural enemies, etc.), national pesticide usage and national pest incidence, all items requested by industry representatives. Updating such databases can be a time-consuming and expensive component of running systems such as DORIS. To minimize these costs and to enhance its cost-effectiveness for users, links have been made to other databases that are remotely updated; for example, information on the latest pest control products can be accessed from LIAISON (Live Interactive Agronomic Information Service on the Net) (Thomas *et al.*, 2000).

A unique feature of DORIS is the integration of chemical and biological control approaches for the management of oilseed rape pests. Industry representatives welcomed the opportunity to gain the environmental and economic advantages of utilizing naturally occurring parasitoids and predators in management systems for the major pests of the crop. However, it was made clear that the requirement that decision-making is ultimately left with the practitioner was essential in this area, owing to concerns centred on the reliability of natural control. In addition, the general instructions relating to minimal data input by the user, and avoidance of running the application on a daily or weekly basis, were stressed as being of particular importance in relation to this component of the DSS. Accordingly, an approach had to be adopted that did not rely on regular sampling of natural enemies by the end user to determine the level of control that could be exerted. The problem was addressed by retaining the current (well-tried and accepted) approach, widely used by farmers/consultants, and then utilizing the potential of the DSS to facilitate accessibility to selected information in a timely and concise fashion.

For example, based on the work described below, the naturally occurring parasitoid *Trichomalus perfectus* is used in conjunction with selected chemicals in the management of cabbage seed weevil. During the calculated risk-window, the system asks the user to take standard beating-tray samples (Walters & Lane, 1994a) to assess the number of cabbage seed weevils in the crop. Combining these data with weather data, crop growth stage and other information, the model then calculates into which of the three decision categories (see above) the field is placed; where appropriate, the model also takes account of tank-mixing (disease control) decisions. If treatment is economically justified, the system automatically presents the user with a list of chemicals, with a sub-set flagged as being compatible with an IPM approach that utilizes *T. perfectus*. In addition, a statement is made concerning the advantages of using the IPM approach and a link is provided to more detailed information, which the user can select if required. Thus, the farmer is asked to do no more than to continue to use the familiar, trusted approach to the control of this pest, but to apply insecticides only where necessary. If treatment is needed, then users are asked to select from a list of products known to have little or no effect on parasitoid populations. Hence, the user maintains complete control of the decision-making process and retains the option of using insecticides whenever he/she thinks it appropriate to do so. Nevertheless, use of the system will result in greater survival of *T. perfectus* on the farm, leading to increased parasitoid populations and a reduced probability of chemical treatment being justified in future years. Continuous use of the system over successive seasons should result in the development of progressively larger populations of *T. perfectus* on the farm. The success of this element of the DSS has resulted in plans to add further IPM components once sufficient information on other natural enemies becomes available.

The development of DORIS was guided by continuous industry-based feedback from the establishment of the initial concept to the completion of the advanced prototype. This approach gave potential users confidence that the DSS addressed their key problems and their concerns relating to the control of oilseed rape pests; also, it was seen to be robust and suitable for use on commercial crops. The output was formatted in a style, and on a time-scale, that enabled its ready integration into day-to-day farm activities. As experienced during the development of PRO_PLANT (Johnen & Meier, 2000), despite the large amount of information that has been published on oilseed rape pests and the availability of extensive data sets collected from manipulative experiments, critical gaps were found in current knowledge and these had to be filled. In particular, industry consultation frequently identified readily achievable components of the management system that would offer substantial advantages for farmers, but which had been overlooked by academic researchers. For example, the need for a more cost-effective assessment technique for cabbage stem flea beetle was raised by several crop consultants; this was addressed by re-analysis of existing data sets, and the revised method immediately adopted and used by the industry (Walters *et al.*, 2001). In addition to new developments stimulated by DSS construction, existing advances that, hitherto, had been too time consuming or complicated for on-farm application were automated at the behest of the industry and made assessable and usable by inclusion in the DSS. An example was the uptake of an improved, streamlined method for assessing the need to control cabbage seed weevil adults to prevent subsequent crop damage by the larvae (Walters & Lane, 1994b). The existing long-term data sets, collected by annual monitoring of the major pests in commercial crops throughout England and Wales, were an essential component of the DSS development.

The consultation group recognized that DSSs provide opportunity for real-time experimentation with different management decisions, to estimate their potential costs and effects; this offers further opportunity to optimize decision making on a field-by-field basis. Moreover, the group considered that the encyclopaedic elements of DORIS were an important overview of the problems faced by growers, and the system itself was a compendium that offered up-to-date solutions that farmers could select to download. Thus, the DSS also has a role in raising awareness of crop protection amongst advisors and farmers and can assist in the introduction of new methods and approaches, such as techniques for enhancing control by naturally occurring parasitoids and predators.

Ultimately, a DSS approach will support the general move towards reduced pesticide use and greater reliance on biological control by naturally occurring predators and parasitoids. It facilitates the integration and collation of a range of factors that need to be considered when operating an IPM system in an expansive, field-scale crop. It can determine when economic action thresholds are exceeded, and can modify them to take account of the presence of natural enemies. It can calculate when tank-mixing insecticides are economically justifiable, thus avoiding the unnecessary and damaging use of insecticides. Timing of essential treatments can be optimized to avoid periods in which adult parasitoids or predators are active in the crop or when bees (and other non-target organisms) may be harmed. Information essential to the correct selection of chemicals for use in IPM systems can be provided in a timely and efficient manner; for example, advice to avoid the use of carbamate-based molluscicides when predators such as ground beetles are present in the field. General advice on crop management can also be transmitted. For example:

- over-cleaning of fields is undesirable, as this reduces the survival of BCAs;
- appropriate management of field margins can encourage natural enemies;
- ploughing might usefully be avoided if post-harvest non-inversion tillage systems that enhance survival of parasitoids or predators are possible.

The success of DORIS has been based on industry guidance throughout its development; in the case of other such systems, this has led to improved initial uptake following its validation and release. Such consultation is an essential component in the development of complex management support systems, and should be widely adopted in any future schemes.

Case Study: An Integrated Management Strategy for Cabbage Seed Weevil in the UK

Cabbage seed weevil is widespread in the UK and the major pest of winter oilseed rape (Alford *et al.*, 1996). Adults fly into the flowering crop during April and May when temperatures rise to 15°C and above (Kjaerpedersen, 1992), attracted by visual (Smart *et al.*, 1997) and chemical (Evans & Allen-Williams, 1992, 1993; Blight *et al.*, 1995) cues from the crop. The adults feed on the crop for three to four weeks, while the ovaries of the female weevils mature, and eggs are eventually deposited within the developing seed pods (Free & Williams, 1979). An oviposition-deterring pheromone secreted by a gland in the female abdomen ensures that, usually, only one egg is laid in each pod (Mudd *et al.*, 1997; Ferguson *et al.*, 1999). Punctures in pod walls produced by adult feeding and egg-laying are often utilized by female brassica pod midges for oviposition (Alford *et al.*, 1991). Cabbage seed weevil eggs hatch after eight to ten days and larvae feed for four to five weeks on the developing seeds before boring an exit hole in the pod wall and pupating in the soil (Williams & Free, 1978). There is a strong correlation between heavy infestations, crop damage and reduced crop yields (Winfield, 1986; Buntin *et al.*, 1995). New-generation adults emerge in August, before harvest, and feed on various brassicaceous plants such as garlic mustard (*Alliaria petiolata*) before hibernating for the winter (Free & Williams, 1978).

Over 20 species of hymenopterous parasitoid have been reported to attack cabbage seed weevil in the UK (Murchie & Williams, 1998a) – for further details see Chapter 6. The ectoparasitoid *Trichomalus perfectus* (Hymenoptera: Pteromalidae) is widely distributed and particularly important for the control of cabbage seed weevil larvae in both continental Europe and the UK (Rosen, 1964; Lerin, 1987; Alford *et al.*, 1995; Murchie & Williams, 1998a). In addition to killing larvae through direct parasitism, *T. perfectus* reduces damage to infested pods by preventing host larvae from eating their full complement of seeds (Murchie & Williams, 1998b); an element of host feeding by the adult wasps (see Chapter 6) also occurs. Cabbage seed weevil tends to congregate at the edges as it migrates into the crop before spreading throughout the crop later in the season; *T. perfectus* adults, which arrive in the crop slightly later than the weevils, are distributed at the edge of the crop only during the earliest phase of their immigration period (Free & Williams, 1979; Ferguson *et al.*, 1999; Murchie *et al.*, 1999). The host-finding behaviour of this parasitoid has been studied in the laboratory by Dmoch (1997). *T. perfectus* females usually lay a single egg on the body surface of third-instar host larvae, and the egg hatches about three days later (Dmoch & Klimek, 1975). Larval and pupal stages last for an average of seven and eight days, respectively, depending on conditions (such as the ambient temperature).

Basis for chemical control decisions

Historically, strategies for managing cabbage seed weevil on oilseed rape in the UK have been based on field walking, and acceptance of an associated set of economic thresholds that determine the economic justification for treatment with an insecticide. The start of weevil migration into crops varies from year to year and from farm to farm. Thus, assessment of population levels during flowering and early pod formation is necessary to assess the risk to crops (Free & Williams, 1979). For control decisions, it is recommended that winter rape should be sampled for adult cabbage seed weevils on three occasions during the flowering period, using a beating-tray method (Walters & Lane, 1994a). More recently, yellow water traps, combined with isothiocyanate/methyl salicylate odour lures (Free & Williams, 1978; Smart *et al.*, 1996; Murchie *et al.*, 1999) and sticky-card traps (Smart *et al.*, 1996), have been developed and can be used to monitor the pest and its parasitoid. However, no clear thresholds for their use in decision-making for insecticide treatments have as yet been established. The use of cube-shaped emergence boxes has also been recommended to monitor overwintering sites of cabbage seed weevil and other pests of oilseed rape (Büchi, 1986).

Oilseed rape plants possess a considerable capacity to compensate for bud and pod damage (including that caused by pests), although late damage may result in the presence of immature pods at harvest (Williams & Free, 1979). Economic treatment thresholds for cabbage seed weevil on modern winter rape cultivars, grown under current farm practices, have been confirmed as two adults per plant during flowering or one adult per plant in those prone to pod midge attack (Lane & Walters, 1993, 1994; Alford *et al.*, 1996) (see also Chapter 2). Shared application costs (including consideration of wheeling damage caused by sprayer machinery) result in the threshold being lowered to one weevil per plant if the insecticide is tank mixed with a fungicide.

Efficacy of *Trichomalus perfectus* for control of cabbage seed weevil

During the mid-1970s, a period during which few insecticide applications were made to winter oilseed rape in the UK, high levels of natural control of cabbage seed weevil were exerted by the naturally occurring pteromalid parasitoid *Trichomalus perfectus*, often in excess of 70% (D. V. Alford, *in litt.*). Since then, insecticide usage on oilseed rape has increased dramatically (Thomas *et al.*, 1997); in 1990, over 390,000 ha of rape was grown and over 546,000 ha received an insecticide treatment (equivalent to over 1.4 treatments/ha). Coincident with this increase in insecticide use, the levels of *T. perfectus* parasitism on commercial crops greatly declined.

Multi-site, large-plot field experiments instigated in the early 1990s (Alford *et al.*, 1996) provided evidence that insecticide use may have contributed to parasitoid decline. In each of the three years studied, parasitism of cabbage seed weevil larvae in unsprayed plots exceeded 50% (Table 3.3), with over 90% parasitism recorded in individual plots. The

Table 3.3 The effect of pesticides on parasitism of cabbage seed weevil (*Ceutorhynchus assimilis*) larvae by *Trichomalus perfectus* (after Alford *et al.*, 1995, 1996).

Tear	Treatment (timing)	Mean % larvae parasitized
1992	Untreated	49
	Alphacypermethrin (during flowering: GS 4.8)	53
	Triazophos (post-flowering: GS 6.1)	19
1993	Untreated	65
	Alphacypermethrin (during flowering: GS 4.8)	71
	Alphacypermethrin (post-flowering: GS 6.1)*	61
	Triazophos (post-flowering: GS 6.1)	11
1994	Untreated	51
	Alphacypermethrin (during flowering: GS 4.8)	49
	Triazophos (post-flowering: GS 6.1)	29

GS = growth stage (codes according to Sylvester-Bradley & Makepeace, 1984).

* Experimental treatment only.

Table 3.4 Decline in the use of triazophos and the level of parasitism of cabbage seed weevil (*Ceutorhynchus assimilis*) larvae by *Trichomalus perfectus* in commercial crops (after Alford *et al.*, 1996).

Year	Area of rape grown ('000 ha)	Area ('000 ha) treated with		Mean level of parasitism in commercial crops (%)
		a pyrethroid	triazophos	
1990	343.8	219.0	48.2	8 ± 15
1991	387.8	247.1	72.7	4 ± 5
1992	362.9	168.2	5.4	5 ± 7
1993	314.8	159.9	2.1	48 ± 31
1994	333.9	165.6	2.7	49 ± 22
1995	301.3	196.5	5.2	58 ± 26

pyrethroid insecticide alphacypermethrin, when applied conventionally during flowering, had no detrimental effect on levels of parasitism. However, post-flowering applications of the organophosphate insecticide triazophos were followed by reduced levels of parasitism in each of the three years. The apparent differences between the reaction to the two insecticides may have been the result of differences in the crop growth stage at which the two are conventionally applied. Accordingly, the effects of flowering treatments targeted to control cabbage seed weevil immediately after its migration to oilseed rape crops may have declined by the time the later-migrating adults of *T. perfectus* arrived in the crop. This possibility was investigated in 1993 by applying a late (experimental) treatment of alphacypermethrin at the same time as the triazophos spray, but the differential parasitism levels following the two products were still apparent.

After 1992, the use of triazophos in commercial crops declined rapidly, offering an opportunity to observe the effect on natural populations of *T. perfectus* on a farm scale. Comparing data on pesticide use collected by the UK Pesticide Usage Survey (Garthwaite *et al.*, 1995) with records from stratified pest and parasitoid monitoring on commercial crops throughout the main rape-growing areas of England and Wales (see Walters *et al.*, 2001), levels of parasitism of cabbage seed weevil larvae by *T. perfectus* were shown to be low during the period of high triazophos use in 1990 and 1991 (Table 3.4). Parasitism rates remained low in 1992, despite the sudden reduction in the amount of the insecticide that was applied; however, after 1993, parasitism rates increased to around 50%, close to the levels recorded in experimental plots. Pest mortality of 50%, resulting from the activities of a naturally occurring parasitoid, would substantially reduce the reliance on chemical insecticides if incorporated into an integrated control strategy.

Integrated management strategy

Probably as a result of the decline in use of insecticides such as triazophos, levels of parasitism of cabbage seed weevil have increased spectacularly and appear to be stabilizing throughout the main oilseed rape growing areas of England and Wales (Alford *et al.*, 1996). The overall effectiveness of *T. perfectus* at controlling cabbage seed weevil is such that it now allows the promotion of a robust biological control strategy in which, irrespective of year-to-year cycles in pest abundance, spraying against cabbage seed weevil could be virtually eliminated. On a crop so attractive to pollinating and other non-target invertebrates, avoiding or at least minimizing the use of insecticides would also have considerable environmental benefits (Alford *et al.*, 1991).

To encourage uptake, it is important that new pest management strategies should, wherever possible, develop rather than replace the existing well-tried and proven methods currently used by farmers. In addition, the strategy should avoid the use of expensive or complicated equipment with which farmers are unfamiliar and it should not require more staff time to operate than current methods. The strategy suggested to take advantage of the increased level of parasitism of cabbage seed weevil by *T. perfectus* meets all these criteria (Alford *et al.*, 1996). As presently recommended, crops of winter rape should be sampled for adult cabbage seed weevils during flowering, using the beating tray method. Only if samples clearly indicate a population in excess of the current economic threshold of two weevils/plant (adjusted downwards, if necessary to take account of brassica pod midge) should an insecticide be applied. However, in recognition of the expected beneficial role of parasitoids, if a spray is required, then a relatively harmless pyrethroid should be used during flowering; post-flowering sprays, and pesticides known to be detrimental to *T. perfectus*, should be totally avoided.

In view of the established beneficial effects of *T. perfectus*, farmers should now have increased confidence when applying thresholds, as any underestimates resulting from errors in pest population assessments are likely to be corrected by the activity of naturally occurring BCAs, thus avoiding the risk of even small losses. Further, since survey data have shown conclusively that, nationally, few crops exceed recommended treatment thresholds and that the exceptions usually do so by only a small margin (Alford *et al.*, 1996), the risk of crop losses from not spraying, even in any expected 'worst cases' situation, is negligible. A fail-safe option of spraying if absolutely necessary is still available. By using a pyrethroid at flowering the farmer will remain in full control of decision-making on pest management, whilst his/her actions will have minimal impact on the basic no-spray strategy. However, cost effectiveness and overall long-term benefits would be lost if the farmer perceived the latter option to be acceptable as routine.

Management of populations of cabbage seed weevil on winter oilseed rape needs to be considered in relation to other spring/summer pests, particularly brassica pod midge and pollen beetle. Brassica pod midge is attacked by various parasitoids (Williams & Walton, 1990) (see also Chapter 7) and, because of the main timing of their flight activity, these will probably be affected adversely by post-flowering applications of insecticides. Pollen beetle is also attacked by various parasitoids (see Chapter 4); these are often abundant on brassica seed crops and are known to be susceptible to insecticides applied in May or early June (Winfield, 1963). Unfortunately, no up-to-date data on oilseed rape are available, but numbers of such parasitoids will probably be reduced when chemicals are applied against spring/summer pests, both during the bud stages and at later growth stages. Thus, parasitoids of pests such as brassica pod midge and pollen beetle need to be considered and safeguarded if an overall holistic biological control strategy for the main UK pest complex on winter oilseed rape in spring/summer is to become a reality. The complications involved with developing an IPM strategy for a series of co-existing pests, rather than for the single pest investigated in this section, illustrate the problems faced by farmers if they are asked to operate complex systems. To assist farmers to apply such systems effectively, and to maximize economic and environmental advantages, modules that provide practical solutions for safeguarding natural enemies (whilst maintaining economic control of pests) have been incorporated into DORIS, as described above. These modules will be further developed as information becomes available, to provide a sound basis for technology transfer.

Evidence is now accruing to indicate that when an integrated management strategy for cabbage seed weevil is followed on a farm over a series of cropping years, numbers of *T. perfectus* increase and weevil populations are progressively suppressed. This reduces the likelihood that damaging pest populations will develop in subsequent years, and is a compelling reason for researching, developing and utilizing the potential of naturally occurring BCAs on offer. Winter oilseed rape, as a major arable crop, offers genuine opportunities for integrated control of an important pest complex by utilizing natural enemies. Detrimental effects of insecticide use on BCAs have been recorded in some situations (Raw, 1989; Ulber *et al*., 1990; Evans & Scarisbrick, 1994). However, these can be overcome by careful selection of product and appropriate timing of application (Umoru *et al*., 1996).

Conclusions

Oilseed rape is a unique and relatively high-input crop in terms of demand for nitrogen and pesticide. There is great scope to reduce such inputs on oilseed rape, and with careful and flexible management strategies it is possible to adopt an ICM approach and to achieve environmentally and economically beneficial reductions in the use of both nitrogen and conventional pesticides.

Naturally occurring biocontrol agents (BCAs) are widely distributed and often common on commercial oilseed rape crops, and these should be treated as a valuable natural resource to be encouraged and safeguarded and, wherever possible, utilized in modern crop protection practices. Farmers and extension workers should be made aware of the potential of biocontrol of oilseed rape pests and of the benefits that can accrue from the enhancement of natural enemies in the farming environment. Wherever appropriate and possible, conventional treatment thresholds advocated for oilseed rape pests should be modified to take account of the presence of natural enemies. In order to maximize the beneficial effects of naturally occurring antagonists in oilseed rape fields, or those associated with oilseed rape crops, insecticides should be applied only when economically justified pest thresholds are reached, and the routine inclusion of insecticides when tank-mixing pesticides should be avoided wherever possible. When treatment is essential, insecticidal sprays should not be applied at times when adult parasitoids or predators are active in the crop, particularly during the flowering period when parasitoids of pollen beetles and ceutorhynchid stem weevils may be harmed. Similarly, insecticides should not be applied at times when bees (which are potential vectors of entomophagous, i.e. beneficial, pathogens as well as useful pollinators) are active in the crop. If it is known that predators such as ground beetles are active in crops, wherever possible use of carbamate-based molluscicides should also be avoided.

Although general information is available on the effects of pesticides on beneficial organisms, their impact on key natural enemies of oilseed rape pests, particularly under field conditions, is less well understood. Data are also wanting on the behavioural traits of natural enemies and on the effects that crop canopy structure and crop architecture have upon them, all factors that can influence their field performance as BCAs.

Other crop husbandry decisions should also be made with BCAs in mind. For example, ploughing should be avoided if other post-harvest non-inversion tillage systems that will enhance survival of parasitoids and predators are possible. Fields should not be over-cleaned and, if feasible, some green cover and crop debris on the soil surface should be maintained between crops, again to enhance survival of BCAs (Powell *et al.*, 1985). Usefully, field margins can be left uncultivated or, alternatively, can be sown with tussock-forming grasses or native perennial wild flowers. These will provide potential overwintering habitats for BCAs or will encourage the development of large populations of natural enemies during the summer. Properly managed, the same areas can significantly increase farmland biodiversity, which is in itself an advantageous side effect.

The potential benefits to be gained from the presence of natural enemies should be recognized in future DSSs. A recent assessment of the overall consequences of phasing out the use of pesticides in Denmark (Bichel, 1999) supports the view that the integration in DSSs of chemical and non-chemical methods (as has been achieved in the DORIS application) could lead to a significant reduction in pesticide use on some crops. The Danish study suggests that, in the absence of pesticides, populations of above-ground insects might be increased by factors of between 1 and 2.5 in the absence of fungicides, by 2 and 4 in the absence of insecticides, and between 2 and 7 in the absence of herbicides. Inevitably, this would include enhancement of populations of natural enemies, especially predators.

In terms of an overall biocontrol strategy for oilseed rape, the whole farm environment (i.e. the on-farm landscape structure) needs to be considered, in order to encourage the survival of naturally occurring beneficial organisms. In this respect the provision of flowering strips on field margins, which can, for example, enhance the parasitization rates of the ichneumonid parasitoids *Phradis* spp. and *Tersilochus heterocerus* on pollen beetle larvae (Büchi, 2002), should be encouraged. Similar encouragement should be given to the establishment and maintenance of mixed on-farm habitats (including refuges) likely to sustain natural enemies throughout the year (Kromp, 1999). Other cultural practices warrant consideration, including the effects of crop rotations and the potential for planting oilseed rape crops as a patchwork rather than in blocks. Trap crops may also have a role to play in crop protection (Büchi, 1990) and these require further attention.

Any proposed ICM system on oilseed rape must be realistic and commercially acceptable. Adoption of ICM on the crop will be complex to manage, despite support from DSSs such as DORIS or PRO_PLANT, and will take time to be implemented. Interactions associated with the adoption of advocated ICM-based practices need careful consideration to ensure satisfactory results, for example non-inversion tillage can exacerbate slug problems (Voss *et al.*, 1998).

Finally, the aims and objectives of conserving and/or enhancing the biocontrol of insect pests in oilseed rape are fully compatible with those of generally accepted ICM strategies. Compared with certain horticultural crops, the development of biocontrol in outdoor field crops remains in its infancy. However, there is great potential to refine ICM practices to promote biocontrol in oilseed rape, the main environmental and agronomic benefits of which lie in minimizing insecticide dependence and use.

Acknowledgements

This work was part-funded by the European Commission Concerted Action: *Minimizing pesticide use and environmental impact by the development and promotion of bio-control for oilseed rape pests, FAIR 96-1314.* Receipt of MAFF (now DEFRA) funds is also acknowledged.

References

Alford, D. V. (1998). BORIS: an EC-funded concerted action on the bio-control of oilseed rape pests. *IOBC/wprs Bulletin* **21**, 183–4.

Alford, D. V., Cooper, D. A. & Williams, I. H. (1991). *Insect Pests of Oilseed Rape.* HGCA Oilseeds Research Review No. OS1. Home-Grown Cereals Authority, Research and Development: London, 130 pp.

Alford, D. V., Williams, I. H., Murchie, A. K. & Walters, K. F. A. (1995). *The Status and Potential of Parasitoids of Seed Weevil and Pod Midge on Winter Oilseed Rape.* HGCA Oilseeds Project Report No. OS14. Home-Grown Cereals Anthority: London, 89 pp.

Alford, D. V., Walters, K. F. A., Williams, I. H. & Murchie, A. K. (1996). A commercially viable low-cost strategy for the management of seed weevil populations on winter oilseed rape in the UK. *Brighton Crop Protection Conference – Pests and Diseases* **2**, 609–14.

Berraondo, P. E. Z., Morgan, D., Walters, K. F. A., Young, J. E. B. & Ellis, S. A. (1998). Spatial recovery of two species of Carabidae following cumulative pesticide applications in winter wheat. *The 1998 Brighton Conference – Pests and Diseases* **2**, 613–8.

Bichel, S. (1999). *The Bichel Committee. The Committee to Assess the Overall Consequences of Phasing Out the Use of Pesticides.* Report from the Main Committee, 148 pp.

Blight, M. M., Pickett, J. A., Wadhams, L. J. & Woodcock, C. M. (1995). Antennal perception of oilseed rape, *Brassica napus*, volatiles by the cabbage seed weevil. *Journal of Chemical Ecology* **21**, 1649–64.

Bommarco, R. (1997). Feeding, reproduction and dispersal of a generalist predator, *Pterostichus cupreus* L. (Coleoptera: Carabidae) related to agricultural landscape composition. *Acta Universitatis Agriculturae Sueciae, Agraria* **59**, 263–71.

Bonnemaison, L. & Jourdheuil, P. (1954). L'altise d'hiver du colza (*Psylliodes chrysocephala* L.). *Annales des Épiphyties* **5**, 345–524.

Büchi, R. (1986). Monitoring of parasitoids in oilseed rape pests during 1989 in Switzerland. *IOBC/wprs Bulletin* **14**, 54–60.

Büchi, R. (1990). Investigation of the use of turnip rape as trap plant to control oilseed rape pests. *IOBC/wprs Bulletin* **13**, 32–9.

Büchi, R. (1995). Combination of trap plants (*Brassica rapa* var. *silvestris*) and insecticide use to control rape pests. *IOBC/wprs Bulletin* **18**, 102–21.

Büchi, R. (2002). Mortality of pollen beetle (*Meligethes* spp.) larvae due to parasitoids and predators in rape fields and the effect of conservation strips. *Agriculture, Ecosystems and Environment* **90**, 255–63.

Büchs, W. & Nuss, H. (2000). First steps to assess the importance of epigaeic active polyphagous predators on oilseed rape insect pests with soil pupating larvae. *IOBC/wprs Bulletin* **23**, 151–63.

Buntin, G. D., McCaffrey, J. P., Raymer, P. L. & Romero, J. (1995). Quality and germination of rapeseed and canola seed damaged by adult cabbage seedpod weevil, *Ceutorhynchus assimilis*. *Canadian Journal of Plant Science* **75**, 539–41.

Copping, L. G. (ed.) (2001). *The Biopesticide Manual.* 2nd edn. British Crop Protection Council: Farnham.

Dmoch, J. (1997). Host-finding behaviour of *Trichomalus perfectus* – a parasitoid of cabbage seedpod weevil *Ceutorhynchus assimilis*. *Biuletyn Warzywniczy* **47**, 49–54.

Dmoch, J. & Klimek, G. (1975). Badania nad pasożytami chowacza podobnika (*Ceuthorrynchus assimilis* Payk.). III. Obserwacje nad biologią *Trichomalus perfectus* (Walker). *Roczniki nauk Rolniczych, Seria E* **5**, 125–36.

Epke, K., Michel, M. & Volk, T. (1996). Entwicklung des wissenbasierten Beratungssystems PRO_PLANT für den umweltschonenden Einsatz von Pflanzenschutzmitteln in der Landwirtschaft. *Forschung und Beratung Reihe* **C 50**, 154.

Evans, K. A. & Allen-Williams, L. J. (1992). Electroantennogram responses of the cabbage seed weevil to oilseed rape volatiles. *Journal of Chemical Ecology* **18**, 1641–59.

Evans, K. A. & Allen-Williams, L. J. (1993). Distant olfactory response of the cabbage seed weevil, *Ceutorhynchus assimilis*, to oilseed rape odour in the field. *Physiological Entomology* **18**, 251–6.

Evans, K. A. & Scarisbrick, D. H. (1994). Integrated insect pest management in oilseed rape crops in Europe. *Crop Protection* **13**, 403–12.

Ferguson, A. W., Solinas, M., Ziesmann, J., Isidoro, N., Williams, I. H., Scubla, P., Mudd, A., Clark, S. J. & Wadhams, L. J. (1999). Identification of the gland secreting oviposition-deterring pheromone in the cabbage seed weevil, *Ceutorhynchus assimilis*, and the mechanism of pheromone deposition. *Journal of Insect Physiology* **45**, 687–99.

Frahm, J., Johnen, A. & Volk, T. (1996). Development of the PRO_PLANT decision support system for plant protection in cereals, sugarbeet and rape. *OEPP/EPPO Bulletin* **26**, 609–22.

Free, J. B. & Williams, I. H. (1978). The responses of the pollen beetle and the seed weevil to oilseed rape and other plants. *Journal of Applied Ecology* **15**, 761–4.

Free, J. B. & Williams, I. H. (1979). The distribution of insect pests on crops of oilseed rape and the damage they cause. *Journal of Agricultural Science* **93**, 139–49.

Garthwaite, D. G., & Thomas, M. R. (1998). *Pesticide Usage Survey Report 159 – Arable Farm Crops in Great Britain 1998.* MAFF: London.

Garthwaite, D. G., Thomas, M. R. & Hart, M. (1995). *Pesticide Usage Survey Report 127 – Arable Farm Crops 1994.* MAFF: London.

Glinwood, R. T., Powell, W. & Tripathi, C. P. M. (1998). Increased parasitization of aphids on trap plants alongside vials releasing synthetic aphid sex pheromone and effective range of the pheromone. *Biocontrol Science and Technology* **8**, 607–14.

Greaves, M. P. & Marshall, E. J. P. (1987). Field margins: definitions and statistics. *British Crop Protection Council Monograph* **35**, 1–10.

Green, D. B., Bennison, J., Emmett, B. & Walters, K. F. A. (1991). Evaluation of alpha-cypermethrin and phorate against cabbage stem weevil in autumn-sown oilseed rape. *Tests of Agrochemicals and Cultivars*, No. 12 (*Annals of Applied Biology* **118**, Supplement), pp. 4–5.

Hardie, J., Nottingham, S. F., Powell, W. & Wadhams, L. J. (1991). Synthetic aphid sex pheromone lures female parasitoids. *Entomologia Experimentalis et Applicata* **61**, 97–9.

Hardwick, N. V. (1998). Disease forecasting. In: D. G. Jones (ed.). *The Epidemiology of Plant Diseases.* Kluwer: Dordrecht.

Hausamman, A. (1996). Strip-management in rape crop: is winter rape endangered by negative impacts of sown weed strips? *Journal of Applied Entomology* **120**, 505–12.

Head, J., Walters, K. F. A. & Langton, S. (2000). The compatibility of the entomopathogenic nematode *Steinernema feltiae* and chemical insecticides for the control of the South American leafminer, *Liriomyza huidobrensis*. *BioControl* **45**, 345–53.

Hokkanen, H. M. T. (1989). Biological and agrotechnical control of the rape blossom beetle *Meligethes aeneus* (Coleoptera, Nitidulidae). *Acta Entomologica Fennica* **53**, 25–9.

Hokkanen, H. M. T. (1991). Trap cropping in pest management. *Annual Review of Entomology* **36**, 119–38.

Hoßfeld, R. (1993). The use of yellow dishes as a decision aid for the control of *Psylliodes chrysocephala* L. *Gesunde Pflanzen* **45**, 291–5.

Johnen, A. & Meier, H. (2000). A weather-based decision support system for managing oilseed rape pests. *The BCPC Conference – Pests and Diseases* **2**, 793–800.

Kjaer, C. & Jepson, P. C. (1995). The toxic effects of direct pesticide exposure for a non-target weed dwelling chrysomelid beetle (*Gastrophysa polygoni*) in cereals. *Environmental Toxicology and Chemistry* **14**, 993–9.

Kjaerpedersen, C. (1992). Flight behaviour of the cabbage seedpod weevil. *Entomologia Experimentalis et Applicata* **62**, 61–6.

Knight, J. D. (1997). The role of decision support systems in integrated crop protection. *Agriculture, Ecosystems and Environment* **64**, 157–63.

Kromp, B. (1999). Carabid beetles in sustainable agriculture: a review on pest control efficacy, cultivation impacts and enhancement. *Agriculture, Ecosystems and Environment* **74**, 187–228.

Lane, A. & Walters, K. F. A. (1993). Recent incidence and cost effective control of pests of oilseed rape in England and Wales. *IOBC/wprs Bulletin* **16** (9), 185–92.

Lane, A. & Walters K. F. A. (1994). Pest control requirements of oilseed rape under CAP reform. *Aspects of Applied Biology* **40**, 171–81.

Lane, A. & Walters, K. F. A. (1995). Prospects for a decision support system for pests of oilseed rape in the UK. *Proceedings of the 9th International Rapeseed Congress, Cambridge, UK*, pp. 1019–21.

Lerin, J. (1987). A short bibliographical review of *Trichomalus perfectus* Walker, a parasite of the seed pod weevil *Ceutorhynchus assimilis* Payk. *IOBC/wprs Bulletin* **10**, 74–8.

Lethmayer, C., Nentwig, W. & Frank, T. (1997). Effect of weed strips on the occurrence of noxious coleopteran species (Nitidulidae, Chrysomelidae, Curculionidae). *Zeitschrift für Pflanzenkrankheiten und Pflanzenschutz* **104**, 75–92.

MacLeod, A. (1999). Attraction and retention of *Episyrphus balteatus* DeGeer (Diptera: Syrphidae) at an arable field margin with rich and poor floral resources. *Agriculture, Ecosystems and Environment* **73**, 237–44.

Meikle, S. M. & Scarisbrick, D. H. (1995). *The profitability of rapeseed production.* In: D. H. Scarisbrick & A. J. Ferguson (eds). *New Horizons for Oilseed Rape*. Semundo: Cambridge.

Morgan, D. & Walters, K. F. A. (1999). Developing innovative computerised technologies to aid rational pest management. *Aspects of Applied Biology* **55**, 67–71.

Morgan, D., Walters, K. F. A., Oakley, J. & Lane, A. (1998). An internet-based decision support system for the rational management of oilseed rape invertebrate pests. *The 1998 Brighton Conference – Pests and Diseases* **1**, 259–64.

Morgan, D., Walters, K. F. A., Oakley, J. N. & Lane, A. (2000). An internet-based decision-support system for insect pests of rape. *OEPP/EPPO Bulletin* **30**, 155–8.

Mudd, A., Ferguson, A. W., Blight, M. M., Williams, I. H., Scubla, P., Solinas, M. & Clark, S. J. (1997). Extraction, isolation and composition of oviposition-deterring secretion of cabbage seed weevil, *Ceutorhynchus assimilis*. *Journal of Chemical Ecology* **23**, 2227–40.

Murchie, A. K. & Williams, I. H. (1998a). A bibliography of the parasitoids of the cabbage seed weevil. *IOBC/wprs Bulletin* **21**, 163–9.

Murchie, A. K. & Williams, I. H. (1998b). Effect of host size on the sex of *Trichomalus perfectus*. *IOBC/wprs Bulletin* **21**, 177–81.

Murchie, A. K., Williams, I. H. & Perry, J. N. (1999). Edge distributions of *Ceutorhynchus assimilis* and its parasitoid *Trichomalus perfectus* in a crop of winter oilseed rape. *BioControl* **44**, 379–90.

Nilsson, C. (1996). Spring turnip rape as a catch crop for pollen beetles in spring rape. *37th Swedish Crop Protection Conference (1996)*, pp. 105–12.

Oakley, J. N., Walters, K. F. A., Ellis, S. A., Green, D. B., Watling, M. & Young, J. E. B. (1996). Development of selective aphicide treatments for integrated control of summer aphids in winter wheat. *Annals of Applied Biology* **128**, 423–36.

Pickett, J. A., Wadhams, L. J., Woodcock, C. M. & Hardie, J. (1992). The chemical ecology of aphids. *Annual Review of Entomology* **37**, 67–90.

Poehling, H. M. (1987). Effect of reduced dose rates of pirimicarb and fenvalerate on aphids and beneficial arthropods in winter wheat. *Bulletin SROP/WPRS* **10**, 75–83.

Powell, W. (1999). *Parasitoid hosts.* In: J. Hardie & A. K. Minks (eds). *Pheromones of Non-Lepidopteran Insects Associated with Agricultural Plants*. CAB International: Wallingford.

Powell, W. (2000). The use of field margins in the manipulation of parasitoids for aphid control in arable crops. *The BCPC Conference – Pests and Diseases* **2**, 579–84.

Powell, W., Dean, G. J. & Dewar, A. (1985). The influence of weeds on polyphagous arthropod predators in winter wheat. *Crop Protection* **4**, 298–312.

Powell, W., Pennachio, F., Poppy, G. M. & Tremblay, E. (1998). Strategies involved in the location of hosts by the parasitoid *Aphidius ervi* Haliday (Hymenoptera: Braconidae: Aphidiinae). *Biological Control* **11**, 104–12.

Raw, K. A. (1989). The effects of summer applications of insecticides on the non-target invertebrate fauna of oilseed rape. *Aspects of Applied Biology* **23**, 295–302.

Rosen, H. von (1964). Untersuchtungen über die Verbreitung und Biologie von zwei Pteromaliden in Rapsschoten (Hym., Chalcidoidea). *Statens vaxtskyddsanstalt Meddelangen* **12**, 453–65.

Singh, S. R., Walters, K. F. A. & Port, G. R. (2001). Behaviour of the adult seven spot ladybird, *Coccinella septempunctata* (Coleoptera: Coccinellidae), in response to dimethoate residue on bean plants in the laboratory. *Bulletin of Entomological Research* **91**, 221–6.

Smart, L. E., Blight, M. M. & Lane, A. (1996). Development of a monitoring trap for spring and summer pests of oilseed rape. *Brighton Crop Protection Conference – Pests and Diseases* **1**, 167–72.

Smart, L. E., Blight, M. M. & Hick, A. J. (1997). Effect of visual cues and a mixture of isothiocyanates on trap capture of cabbage seed weevil, *Ceutorhynchus assimilis*. *Journal of Chemical Ecology* **23**, 889–902.

Sterk, G., Hassan, S. A., Baillod, M., Bakker, F., Bigler, F., Blumel, S., Bogenschutz, H., Boller, E., Bromand, B., Brun, J., Calis, J. N. M., Coremans-Pelseneer, J., Duso, C., Garrido, A., Grove, A., Heimbach, U., Hokkanen, H., Jacas, J., Lewis, G., Moreth, L., Polgar, L., Roversti, L., Samsoe-Petersen, L., Sauphanor, B., Schaub, L., Staubli, A., Tuset, J. J., Vainio, A., van de Veire, M., Viggiani, G., Vinuela, E. & Vogt, H. (1999). Results of the seventh joint pesticide testing programme carried out by the IOBC/WPRS-Working Group 'Pesticides and Beneficial Organisms'. *BioControl* **44**, 99–117.

Sylvester-Bradley, R. & Makepeace, R. J. (1984). A code for stages of development in oilseed rape (*Brassica napus* L.). *Aspects of Applied Biology* **6**, 399–419.

Thiele, H. U. (1977). *Carabid Beetles and Their Environments.* Springer: Berlin.

Thies, C. & Tscharntke, T. (1999). Landscape structure and biological control in agroecosystems. *Science* **285**, 893–5.

Thies, C., Denys, C., Tscharntke, T. & Ulber, B. (1997). Der Einfluß von Saumbiotopen und Ackerbrachen auf Schadlings-Nutzlings-Interaktionen im Raps (*Brassica napus* spp. *oleifera*). *Verhandlungen der Gesellschaft für Okologie* **27**, 393–8.

Thomas, C. F. G. (1996). Field margin terminology and definitions: towards a unified semantic perspective. *Field Margins Newsletter* **6/7**, 35–7.

Thomas, M. R., Garthwaite, D. G. & Banham, A. R. (1997). *Pesticide Usage Survey Report 141. Arable Farm Crops in Great Britain 1996.* MAFF: London.

Thomas, M. R., Turner, J. A., Wardman, O. L. & Maidment, E. (2000). LIAISON – a broadly based agronomic information system serving UK agriculture, horticulture and the food industry. *The BCPC Conference – Pests and Diseases* **3**, 1285–90.

Titi, A. El, Boller E. F. & Gendrier J. P. (1993). Integrated production – principles and technical guidelines. *IOBC/wprs Bulletin* **16** (1), 1–97.

Ulber, B., Stippich, G. & Wahmhoff, W. (1990). Möglichkeiten, Grenzen und Auswirkungen des gezielten Pflanzenschutzes im Ackerbau: III Auswirkung unterschiedlicher Intensität chemischen Pflanzenschutzes auf epigäische Raubarthropoden in Winterweizen, Zuckerrüben und Winterraps. *Zeitschrift für Pflanzenkrankheiten und Pflanzenschutz* **97**, 263–83.

Umoru, P. A., Powell, W. & Clark, S. J. (1996). Effect of pirimicarb on the foraging behaviour of *Diaeretiella rapae* (Hymenoptera: Braconidae) on host-free and infested oilseed rape plants. *Bulletin of Entomological Research* **86**, 193–201.

Vet, L. E. M. & Dicke, M. (1992). Ecology of infochemical use by natural enemies in a tritrophic context. *Annual Review of Entomology* **37**, 141–72.

Voss, M. C., Ulber, B. & Hoppe, H.-H. (1998). Impact of reduced and zero tillage on activity and abundance of slugs in winter oilseed rape. *Journal of Plant Disease and Protection* **105**, 632–40.

Walters, K. F. A. & Hardwick, N. V. (2000). *Principles of pest and disease management.* In: D. V. Alford (ed.). *Pest and Disease Management Handbook.* Blackwell Science: Oxford.

Walters, K. F. A. & Lane, A. (1992). Incidence of pollen beetles in winter oilseed rape and evaluation of thresholds for control. *Brighton Crop Protection Conference – Pests and Diseases* **2**, 545–50.

Walters, K. F. A. & Lane, A. (1994a). Sampling procedures for pests of winter oilseed rape: meeting the needs of the crop consultant. *Aspects of Applied Biology* **37**, 83–94.

Walters, K. F. A. & Lane, A. (1994b). *The development of an improved management strategy for cabbage seed weevil on oilseed rape.* In: S. R. Leather, K. F. A. Walters, N. Mills & A. D. Watts (eds). *Individuals, Populations and Patterns in Ecology.* Intercept Press: Andover.

Walters, K. F. A., Turner, J. A., Elcock, S. J. & Alford, D. V. (2000). A European database on biological control of oilseed rape pests. *IOBC/wprs Bulletin* **23** (6), 141–50.

Walters, K. F. A., Lane, A., Cooper, D. A. & Morgan, D. (2001). A commercially acceptable assessment technique for improved control of cabbage stem flea beetle feeding on winter oilseed rape. *Crop Protection* **20**, 907–12.

Williams, I. H. & Free, J. B. (1978). A survey of the damage caused to crops of oilseed by insect pests in south-central England and their effect on seed yield. *Journal of Agricultural Science* **90**, 417–24.

Williams, I. H. & Free, J. B. (1979). Compensation of oilseed rape plants in response to injury to their flower buds and pods. *Journal of Agricultural Science* **92**, 53–9.

Williams, E. C. & Walters, K. F. A. (2000). Foliar application of the entomopathogenic nematode, *Steinernema feltiae* against leafminers on vegetables. *Biocontrol, Science and Technology* **10**, 61–70.

Williams, I. H. & Walton, M. (1990). A bibliography of the parasitoids of the brassica pod midge (*Dasineura brassica* Winn.). *IOBC/wprs Bulletin* **13**, 46–52.

Winfield, A. L. (1963). A study on the effects of insecticides on parasites of larvae of blossom beetles. *Entomologia Experimentalis et Applicata* **6**, 309–18.

Winfield, A. L. (1986). *Field pests of oilseed rape.* In: D. H. Scarisbrick & R. W. Daniels (eds). *Oilseed Rape.* Collins: London.

Wratten, S. D. & Thomas, C. F. G. (1990). *Farm-scale spatial dynamics of predators and parasitoids in agricultural landscapes.* In: G. H. Bunce & D. C. Howard (eds). *Species Dispersal in Agricultural Habitats.* Belhaven Press: London.

4

Parasitoids of Pollen Beetles

Christer Nilsson
Swedish University of Agricultural Sciences, Department of Crop Science, S-23053 Alnarp, Sweden

Introduction

The parasitoids of pollen beetles (*Meligethes* spp., notably *M. aeneus*) are all larval endoparasitoids, and all are hymenopterans; there are at least nine European species (Table 4.1). A few additional parasitoid species are mentioned in the literature, but these 'records' have not been confirmed by other workers and are omitted from this review. Egg parasitoids are unknown (Hokkanen *et al.*, 1988; Nilsson, 1989). Nor have parasitoids been found to attack the adults, at least of *Meligethes aeneus* (Kaufmann, 1923; Jourdheuil, 1960), although Fritzsche (1957) observed that caged females of *Phradis interstitialis* sometimes could attack adults of *Meligethes picipes* and *M. viridescens*.

Nomenclature of the Parasitoids

There is considerable confusion in the literature concerning the nomenclature of parasitoids of pollen beetles, particularly within the superfamily Ichneumonoidea.

Genus *Phradis*

There are very few papers prior to 1970 that mention *Phradis interstitialis* (Fig. 4.1) as a parasitoid of pollen beetles. In his revision of the European Tersilochines in 1971, and in an article published in 1967, Horstmann was the first to give a useful key for the separation of *P. morionellus* from *P. interstitialis*. Articles written after 1971 have generally separated these two species, and several authors have also found that *P. interstitialis* is, in fact, very common. Many of the older reports on *P. morionellus* actually deal with *P. morionellus* and/or *P. interstitialis*. The *Isurgus* species 'C' of Börner & Blunck (1920) and the *Thersilochus* sp. of Osborne (1960) and Winfield (1963) are certainly all *P. interstitialis*, whereas the other unidentified species in Osborne's paper are probably all *P. morionellus*.

Genus *Blacus*

Blacus nigricornis was first described by Haeselbarth in 1973 and, according to him, had previously been cited as *Blacus hastatus* Haliday, e.g. by Jourdheuil (1960). However, there

Table 4.1 Systematic list of parasitoids of pollen beetles (*Meligethes* spp.). For further details see Fitton *et al.* (1978) and Goulet & Huber (1993).

Order HYMENOPTERA

Superfamily ICHNEUMONOIDEA
Family ICHNEUMONIDAE (Subfamily TERSILOCHINAE)
Aneuclis Förster
– *incidens* (Thomson)†
Phradis Holmgren
– *interstitialis* (Thomson)†
– *morionellus* (Holmgren)†
Tersilochus Holmgren (= ***Thersilochus*** Holmgren)
– *heterocerus* Thomson†
Family BRACONIDAE (Subfamily HELCONINAE)
Blacus Nees
– *nigricornis* Haeselbarth†
Diospilus Haliday
– *capito* (Nees) (= *oleraceus* Haliday)† [Note 1]
Eubazus Nees (= ***Calyptus*** Haliday)
– *sigalphoides* (Marshall)†

Superfamily CHALCIDOIDEA
Family ENCYRTIDAE
Cerchysiella Girault (= ***Zeteticontus*** Silvestri)
– *planiscutellum* (Mercet)†

Superfamily PROCTOTRUPOIDEA
Family PROCTOTRUPIDAE
Brachyserphus Hellén
– *parvulus* (Nees)†

† Species included in the key to adults (Chapter 11).

Note 1. The name *Diospilus oleraceus* Haliday, 1835, appears in the literature as a synonym of *D. capito* but is a source of confusion since *Diospilus oleraceus* Haliday, 1933, is a quite separate species. Use of the name *Diospilus oleraceus* elsewhere in this monograph is restricted to the latter.

should be no confusion because *Blacus hastatus* is a parasitoid of beetles living off fungi on wood and usually found in forest areas.

Genus *Cerchysiella*

Osborne (1960) bred a species of *Cerchysiella* (cited as *Zeteticontus*) new to the British fauna, from both *Meligethes aeneus* and *M. viridescens*; it is highly probable that this parasitoid was *Cerchysiella planiscutellum*.

Distribution of the Parasitoids

The majority of the parasitoid species are found over most of Europe, from the southern part of Sweden and down to France, Italy and Hungary (Delucchi, 1956; Aubert &

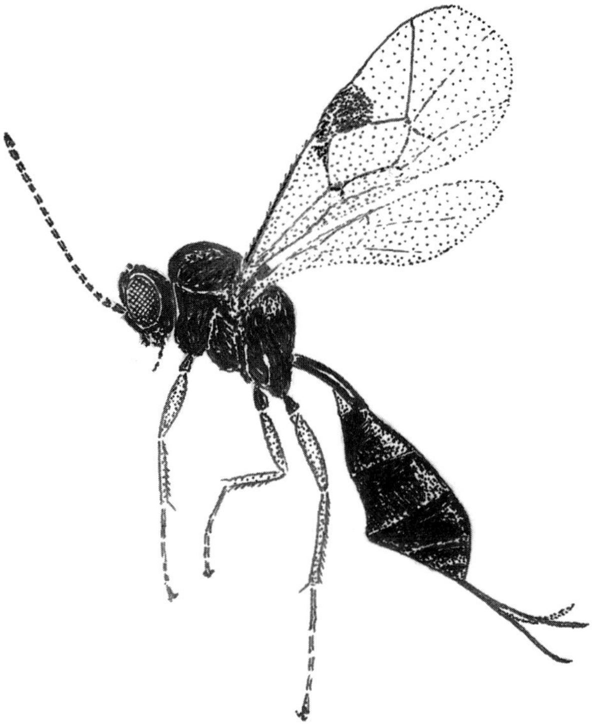

Fig. 4.1 Female *Phradis interstitialis* – family ICHNEUMONIDAE (×18).

Jourdheuil, 1958; Hellén, 1958; Blazejewska, 1960; Jourdheuil, 1960; Osborne, 1960; Herrström, 1964; Lehmann, 1965a, 1965b, 1969; Miczulski, 1967, 1968; Horstmann, 1971; Haeselbarth, 1973; Šedivý, 1983; Nilsson & Andreasson, 1987; Hokkanen, 1989; Büchi & Roos-Humbel, 1991; Nissen 1999). There are no records from most of the Balkan countries, but probably all central European species also occur here. *Diospilus capito*, *Phradis morionellus*, *P. interstitialis* and *Tersilochus heterocerus* have all been recorded at the northern distribution limit of oilseed rape, in Finland (Hellén, 1958; Hokkanen, 1989). *Eubazus sigalphoides* is recorded as a parasitoid of pollen beetles only from France and *Cerchysiella planiscutellum* only from southern England, France and Italy, but both species are very rare. *Aneuclis incidens* seems to have a more southerly distribution than the other cited parasitoids; only a few specimens of this species have been found in Scandinavia, but it is known from Turkey; the developmental time at moderate summer temperatures is also longer compared with that of the other multivoltine species (Jourdheuil, 1960).

The occurrence and abundance of the different species depends on factors such as the local climate, the crops grown during previous years and the cultivation techniques used. Usually, the most abundant species are one or more of *P. interstitialis*, *P. morionellus* or *T. heterocerus* in winter rape and the two last-mentioned in spring rape. *T. heterocerus* often dominates in winter rape and *P. morionellus* in spring rape. In addition, in spring rape, *Brachyserphus parvulus*, *Blacus nigricornis* and *D. capito* can be quite common in certain situations.

Life History and Biology on the Crop

Species of the genus *Phradis* and *Tersilochus* are univoltine, whereas all the other pollen beetle parasitoids are multivoltine.

Univoltine parasitoids

This group consists of *Phradis interstitialis*, *P. morionellus* and *Tersilochus heterocerus*, all of which have just one generation a year. Most parasitoid larval growth takes place in the host pupal chamber, made in the soil by the pollen beetle larva. About a month after the parasitoid pupates it develops into an adult, but the parasitoid stays in diapause in the cocoon until the following spring or summer. If the host beetle larva completed its development during the flowering period of winter rape, the parasitoid will emerge during the flowering stage of winter rape in the following year; similarly, if the parasitoid eggs were laid in pollen beetle larva developing on spring rape, the parasitoid will emerge during flowering of spring rape in the following year. The variation in emergence times for parasitoids that develop between these two extremes is considerable (Jourdheuil, 1960; Nilsson, 1985). In Finland, Hokkanen *et al.* (1988) showed that *P. morionellus* emergence from a spring turnip rape field of the previous year started at 280 degree days (DD) and peaked at 330 DD (base 5°C). A few days after emergence the parasitoids appear in the current year's oilseed rape crop. The females are sexually mature on emergence from hibernation, and can start laying eggs immediately when they arrive on the crop, searching the buds and flowers of the rape crop for suitable hosts. The adult female parasitoids live for less than a week at summer temperatures (but for up to a month at 5°C). Most of the males emerge before the females and they live for a much shorter time. They swarm above the crop, where they can be caught in large numbers, giving a misleading impression of the sex ratio. Sweep-net catches will often be poorly correlated to parasitization levels (Šedivý, 1983; Wyrostkiewicz & Blazejewska, 1985), possibly because of differences in male and female behaviour. This can also be part of the reason why reported sex ratios vary greatly between investigations.

The females of *P. morionellus* and *T. heterocerus* prefer to oviposit in large, second-instar larvae (Nilsson & Andreasson, 1987), and mainly search for hosts in open flowers, where such larvae are more numerous (Winfield, 1963; Lehmann, 1965b); however, the time during flowering that the parasitoids enter the field varies from year to year (Nilsson, 1985). The coincidence between crop development and parasitoid appearance is not precise. This means that part of the host larval population will escape parasitization in some years. In years with the presence of a very large pollen beetle population, many immature pollen beetle larvae and eggs will drop to the ground following bud abortion and loss of petals. Parasitoid eggs in these larvae will be lost, i.e. they will not develop further (Nilsson, 1989).

Adults of *P. morionellus* and *T. heterocerus* appear on the crop at the same time, usually at the beginning of flowering, although in some years one or the other can be earlier (Jourdheuil, 1960; Nilsson, 1985). Adults of *P. interstitialis*, on the other hand, normally emerge more than a week before the two other species and can reach the rape crop well before flowering. This species is reported, also, to search the buds for hosts (Börner *et al.*, 1921; Winfield, 1963); it can oviposit in first-instar larvae and eggs of pollen beetle, through the bud wall (as indicated by the long, slender form of its ovipositor) (Börner *et al.*, 1921; Osborne, 1960; Winfield, 1963). The ovipositor of *P. interstitialis* measures approximately 1.7 mm, that of *Diospilus capito* and of *P. morionellus* 1.6 mm, and that of *T. heterocerus* only 1.2 mm, reflecting the differences in their searching behaviour.

When there is competition between *P. interstitialis* and *T. heterocerus* in a host larva, the former species has an advantage (Nitzsche & Ulber, 1998), probably because its eggs hatch before those of the latter. Possibly as a result of this, compared with the relative quantitative distribution of parasitization by the two species in the preceding year, a proportionally larger number of adults of *P. interstitialis* have been observed to emerge in the spring (Nitzsche & Ulber, 1998); however, such a difference could at least in part be a result of greater egg mortality in *T. heterocerus*, as unhatched eggs of this species are often found in the fat bodies of adult pollen beetles (Osborne, 1960).

First-instar parasitoid larvae hatch shortly before the host larvae are fully fed and ready to enter the soil to pupate. Descriptions of eggs, first- and final-instar larvae of these parasitoids are given by Osborne (1960). The eggs of *T. heterocerus* are black and can readily be seen through the skin of the host larvae. The larva is pigmented in the caudal part of each segment, and appears striped. The egg breaks opens with a transverse cut, but the larva will not leave the egg shell completely until the next ecdysis. In the soil, the parasitoid larva completes its feeding and pupates outside the host. The cocoon consists of two layers of paper-like silk. On average, parasitized final-instar (second-instar) host larvae have smaller head-capsules than unparasitized ones, but the significance of this is not known (Zill, 1994; Nissen, 1997).

Univoltine parasitoids can reach parasitization levels of more than 50%. Owing to multi-parasitism, the number of host larvae that are subjected to successful oviposition is greater than the number of larvae that are actually parasitized, as the parasitoid females lay their eggs at random. At higher parasitization levels, it is not unusual to find more than one egg, sometimes five or more, in the same host larva (Jourdheuil, 1960; Nilsson & Andreasson, 1987; Nissen, 1997). Consequently, parasitization efficiency is quite low and will also be limited by the short life-span of the parasitoid female. Fritzsche (1957) reported that *T. heterocerus* females laid 40 eggs and *P. interstitialis* only 24. Dissections, however, have shown the presence of 200 ovarioles (Jourdheuil, 1960), so the potential fecundity is regularly not utilized.

Multivoltine parasitoids

The remaining parasitoid species are multivoltine, of which *Brachyserphus parvulus*, *Blacus nigricornis* and *Diospilus capito* are of greatest importance, at least in northern and central Europe. They have several generations each year and they seem to have no diapause, apart from a resting period determined by temperature. Consequently, the number of generations will be two or three in northern Europe and may be five or more in central and southern Europe. Osborne (1960) has described the egg, first- and final-instar larvae of *B. parvulus* and *D. capito*, and Čapek (1983) has described the larva of *B. parvulus*.

Diospilus capito

Females of *Diospilus capito* oviposit in host larvae of different sizes (including through the bud wall, as the female has a relatively long ovipositor) (Börner & Blunck, 1920; Osborne, 1960). The first-instar larva has a distinctive head capsule, with strong mandibles, which is visible through the cuticle of the host larva. Later instars have a less sclerotized head.

The larva of *D. capito* can kill other parasitoid larvae that might be present in the same host body. Jourdheuil (1960) observed that *D. capito* was superior in most situations where there was interspecific competition.

When the host larva enters the soil to pupate, the larva of the parasitoid (i.e. of *D. capito*) is in its intermediate instar. The host larval skin bursts on the ventral side and the parasitoid larva wriggles out head first, and it completes its feeding after leaving the host. When fully grown, the parasitoid larva spins a cocoon in the soil cavity that its host had already prepared for its own pupation. The parasitoid cocoon is greyish and consists of two separate layers: an outer layer of loosely woven yellow silken threads and an inner paper-like layer composed of yellow silken threads (Börner & Blunck, 1920; Osborne, 1960).

Several authors have claimed that *D. capito* alternates between *Meligethes* spp. in the summer and cabbage stem flea beetle (*Psylliodes chrysocephala*) or *Ceutorhynchus* spp. in winter (Börner *et al.*, 1921; Meuche, 1940; Kaufmann, 1941; Börner, 1942). In these reports there is sometimes a confusion between two species of *Diospilus*: namely, *D. capito* and *D. oleraceus* (Börner, 1942). Jourdheuil (1960) concludes that, in the Paris area, *D. capito* does not seem to attack hosts other than those of *Meligethes* developing on brassicaceous plants. Jourdheuil found several parasitized pollen beetle larvae very late in the autumn; he concluded that the low temperature in the autumn did not allow completion of their development until next spring and that this was the way in which *D. capito* overwintered. Although not likely, *Ceutorhynchus leprieuri* Brisout has been suggested to provide a winter host for *D. capito*. Cabbage stem flea beetle is very rare in central Sweden and Finland, and *C. leprieuri* is not known from Sweden, where *D. capito* can be found and where it sometimes achieves high parasitization levels in *Meligethes* larvae (Herrström, 1964; Hokkanen *et al.*, 1988).

Usually, few individuals of *D. capito* survive the winter. The first adults typically appear during flowering, often during the final stages of flowering in winter rape, but usually in very low numbers. The next generation can be quite numerous, adults of which appear during the flowering stage of spring oilseed rape. The following one or two generations during the summer can sometimes also be very numerous (Börner *et al.*, 1921; Kaufmann, 1923; Miczulski, 1967).

Other multivoltine species

The biology of multivoltine species other than *Diospilus capito* is poorly known.

Blacus nigricornis is considered to be nocturnal. The males swarm at the emergence sites of females and copulation follows soon after the latter appear (Haeselbarth, 1973; Čapek, 1983).

As in the case of *D. capito*, the first-instar larva of *Brachyserphus parvulus* is pigmented and sclerotized. However, the larva of *B. parvulus* completes its feeding within the host larval skin, and pupates without forming a cocoon, partly hidden within the host larval skin. Emergence of the adult takes place a couple of weeks later. Mating and oviposition start shortly afterwards. The ovipositor is very short, but the hind part of the female's abdomen is highly flexible. Using her antennae, the female locates a host larva in an open flower. She then, at high speed, repeatedly stabs it with her ovipositor. Second-instar host larvae are preferred (Osborne, 1960; Nissen, 1997). *B. parvulus* has also been recorded as a parasitoid of other beetle species (Pschorn-Walcher, 1971).

In the case of *Cerchysiella planiscutellum*, pupation occurs within the host larval skin, which becomes swollen and reddish brown, resembling a dipterous puparium (Osborne, 1960).

Host Selection, Host Defence Reactions and Parasitoid Mortality

The parasitoids *Brachyserphus parvulus* and *Diospilus capito* lay their eggs at random and do not discriminate between a parasitized and a non-parasitized host (Lehmann, 1969; Wyrostkiewicz & Blazejewska, 1985; Nilsson & Andreasson, 1987; Nitzsche & Ulber, 1998). The females lay one egg at a time. The lack of discrimination ability will markedly lower the efficiency of each female at low host: parasitoid (H: P) ratios. It can take 460 eggs to parasitize 99 out of 100 host larvae (Jourdheuil, 1960). This is a considerable wastage of eggs and is most pronounced at parasitization levels greater than 50–60%. This probably happens when beetle density is much lower than it was the previous year, thereby lowering the H: P ratio, and increasing parasitoid intraspecific competition.

The mechanisms behind host finding in these parasitoids have not been studied. Visual attraction could be important, as they are caught in large numbers in yellow traps. Odours are probably also of importance, at least within the field, but whether the odours of the rape plant, its flowers or just the pollen beetle larvae (or the damage made to the stamens and buds by the larvae) have any importance is not known. Jourdheuil (1960) observed that some of the multivoltine parasitoids were found in greater numbers the more intensive the flowering of the plants, irrespective of the larval population density.

The ichneumonid parasitoids of pollen beetles can provoke defence reactions in their hosts. Eggs of both *Phradis morionellus* and *Tersilochus heterocerus* are often encapsulated by the pollen beetle larva. Haemocytes aggregate on the parasitoid egg within hours after oviposition, starting at the micropyle. Through autolysis over a four-day period, the haemocytes are transformed into a thin, hard, dark-brown capsule. Also, first-instar larvae can be totally or partially encapsulated (Jourdheuil, 1960). If the vitality of the host is lowered by multiple egg laying this can suppress the encapsulation and thereby increase the survival of the parasitoid eggs (Jourdheuil, 1960; Salt, 1968). Thus, the wastage of parasitoid eggs increases the survival of the progeny. The interspecific competition between *P. morionellus* and *T. heterocerus* could even increase the survival of both of the species at intermediate H: P ratios, when the collective oviposition of both species is enough to avoid encapsulation, and it is then a random event which of the two survives the competition.

This encapsulation could be a considerable source of error in calculations of mortality. Osborne (1960) found 'numerous examples of overwintered *Meligethes* adults with one to three black egg-shells embedded in the fat-body'. An unknown proportion of the parasitized pollen beetle larvae will, at least under certain conditions, be able to survive. Estimates of mortality due to parasitoids (based on egg density) can, consequently, give too low an estimate of mortality of pollen beetle pupae and too high an estimate of parasitoid mortality in winter if this is estimated through spring emergence.

Mortality factors of parasitoid larvae and pupae

The loss of parasitoid eggs and young larvae as a result of bud abortion (owing to pollen beetle oviposition in small buds) can be substantial (Nilsson, 1989). The number of parasitoid eggs lost simultaneously is not known, but the earliest-appearing species, such as *Phradis interstitialis*, could be particularly affected. Some first-instar and some immature second-instar pollen beetle larvae involuntarily fall to the soil surface at petal fall, probably because they cling to petals. This mortality is mainly dependent on differences in the rate of (a) bud and flower growth and (b) pollen beetle larval development. In some years the loss of larvae in this way can be considerable: more than a third of the population. Thus, depending on the parasitization level, a corresponding proportion of the parasitoid population is also lost (Nilsson, 1989).

Pollen beetles have a high and virtually unexplained mortality during the pupal stage. This is also true at least for some of the parasitoids (Jourdheuil, 1960; Hokkanen *et al.*, 1988; Nitzsche & Ulber, 1998). This difference between parasitization of larvae and emergence of adult parasitoids may be explained by host reaction, by differential survival of parasitoid larva in multi-parasitized hosts and by other, earlier mentioned, factors. It can also be a result of unquantified mortality factors, such as predators or insect pathogens in the soil.

Parasitoid larvae of all species complete their development in the soil and pupate in the pollen beetle pupation chamber. Univoltine parasitoids also hibernate in the former rape field. The cultivation technique used to establish the crop following rape will greatly influence the mortality of the parasitoids (Nilsson, 1985; Hokkanen *et al.*, 1988; Klingenberg & Ulber, 1994; Nitzsche & Ulber, 1998; Thies & Tscharntke, 1999; Wahmhoff *et al.*, 1999). Experiments done in Finland, Germany and Sweden, in both spring and winter rape, show that ploughing or any other soil tillage will reduce the parasitoid population by 50–75% (Nilsson, 1985; Hokkanen *et al.*, 1988; Nitzsche & Ulber, 1998; Wahmhoff *et al.*, 1999). Parasitoid population density is less affected if the crop that follows rape is established through direct drilling. Nitzsche & Ulber (1998) showed that loosening the soil with a chisel plough before drilling had only a minor effect on parasitoid emergence in the following spring.

Population dynamics

Multivoltine parasitoids, such as *Diospilus capito*, are favoured by the presence of both spring and winter rape crops in the same area, and even more so if there is a difference in the phenology of the crops, as when both rape and turnip rape are grown in the same region (Jourdheuil, 1960; Karltorp & Nilsson, 1981). Growing all four oilseed rape crops in the same area creates increased problems with pests such as brassica pod midge (*Dasineura brassicae*) and cabbage aphid (*Brevicoryne brassicae*) and, therefore, is not recommended. However, it is also doubtful whether there would be a similar influence on pollen beetle populations, as the population of multivoltine parasitoids in winter rape is almost always very low. Conditions for breeding during late summer and autumn, and for hibernation, seem to be of greater importance (Jourdheuil, 1960). Prior to the use of pesticides in rape crops, and the intensive use of herbicides in agriculture in general, *D. capito* was reported to attain high levels of parasitization (Börner *et al.*, 1921; Kaufmann, 1923). Kaufmann claimed that if *D. capito* had arrived earlier in the season it would have totally suppressed its host. Apparently, he was dealing with the second generation of *D. capito*, which can be very numerous if the appropriate multiplication conditions are met during the spring. In Jourdheuil's (1960) investigations (from 1952 to 1956), this species was the most common of all parasitoids and far more common than the various Tersilochinae – attaining levels of parasitization of 5–45%, compared with 2–12% for the latter. In more recent studies, *D. capito* seems to be rare (Winfield, 1963; Lehmann, 1965b; Karltorp & Nilsson, 1981) or highly fluctuating in its occurrence (Nilsson & Andreasson, 1987). As concluded above, it is highly probable that late-breeding pollen beetles are more important to the population of parasitoids in the next (i.e. following) spring than alternative hosts. If this is true, brassicaceous weeds in the field margins and on fallow land could be of great importance, and the observed lower abundance of *D. capito* a result of the present intensive herbicide use. A closer examination of the biology and hibernation conditions for the multivoltine parasitoids of pollen beetle, particularly *D. capito*, could be rewarding.

Both interspecific and intraspecific competition is high at low H : P ratios. As a rule only one of the parasitoid larvae will survive, killing any of its competitors. Larvae of *D. capito* have been observed to be superior to those of *Phradis* or *Tersilochus* (Jourdheuil, 1960). It is not subjected to encapsulation reactions from the host and, therefore, will have its greatest impact on the beetle population through the greater survival at high H : P ratios, when multiple egg laying is low and encapsulation of the ichneumonid eggs is most common.

The highest level of biocontrol of pollen beetles should be attained where several parasitoid species are present, with peak activities during different parts of the host larval development. A low correlation is often reported between the abundance of parasitoids and the frequency of parasitization. The density of a parasitoid population is a function of the density of the pollen beetle population the previous year. However, the density of the pollen beetle population is, to a large extent, a random effect. The total beetle population is, of course, dependent also on the parasitoid population density, but wind direction and other weather factors will strongly influence the distribution of the beetles among their breeding sites. Thus, highly fluctuating population densities are to be expected, and also a highly fluctuating parasitization frequency.

Parasitoids probably cannot migrate over long distances, because of their limited life and because their hibernation strategy is to stay in the place where the host was found in the previous year. For these reasons, it could be argued that the area of rape in a region should not be allowed to fluctuate widely between years. An increased area of oilseed rape should be partitioned over several years, to give parasitoid populations the opportunity to spread out and multiply. Ichneumonids would have to migrate from the previous year's fields and migration losses will, of course, be minimal if crops are as close together as possible (Hokkanen *et al.*, 1988).

Effects of Insecticides

Phradis interstitialis is usually the first parasitoid to enter winter rape fields, followed by the other species, usually during flowering or late flowering of the crop. Consequently, *P. interstitialis* can be killed by pre-flowering insecticide treatments directed against pollen beetle. If chemical control measures on commercial crops could be decreased, an increase in the parasitization from *P. interstitialis* would probably result. To spray only during early bud growth is one way to achieve this. Neither *Phradis morionellus* nor *Tersilochus heterocerus* seems to be harmed by sprays applied during the bud stages (Lehmann, 1965b; Goos & Klein, 1971; Nilsson & Andreasson, 1987; Büchi, 1995). It can be argued that if part of the pollen beetle population can be killed without harming the parasitoid population this will create a more favourable H : P ratio and increase the level of control (Hokkanen *et al.*, 1988). The use of insecticides during the later part of bud development or during flowering (which has been proposed, or is used, for control of other pests – see Chapter 2) should be avoided wherever possible, as this will cause great harm to many of the parasitoid species (Winfield, 1963; Hokkanen *et al.*, 1988; Nissen, 1999).

Some parasitoids of pollen beetles migrate from winter rape to spring rape, if this is possible. However, they will often arrive before spring rape anthesis, and could be killed by the insecticides that are used against the pest. At present, where insecticides must be used, such losses are unavoidable, which limits the advantage to the parasitoids of growing winter and spring rape crops in the same area. The use of non-selective organophosphate insecticides will kill more parasitoids than will the use of pyrethroids (Nilsson, 1980; Hokkanen *et al.*, 1988).

When univoltine parasitoids emerge from the previous year's spring rape fields in mid-June, this can (if the crop that follows after spring rape is a spring cereal) coincide with a combined fungicide/prophylactic insecticide spray treatment, killing parts of the parasitoid population (Nilsson, 1994; Nissen, 1999). It is of great importance, therefore, that control thresholds for aphid control in spring cereal crops are adopted and followed. In barley, an insecticide application is normally not required before the end of June, if at all; moreover, if application of an insecticide is necessary, then pirimicarb (which is fairly selective) can be used, the application of which is far preferable to that of broader-spectrum alternatives.

Landscape Structure

Hokkanen *et al.* (1988) and Hokkanen (1989) noted that parasitization of *Diospilus capito* and *Phradis morionellus* was lowest in areas where spring oilseed rape cultivation had occurred over the largest uniform areas. Correspondingly, the rates tend to be highest where the fields were more scattered among other crops. In Germany, Thies & Tscharntke (1999), working only with univoltine species (*Phradis interstitialis*, *Tersilochus heterocerus* and the very rare *Phradis morionellus*), made similar observations.

The characteristics of field margins also greatly influenced parasitization levels in the middle of rape crops, which was enhanced by old field margins – increasing from 20% to 60%. Other investigations have also reported that parasitization under current growing conditions drops in the centres of large fields (Šedivý, 1983; Hausammann, 1996). Evidently, the influence of field size is only indirect, and other factors (such as parasitoid population density) are more important. Old and unploughed strips with natural vegetation enabled parasitoid populations to build up and to disperse into the fields, indicating losses during migration. Old fallow had an even higher effect: age and size of non-crop areas (previous crop) were the major factors that enhanced the parasitization in winter rape fields or in spring rape plots. Rates of parasitism were not related to host density or to the relative density of rape fields in the area. When the non-crop area dropped below approximately 20%, parasitism dropped below the suggested threshold for successful biological control (i.e. below one-third parasitized host larva) (Hawkins & Cornell, 1994). The surrounding landscape can be seen as functioning as a source and a sink for parasitoid populations (Pulliam, 1988), and the composition of different landscape elements contributed greatly to the degree of biocontrol achieved in nearby crops. In the investigation of Thies & Tscharntke (1999), flowers in the field margins were not shown to have any importance. However, the univoltine parasitoids, in contrast to *D. capito*, are mature at emergence and do not undergo any maturation feeding (Jourdheuil, 1960). The conclusion could have been different if multivoltine parasitoids were also considered, as for them maturation feeding from pollen and nectar is mandatory. Thus, for them, flowering field margins are needed, but they may also be of importance as food sources to univoltine parasitoids, enabling them to maintain adequate food reserves to migrate from hibernation sites to oilseed rape crops.

Acknowledgements

This work was part-funded by the European Commission Concerted Action: *Minimizing pesticide use and environmental impact by the development and promotion of bio-control for oilseed rape pests, FAIR 96-1314*.

References

Aubert, J.-F. & Jourdheuil, P. (1958). Nouvelle description et biologie de quelques Ichneumonides appartenant aux genres *Aneuclis* Först., *Isurgus* Först. et *Thersilochus* Holm. *Revue de Pathologie Végétale et d'Entomologie Agricole de France* **37**, 175–93.

Blazejewska, A. (1960). Studia nad biologia slodyszka rzepakowego (*Meligethes aeneus* Fabr.). *Studia Societatis Scientiarum Torunensis, Sectio E* **4**, 1–70.

Börner, C. (1942). Wirtswechsel der Schlupfwespe *Diospilus capito* zwischen den Larven von Rapsglanzkäfer und Kohlblattrüßler. *Zeitschrift für Pflanzenkrankheiten* **52**, 107–13.

Börner, C. & Blunck, H. (1920). Zur Lebensgeschichte des Rapsglanzkäfers. *Mitteilungen aus der Biologischen Bundesanstalt für Land- und Forstwirtschaft, Berlin-Dahlem* **18**, 91–109.

Börner, C., Blunck, H., Speyer, W. & Dampf, A. (1921). Beiträge zur Kenntnis vom Massenwechsel (Gradation) schädlicher Insekten. *Arbeiten aus der Biologischen Reichsanstalt für Land- und Forstwirtschaft, Berlin* **10**, 405–66.

Büchi, R. (1995). Combination of trap plants (*Brassica rapa* var. *silvestris*) and insecticide use to control rape pests. *Bulletin OILB/SROP* **18**, 102–21.

Büchi, R. & Roos-Humbel, S. (1991). Nützlinge reduzieren die Zahl der Rapsschädlinge. *Landwirtschaft Schweiz* **4**, 69–73.

Čapek, M. (1983). Relationships and taxonomic placement of the genus *Blacus* (Hym., Braconidae). *Contribution of the American Entomological Institute* **20**, 119–24.

Delucchi, V. (1956). Liste d'identification No. 1. *Entomophaga* **1**, 113–27.

Fitton, M. G., Graham, M. W. R. de V., Bouček, Z. R. L., Fergusson, N. D. M., Huddleston, T., Quinlan, J. & Richards, O. W. (1978). A check list of British insects. Part 4: Hymenoptera. *Handbooks for the Identification of British Insects* **11** (4), 1–159.

Fritzsche, R. (1957). Beeinflußung der Populationsdichte verschiedener *Meligethes*-Arten von gleichen Wirtspflanzen durch Parasiten. *Bericht über die Hundertjahrfeier der Deutschen Entomologischen Gesellschaft Berlin*, pp. 141–45.

Goos, A. & Klein, K. (1971). Einfluß der Rapsglanzkäferspritzung auf die Entomofauna des Winterrapses. *XIII International Kongress für Entomologie, Moskva 1968*, pp. 87–8.

Goulet, H. & Huber, J. T. (eds) (1993). Hymenoptera of the world: an identification guide to families. *Agriculture Canada, Research Branch, Publication 1894/E*, 668 pp.

Haeselbarth, E. (1973). Die *Blacus*-Arten Europas und Zentral-Asiens (Hymenoptera, Braconidae). *Veröffentlichungen der Zoologischen Staatssammlung München* **16**, 69–170.

Hausammann, A. (1996). Strip-management in rape crop: is winter rape endangered by negative impacts of sown weed strips? *Journal of Applied Entomology* **120**, 505–12.

Hawkins, B. A. & Cornell, H. V. (1994). Maximum parasitism rates and successful biological control. *Science* **266**, 1886.

Hellén, W. (1958). Die Tersilochinen Finnlands (Hym., Ichn.). *Notulae Entomologicae* **38**, 4–23.

Herrström, G. (1964). Untersuchungen über Parasiten von Ölfruchschädlingen in Schweden. *Statens växtskyddsanstalt Meddelanden* **12**, 437–48.

Hokkanen, H. (1989). Biological and agrotechnical control of the rape blossom beetle *Meligethes aeneus* (Col., Nitid.). *Acta Entomologica Fennica* **53**, 25–9.

Hokkanen, H., Husberg, G.-B. & Söderblom, M. (1988). Natural enemy conservation for the integrated control of the rape blossom beetle *Meligethes aeneus* F. *Annales Agriculturae Fenniae* **27**, 281–94.

Horstmann, K. (1967). Bemerkungen zur Taxonomie der Tersilochinen (Hym., Ichneumonidae). *Opuscula Entomologica* **32**, 123–30.

Horstmann, K. (1971). Revision der europäischen Tersilochinen I (Hym., Ichneumonidae). *Veröffentlichungen der zoologischen Staatssammlung (München)* **15**, 45–138.

Jourdheuil, P. (1960). Influence de quelques facteurs écologiques sur les fluctations de population d'une biocénose parasitaire: étude relative à quelque Hyménoptères (Ophioninae, Diospilinae, Euphorinae) parasites de divers Coléoptères inféodès aux crucifères. *Annales des Épiphyties* **11**, 445–658.

Karltorp, M. & Nilsson, C. (1981). Rapsbaggar i mellansvenska vårrapsodlingar. *Växtskyddsnotiser* **45**, 146–54.

Kaufmann, O. (1923). Beobachtungen und Versuche zur Frage der Überwinterung und Parasitierung von Ölfruchtschädlingen aus den Gattungen *Meligethes*, *Phyllotreta*, *Psylliodes* und *Ceutorrhynchus*. *Arbeiten aus der Biologischen Reichsanstalt für Land- und Forstwirtschaft, Berlin* **12**, 109–69.

Kaufmann, O. (1941). Untersuchungen am Rapserdfloh (*Psylliodes chrysocephala* L.) in Ostholstein. *Zeitschrift für Angewandte Entomologie* **27**, 464–95.

Klingenberg, A. & Ulber, B. (1994). Untersuchungen zum Auftreten der Tersilochinae (Hym., Ichneumonidae) als Larvalparasitoide einiger Rapsschädlinge im Raum Göttingen 1990 und 1991 und zu deren Schlupfabundanz nach unterschiedlicher Bodenbearbeitung. *Journal of Applied Entomology* **117**, 287–99.

Lehmann, W. (1965a). Einfluß chemischer Bekämpfungsmaßnahmen auf einige Rapsschädlinge und ihre Parasiten. II. Knospen- und Stengelschädlinge. *Archiv für Pflanzenschutz* **1**, 209–19.

Lehmann, W. (1965b). Der Einfluß chemischer Bekämpfungsmaßnahmen auf einige Parasiten von Rapsschädlingen. *Mitteilungen aus der Biologischen Bundesanstalt für Land- und Forstwirtschaft Berlin-Dahlem* **115**, 35–42.

Lehmann, W. (1969). Beiträge zur Parasitenfauna in Rapsbeständen. *Deutsche Akademie der Landwirtschaftswissenschaften zu Berlin. Tagungsberichte* **80**, 513–28.

Meuche, A. (1940). Untersuchungen am Rapserdfloh (*Psylliodes chrysocephala* L.) in Ostholstein. *Zeitschrift für Angewandte Entomologie* **27**, 464–95.

Miczulski, B. (1967). Blonkówki (*Hymenoptera*) w biocenozie upraw rzepaku. Czesc III. Meczelkowate (Braconidae) i mszycarzowate (Aphidiidae). *Polskie Pismo Entomologiczne* **37**, 167–91.

Miczulski, B. (1968). Blonkówki (*Hymenoptera*) w biocenozie upraw rzepaku. Czesc VII. Ogólne podsumowanie wyników. *Polskie Pismo Entomologiczne* **38**, 475–95.

Nilsson, C. (1980). Effekter av syntetiska pyretroider på nyttoinsekter. *Växtskyddsrapporter* **12**, 76–82.

Nilsson, C. (1985). Impact of ploughing on emergence of pollen beetle parasitoids after hibernation. *Zeitschrift für Angewandte Entomologie* **100**, 302–8.

Nilsson, C. (1989). The pollen beetle (*Meligethes aeneus* F.) in winter and spring rape at Alnarp 1976–1978. III. Mortality factors. *Växtskyddsnotiser* **52**, 145–50.

Nilsson, C. (1994). Pollen beetles (*Meligethes* spp.) in oil seed rape crops (*Brassica napus* L.): biological interactions and crop losses. *Department of Plant Protection Sciences, SLU. Dissertations* **1**, 39 pp.

Nilsson, C. & Andreasson, B. (1987). Parasitoids and predators attacking pollen beetles (*Meligethes aeneus* F.) in spring and winter rape in southern Sweden. *IOBC/wprs Bulletin* **10**, 64–73.

Nissen, U. (1997). *Ökologische Studien zum Auftreten von Schadinsekten und ihre Parasitoiden an Winterraps norddeutscher Anbaugebiete.* Dissertation, Universität Kiel, Germany.

Nissen, U. (1999). Natürliche Gegenspieler von Rapsschädlingen. *Raps* **17**, 79–82.

Nitzsche, O. & Ulber, B. (1998). Einfluß differenzierter Bodenbearbeitungssysteme nach Winterraps auf die Mortalität einiger Parasitoiden des Rapsglanzkäfers (*Meligethes* spp.). *Zeitschrift für Pflanzenkrankheiten und Pflanzenschutz* **105**, 417–21.

Osborne, P. (1960). Observations on the natural enemies of *Meligethes aeneus* (F.) and *viridescens* (F.) [Coleoptera: Nitidulidae]. *Parasitology* **50**, 91–110.

Pschorn-Walcher, H. (1971). Heloridae et Proctotrupidae. *Insecta Helvetica* **4**, 63 pp.

Pulliam, H. R. (1988). Sources, sinks, and population regulation. *American Naturalist* **132**, 652–61.

Salt, G. (1968). The resistance of insect parasitoids to the defence reactions of their hosts. *Biological Reviews* **43**, 200–32.

Šedivý, J. (1983). Tersilochinae as parasitoids of insect pests of winter rape (Hymenoptera: Ichneumonidae). *Contributions of the American Entomological Institute* **20**, 266–76.

Thies, C. & Tscharntke, T. (1999). Landscape structure and biological control in agroecosystems. *Science* **285**, 893–5.

Wahmhoff, W., Hedke, K., Tiedemann, A. von, Nitzsche, O. & Ulber, B. (1999). Zum Einfluß von Fruchtfolge und Bodenbearbeitung auf die wichtigen Schaderreger des Winterrapses. *Zeitschrift für Pflanzenkrankheiten und Pflanzenschutz* **106**, 57–73.

Winfield, A. L. (1963). A study on the effects of insecticides on parasites of larvae of blossom beetles (*Meligethes aeneus* F., Coleoptera: Nitidulidae). *Entomologia Experimentalis et Applicata* **6**, 309–18.

Wyrostkiewicz, K. & Blazejewska, A. (1985). *Isurgus heterocerus* Thoms. i *I. morionellus* Holmgr. (Hym., Ichneumonidae) parazytoidy larw slodyszka rzepakowego – *Meligethes aeneus* F. (Col., Nitidulidae). *Polskie Pismo Entomologiczne* **55**, 391–404.

Zill, H. (1994). Studien zur Parasitierung des Rapsglanzkäfers *Meligethes aeneus* F. (Col., Nitidulidae) durch *Tersilochus heterocerus* Thoms. (Hym., Ichneumonidae). *Semesterarbeit Universität Kiel*, 7 pp.

5

Parasitoids of Ceutorhynchid Stem Weevils

Bernd Ulber
Institut für Pflanzenpathologie und Pflanzenschutz, Universität Göttingen, Grisebachstraße 6, D-37077 Göttingen, Germany

Introduction

This chapter deals with parasitoids of stem-boring ceutorhynchid weevils (*Ceutorhynchus* spp.), whose larvae tunnel within the petioles and stems of oilseed rape. The three most important species are: rape stem weevil (*C. napi*); cabbage stem weevil (*C. pallidactylus*) and rape winter stem weevil (*C. picitarsis*). For details of these pests see Chapter 2.

All recorded parasitoids of stem-boring ceutorhnychid weevils are hymenopterans, with representatives from three families: Ichneumonidae, Braconidae and Pteromalidae (Table 5.1).

Parasitism of Rape Stem Weevil

Two ichneumonid species of the subfamily Tersilochinae have been reported as parasitoids of rape stem weevil larvae: *Tersilochus fulvipes* (Kazda, 1956; Aubert & Jourdheuil, 1958; Jourdheuil, 1960; Horstmann, 1971; Šedivý, 1983; Ulber, 2000) and *Tersilochus moderatus* (Dosse, 1951). In addition, the pteromalid *Stenomalina gracilis* has been reared recently from larvae of rape stem weevil in Poland (Klukowski & Kelm, 2000).

The most abundant parasitoid (*T. fulvipes*) has been found in Austria (Kraus & Kromp, 2002), France (Jourdheuil, 1960), Germany (Buhl, 1952; Lehmann, 1965; Nitzsche, 1998), Hungary (Horstmann, 1981) and Switzerland (Günthardt, 1949), and in the former Czechoslovakia (Kazda, 1956; Šedivý, 1983). *T. moderatus* has been reported only from southern Germany (Dosse, 1951).

Life history of *Tersilochus fulvipes*

The most extensive information on the life histories of the Tersilochinae is given by Jourdheuil (1960). *Tersilochus fulvipes* is a solitary, univoltine endoparasitoid of rape stem weevil. There is no information on alternative host species. The adults colonize rape crops in April or May, usually shortly before or at the beginning of flowering. Peak

Table 5.1 Systematic list of parasitoids of stem weevils (*Ceutorhynchus* spp.). For further details see Fitton *et al.* (1978) and Goulet & Huber (1993).

Order HYMENOPTERA
  Family ICHNEUMONIDAE
    Subfamily PHYGADEUONTINAE
      Stibeutes Förster
      – *curvispina* (Thomson)†
    Subfamily TERSILOCHINAE
      Tersilochus Holmgren (= ***Thersilochus*** Holmgren)
      – *exilis* Holmgren† [Note 1]
      – *fulvipes* (Gravenhorst) (= *gibbus* Holmgren)† [Note 2]
      – *moderatus* Linnaeus
      – *obscurator* Aubert† [Note 3]
      – *stenocari* (Gregor)†
      – *tripartitus* Brischke (= *melanogaster* Thomson)†
  Family BRACONIDAE
      Diospilus Haliday
      – *oleraceus* Haliday
      Microctonus Wesmael
      – *melanopus* Ruthe† [Note 4]
      Sigalphus Latreille
      – *obscurellus* Nees†
  Family PTEROMALIDAE
      Stenomalina Ghesquière
      – *gracilis* (Walker) (= *muscarum* misidentification)†
      Trichomalus Thomson
      – *lucidus* (Walker)†

† Species included in the key to adults (Chapter 11).

Note 1. Also cited in the literature as *Probles exilis* (Holmgren).
Note 2. Also cited in the literature as *Porizon fulvipes* (Gravenhorst) and as *Thersilochus fulvipes* (Gravenhorst) ssp. *gallicator* Aubert.
Note 3. Also cited in the literature as *Thersilochus tripartitus* Brischke spp. *obscurator* Aubert.
Note 4. Also cited in the literature as *Perilitus melanopus* Ruthe.

activity occurs when the crop is at full bloom and declines at the end of flowering (Jourdheuil, 1960; Lehmann, 1965; Nitzsche, 1998). Under laboratory conditions, the longevity of two *T. fulvipes* females, provided with rape flowers and water, was 48 and 58 days, respectively (Nitzsche, 1998).

Females forage on stems, with antennation of the stem surface and ovipositor probing close to infested parts, suggesting that host microhabitat location and host recognition are assisted by contact chemosensory cues originating from the host plant or host (B. Ulber, *unpublished*). Females lay their eggs through the stem tissue, singly, into host larvae while these are mining within the stem pith. After hatching, the parasitoid persists within the host in its primary larval instar until maturation of the weevil larva (Jourdheuil, 1960). During this period, the development of the host is apparently unaffected. However, when the full-grown host larva has left the plant to pupate in the soil, the parasitoid develops rapidly and kills the host prepupa (koinobiont parasitoid). The adult parasitoid diapauses in its pupal cocoon within the earthen cell formed by the host larva. Parasitoids emerge in the following spring and then migrate to new oilseed rape crops.

Status and importance of *Tersilochus fulvipes*

High levels of parasitism of rape stem weevil have been reported frequently from various central European countries. In France, Jourdheuil (1960) observed an increase in levels of parasitism of rape stem weevil by *Tersilochus fulvipes* from 22% to 33% in 1952 to 1973, and 95% in 1956. However, from 1987 to 1990, Y. Ballanger (*pers. comm.*), also in France, obtained only 22–30% of unidentified parasitoid cocoons from extensive rearings of rape stem weevil larvae. In the former Czechoslovakia, levels of parasitism of rape stem weevil increased from 26% in 1949 to 95% in 1954 (Kazda, 1956); when Šedivý (1983) studied the parasitism of the same host from 1951 to 1982 in this country, he found 79–98% of the larvae parasitized by *T. fulvipes*. As in France (Jourdheuil, 1960), increasing levels of parasitism were accompanied by decreasing levels of infestation of oilseed rape by rape stem weevil, which led to the conclusion that *T. fulvipes* had a significant effect on the population dynamics of this pest species.

In Switzerland, in 1945, 30–40% of the prepupae of rape stem weevil were parasitized by *Tersilochus fulvipes* (Günthardt, 1949); however, several decades later (in 1989 and 1990), in the same region of Switzerland, Büchi & Roos-Humbel (1991) could not find any parasitism of this pest; further, there was only 6.4% of parasitism in 1993 (R. Büchi, *pers. comm.*). In rape fields around Göttingen in Germany, no parasitism of rape stem weevil was observed from 1990 to 1994, in spite of regularly high levels of infestation of oilseed rape crops by the pest and extensive sampling programmes (Klingenberg & Ulber, 1994; Nitzsche, 1998). However, in 1995, 1996 and 1997, *T. fulvipes* accounted for 1%, 21% and 6% of larval parasitism, respectively, in this region (Nitzsche, 1998; Ulber, 2000). By contrast, in Austria, the level of parasitism of rape stem weevil ranged from 60% to 81% in 1996 and from 0% to 26% in 1999 (Kraus & Kromp, 2002); the parasitoid species was determined as *T. fulvipes*.

The parasitism of rape stem weevil by *T. moderatus* has not been assessed quantitatively (Dosse, 1951).

Factors affecting the efficacy of *Tersilochus fulvipes*

The immigration of *Tersilochus fulvipes* females into oilseed rape crops is usually synchronized with the early development of rape stem weevil larvae in the plants. However, investigations in northern Germany provided some evidence that phenological asynchrony between *T. fulvipes* females and host larval stages suitable for parasitization occasionally might result in low levels or complete failure of parasitism at these locations (Nitzsche, 1998). In spite of consistently high levels of abundance in 1996 and 1997 of both adult parasitoids and host larvae, the parasitism rate of rape stem weevil by *T. fulvipes* was considerably higher in 1996, when the development of weevil larvae was delayed by low temperatures in spring. Accordingly, the effectiveness of *T. fulvipes* in reducing host populations is obviously biased by a close temporal synchronization of the parasitoid with the occurrence of suitable host larvae within the plants. Host-instar selection by *T. fulvipes* has not been studied in detail.

A significant correlation has been observed between the level of parasitization of rape stem weevil by *Tersilochus fulvipes* and the diameter of the rape stem (Nitzsche & Ulber, 2002); both at the stem base and at 50 cm up from the stem base, a greater proportion of larvae was parasitized when stems were less than 8–9 mm in diameter than when they were more than 8–9 mm in diameter (Fig. 5.1). By comparing the ovipositor length of *T.*

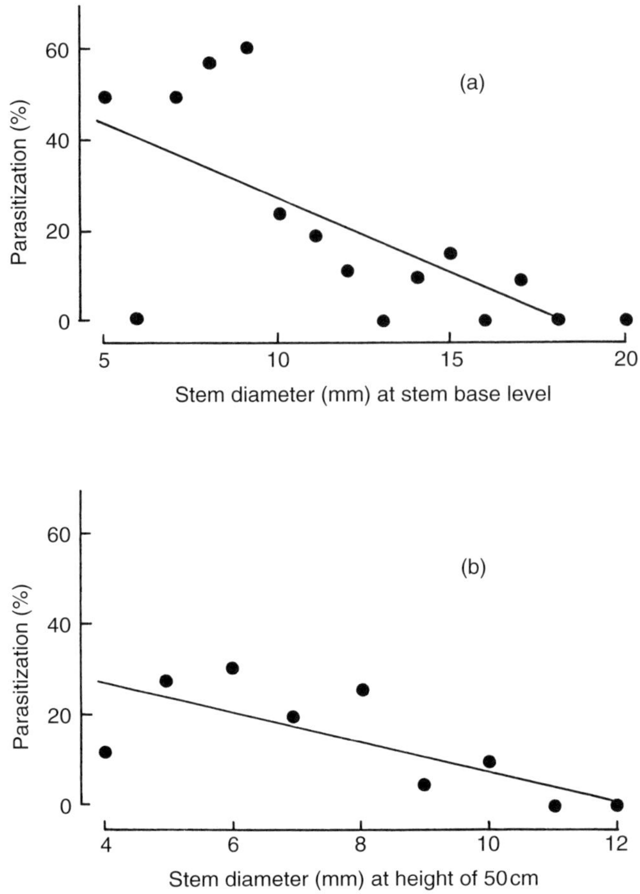

Fig. 5.1 Effect of stem diameter of oilseed rape plants on the extent of parasitism of rape stem weevil (*Ceutorhynchus napi*) by *Tersilochus fulvipes*. (a) At soil (stem base) level: $y = 60.78 - 3.35x$, $r^2 = 0.46$, $n = 15$, $P < 0.01$; (b) at 50 cm above the stem base: $y = 40.26 - 3.19x$, $r^2 = 0.51$, $n = 9$, $P < 0.05$.

fulvipes females with the diameter of rape stems, it was demonstrated that thick stems can provide a structural refuge for host larvae. Owing to its shortness (4.21 mm ± 0.25 SD), the ovipositor of *T. fulvipes* females is incapable of reaching the host larvae when these are feeding in the centre of thick stems.

Parasitism of Cabbage Stem Weevil

Four species of ichneumonid parasitoid have been found to attack the larvae of cabbage stem weevil: *Tersilochus obscurator* (Aubert & Jourdheuil, 1958; Jourdheuil, 1960; Lehmann, 1965; Horstmann, 1971; Šedivý, 1983; Klingenberg & Ulber, 1994; Nissen, 1997; Nitzsche, 1998; Ulber, 2000; Kraus & Kromp, 2002), *Tersilochus tripartitus* (Günthardt, 1949; Jourdheuil, 1960), *Tersilochus exilis* (Herrström, 1964) and *Stibeutes curvispina* (Nissen, 1997). In addition, the pteromalid *Trichomalus lucidus* has been reared recently from larvae of cabbage stem weevil by Nissen (1997).

Adults of cabbage stem weevil have been reported to be parasitized by the braconid *Microctonus melanopus* (Speyer, 1925; Günthardt, 1949; Jourdheuil, 1960). This parasitoid species has also been reared from adults of some other species of *Ceutorhynchus* and also from adults of cabbage stem flea beetle (*Psylliodes chrysocephala*) (Jourdheuil, 1960).

The most widely distributed parasitoid of cabbage stem weevil (namely, *Tersilochus obscurator*) has been reared from larvae in the former Czechoslovakia (Šedivý, 1983) as well as in Austria (Kraus & Kromp, 2002), France (Jourdheuil, 1960), Germany (Lehmann, 1965; Klingenberg & Ulber, 1994; Nissen, 1997; Nitzsche, 1998) and Switzerland (R. Büchi, *pers. comm.*). It has also been collected from oilseed rape crops by sweep netting and captured in water trays in Germany (Lehmann, 1965) and in omniflight traps in the UK (I. H. Williams, *pers. comm.*), and has been found in Hungary, Ireland, Sweden and the Ukraine (Horstmann, 1981). *Tersilochus tripartitus* has been observed in Austria, Denmark, Ireland (Horstmann, 1981), France (Jourdheuil, 1960) and Switzerland (Günthardt, 1949). *Tersilochus exilis* was reared from larvae of cabbage stem weevil collected from spring rape in southern Sweden by Herrström (1964). *Stibeutes curvispina* and *T. lucidus* have been reported only recently, both from northern Germany (Nissen, 1997).

Life histories of the main parasitoids

Tersilochus obscurator

The life history and phenology of the solitary koinobiont *Tersilochus obscurator* correspond very closely to that of *T. fulvipes* (Jourdheuil, 1960). Adult parasitoids colonize rape fields in April or May, coinciding with the occurrence of early larval instars of cabbage stem weevil within petioles. The parasitoid females preferentially oviposit within young host larvae, possibly because they mainly search leaf petioles where, compared with stems, they can more reliably locate hosts within the plant tissue. Observations by Jourdheuil (1960) and Nissen (1997) provide some evidence that *T. obcurator* females can parasitize host larvae only as long as these are feeding within petioles; once the weevil larvae have bored into the stem, they appear to be inaccessible to the short ovipositor of *T. obscurator* females, which is only 2.6 mm ± 0.17 SD in length (Nitzsche, 1998). In contrast to these findings, Šedivý (1983) reported that *T. obscurator* females prefer to parasitize the third-instar larvae of cabbage stem weevil. As with *T. fulvipes*, consumption and killing of the host larvae is delayed until these have entered the prepupal stage in the ground. Overwintering of the adult stage of *T. obscurator* occurs within their silken pupal cocoon.

Trichomalus lucidus

The ectoparasitic idiobiont *Trichomalus lucidus* parasitizes the larvae of cabbage stem weevil, mostly in May; females lay their eggs through the stem tissue onto final-instar host larvae (Nissen, 1997). After consumption of the host, the parasitoid pupates within the plant in mid-June. The newly emerged adults leave the stem (Nissen, 1997). *T. lucidus* has also been reared from larvae of cabbage stem flea beetle (Nissen, 1997) – see Chapter 8.

Stibeutes curvispina

Stibeutes curvispina, another ectoparasitic solitary idiobiont, attacks the prepupae and pupae of cabbage stem weevil, and probably of other species of *Ceutorhynchus* in the soil of oilseed rape fields (Nissen, 1997). There is, however, only sparse information in the literature on the biology, phenology and host range of this, or the previous, parasitoid species.

Status and importance of parasitoids

Studies on the level of parasitism of cabbage stem weevil have shown *Tersilochus obscurator* to be the most abundant parasitoid of this pest in Europe. In France, 35% and 54% of larvae were parasitized in 1953 and 1956, respectively (Jourdheuil, 1960). In the former Czechoslovakia, in 1951 to 1982, the parasitism of larvae ranged from 6% to 69% (Šedivý, 1983). Günthardt (1949) found that, in 1945, from 35% to 75% of the prepupae of cabbage stem weevil were parasitised by *T. tripartitus* in the soil of rape crops in Switzerland. However, when Büchi (1991; *pers. comm.*) studied the parasitism of cabbage stem weevil in the same area from 1989 to 1996, the percentage of larvae parasitized by *T. obscurator* ranged from 0% to 20%. In eastern Austria, in 1996, 1999 and 2000, 40–80% of larvae were parasitized by *T. obscurator* (Kraus & Kromp, 2002). In Sweden, in 1951, from 0% to 30% of larvae were parasitized by *T. exilis* in spring rape (Herrström, 1964). In the northern part of Germany, the level of parasitism of cabbage stem weevil larvae by *T. obscurator* in the period 1990–1997 ranged from 22% to 52% (Klingenberg & Ulber, 1994; Thies *et al.*, 1997; Nitzsche, 1998; Ulber, 2000). Similar levels were found by Nissen (1997) in 1993 and 1994 in northern coastal areas of Germany, with levels of parasitism at 34% and 51%, respectively.

Factors affecting efficacy of parasitoids

Dissections of cabbage stem weevil larvae sampled from winter rape crops in Germany indicated that superparasitism by *Tersilochus obscurator* occurs commonly and at various levels of parasitization: B. Ulber & C. Schulz (*unpublished data*) found that 63.3%, 26.0%, 6.3%, 2.5% and 1.9% of the larvae contained one, two, three, four and five eggs and/or larvae of *T. obscurator*, respectively, when the level of parasitism in different plots of a winter rape field ranged from 4.1% to 27.2%. In coastal areas, 51.2% of larvae were parasitized by *T. obscurator*, and 44.9% of them were superparasitized, with up to seven eggs and/or larvae per host (Nissen, 1997). High levels of superparasitism presumably result from relatively short periods of vulnerability to parasitization of the host larvae within petioles.

The density dependence of parasitism by *Tersilochus obscurator* in an oilseed rape field was determined by B. Ulber & C. Schulz (*unpublished data*). A range of various host densities was established by releasing different numbers of cabbage stem weevil adults into field cages. After oviposition, the cages were removed and weevil larvae exposed to natural parasitization. The proportion of larvae parasitized by *T. obscurator* increased from 17.9% to 27.2% as host densities increased from 140 to 345 larvae per 50 plants. However, the statistical analysis did not prove a significant density-dependent response of parasitism by *T. obscurator* to increasing host densities.

Parasitism of Rape Winter Stem Weevil

Three species of endoparasitoids have been reported attacking larvae of rape winter stem weevil in France: *Tersilochus stenocari*, *Sigalphus obscurellus* and *Diospilus oleraceus*; in addition, *Microctonus melanopus* is known as a parasitoid of the adults (Jourdheuil, 1960).

Tersilochus stenocari attacks the larvae of rape winter stem weevil while these are mining within petioles or stems of winter rape. A high percentage of larvae was parasitized in France in 1956 (70%) (Jourdheuil, 1960). In Slovakia, *T. stenocari* has also been found to parasitize the larvae of the weevil *Stenocarus fuliginosus* (Rozsypal, 1941). In Switzerland, in 1992, from 42% to 61% of rape winter stem weevil larvae were parasitized by an undetermined parasitoid species (R. Büchi, *pers. comm.*).

The plurivoltine braconids *S. obscurellus* and *D. oleraceus* parasitize the larvae of rape winter stem weevil as well as those of turnip gall weevil (*Ceutorhynchus pleurostigma*) in autumn and winter, whilst in summer they can be found in larvae of cabbage seed weevil and in other hosts (Jourdheuil, 1960). Only low levels of parasitism of rape winter stem weevil by *S. obscurellus* have been reported.

Effects of Abiotic Factors on Parasitoids

Effects of soil tillage on population dynamics of parasitoids

Both reduced tillage and direct drilling after harvest of oilseed rape have been shown to reduce significantly the overwintering mortaliy of *Tersilochus fulvipes* and *T. obscurator* (Klingenberg & Ulber, 1994; Nitzsche, 1998). Following ploughing, rotary harrowing, heavy cultivation or direct drilling in autumn the mean density of *T. fulvipes* adults emerging from soil in spring was 4.8/m^2, 9.2/m^2, 11.3/m^2 and 12.8/m^2, respectively. The effects of tillage on *T. obscurator* were similar to those on *T. fulvipes*: on average, 43.0/m^2, 70.4/m^2, 78.3/m^2 and 84.3/m^2 of adult *T. obscurator* emerged in spring, when the soil was cultivated in autumn by ploughing, rotary harrowing, heavy cultivation or direct drilling, respectively. Nearly two or three times as many parasitoids emerged from direct-drilled plots, as opposed from plots cultivated by a conventional plough, suggesting that disturbance of the pupae and burying of overwintering adults into the ground by deep ploughing caused high mortalities of these parasitoid species.

Effects of insecticides on parasitoids

Immigration of adults of the parasitoids *Tersilochus fulvipes* and *T. obscurator* occurs mostly at the beginning of flowering. It follows that early applications of insecticide sprays to control adults of cabbage stem weevil and rape stem weevil before oviposition or to prevent feeding damage by pollen beetle (*Meligethes aeneus*) at the early bud stage, rather than later, would provide some conservation of these parasitoids (Lehmann, 1965; Nitzsche, 1998). For further details, see Chapter 2.

Acknowledgements

This work was part-funded by the PhD graduate programme of the Deutsche Forschungsgemeinschaft and the European Commission Concerted Action: *Minimizing pesticide use and environmental impact by the development and promotion of bio-control strategies for oilseed rape pests, FAIR 96-1314.*

References

Aubert, J.-F. & Jourdheuil, P. (1958). Nouvelle description et biologie de quelques Ichneumonides appartenant aux genres *Aneuclis* Först., *Isurgus* Först. et *Thersilochus* Holm. *Revue de Pathologie Vegetale et d'Entomologie Agricole de France* **37**, 175–93.

Büchi, R. (1991). Monitoring of parasitoids in oilseed rape pests during 1989 in Switzerland. *IOBC/wprs Bulletin* **19** (6), 54–60.

Büchi, R. & Roos-Humbel, S. (1991). Nützlinge reduzieren die Zahl der Rapsschädlinge. *Landwirtschaft Schweiz* **4**, 69–73.

Buhl, C. (1952). Der Grosse Kohltriebrüssler (*Ceutorhynchus napi* Gyll.), ein bisher im Glückstaedter Gemueseanbaugebiet unbekannter Schädling. *Zeitschrift für Pflanzenkrankheiten und Pflanzenschutz* **59**, 326–34.

Dosse, G. (1951). Der Große Kohltriebrüßler (*Ceuthorrhynchus napi* Gyll.). Biologie, Schadauftreten und Bekämpfung unter besonderer Berücksichtigung der Gallbildung an Kohlpflanzen. *Zeitschrift für Angewandte Entomologie* **32**, 489–566.

Fitton, M. G., Graham, M. W. R. de V., Bouček, Z. R. L., Fergusson, N. D. M., Huddleston, T., Quinlan, J. & Richards, O. W. (1978). A check list of British insects. Part 4: Hymenoptera. *Handbooks for the Identification of British Insects* **11** (4), 1–159.

Goulet, H. & Huber, J. T. (eds) (1993). Hymenoptera of the world: an identification guide to families. *Agriculture Canada, Research Branch, Publication 1894/E*, 668 pp.

Günthardt, E. (1949). Beiträge zur Lebensweise und Bekämpfung von *Ceuthorrhynchus quadridens* Panz. und *Ceuthorrhynchus napi* Gyll., mit Beobachtungen an weiteren Kohl- und Rapsschädlingen. *Mitteilungen der Schweizerischen Entomologischen Gesellschaft* **22**, 441–591.

Herrström, G. (1964). Untersuchungen über Parasiten von Ölfruchtschädlingen in Schweden. *Statens Växtskyddsanstalt. Meddelanden* **12**, 437–48.

Horstmann, K. (1971). Revision der europäischen Tersilochinen (Hym.: Ichneumonidae). Teil I. *Veröffentlichungen der Zoologischen Staatssammlung München* **15**, 47–138.

Horstmann, K. (1981). Revision der europäischen Tersilochinae II (Hymenoptera, Ichneumonidae). *Spixiana, Supplementum* **4**, 1–76.

Jourdheuil, P. (1960). Influence de quelques facteurs écologiques sur les fluctuations de population d'une biocénose parasitaire: étude relative à quelques Hyménoptères (Ophioninae, Diospilinae, Euphorinae) parasites de divers Coléoptères inféodés aux Crucifères. *Annales des Épiphyties* **11**, 445–658.

Kazda, V. (1956). Die in der Tschechoslowakei an den Kohl- und Rapspflanzen schädlichen Rüsselkäferarten unter besonderer Berücksichtigung des *Ceuthorrhynchus napi* Gyll. *Meddedelingen Landbouwhogeschool Opzoekingsstation Gent* **21**, 411–20.

Klingenberg, A. & Ulber, B. (1994). Untersuchungen zum Auftreten der Tersilochinae (Hym., Ichneumonidae) als Larvalparasitoide einiger Rapsschädlinge im Raum Göttingen 1990 und 1991 und zu deren Schlupfabundanz nach unterschiedlicher Bodenbearbeitung. *Journal of Applied Entomology* **117**, 287–99.

Klukowski, Z. & Kelm, M. (2000). *Stenomalina gracilis* (Walker), a new parasitoid reared from *Ceutorhynchus napi* Gyll. in Poland. *IOBC/wprs Bulletin* **23** (6), 135–8.

Kraus, P. & Kromp, B. (2002). Parasitization rates of the oilseed rape pests *Ceutorhynchus napi, Ceutorhynchus pallidactylus* (Coleoptera, Curculionidae) and *Meligethes aeneus* (Coleoptera, Nitidulidae) by Ichneumonids in several localities of eastern Austria. *IOBC/wprs Bulletin* **25** (2), 117–22.

Lehmann, W. (1965). Einfluss chemischer Bekämpfungsmassnahmen auf einige Rapsschädlinge und ihre Parasiten. II. Knospen- und Stengelschädlinge. *Archiv für Pflanzenschutz* **1**, 209–19.

Nissen, U. (1997). *Ökologische Studien zum Auftreten von Schadinsekten und ihren Parasitoiden an Winterraps norddeutscher Anbaugebiete*. PhD thesis Christian-Albrechts-University of Kiel, Germany.

Nitzsche, O. (1998). *Auftreten und Effizienz von Parasitoiden als natürliche Gegenspieler von Schadinsekten im Winterraps unter besonderer Berücksichtigung unterschiedlicher Bodenbearbeitungsmaßnahmen nach Winterraps*. PhD thesis, Georg-August-University of Göttingen, Germany.

Nitzsche, O. & Ulber, B. (2002). Studies on parasitism of rape stem weevil, *Ceutorhynchus napi* Gyll. (Col.: Curculionidae), by *Tersilochus fulvipes* Grav. (Hym.: Ichneumonidae) in oilseed rape. (Submitted).

Rozsypal, J. (1941). Ein Beitrag zur Kenntnis des Mohnwurzelrüsslers (*Stenocarus fuliginosus* MARSH.). *Folia Entomologica Brno* **4**, 43–60.

Šedivý, J. (1983). Tersilochinae as parasitoids of insect pests of winter rape (Hymenoptera: Ichneumonidae). *Contributions of the American Entomological Institute* **20**, 266–76.

Speyer, W. (1925). *Perilitus melanopus* Ruthe (Hym. Braconidae) als Imaginalparasitoid von *Ceuthorrhynchus quadridens* Panz. *Zeitschrift für Angewandte Entomologie* **11**, 132–46.

Thies, C., Denys, C., Tscharntke, T. & Ulber, B. (1997). Der Einfluß von Saumbiotopen und Ackerbrachen auf Schädlings-Nützlings-Interaktionen im Raps (*Brassica napus* spp. *oleifera*). *Verhandlungen der Gesellschaft für Ökologie* **27**, 393–8.

Ulber, B. (2000). Bibliography of parasitoid species and levels of parasitism of rape stem weevil *Ceutorhynchus napi* Gyll. and cabbage stem weevil *Ceutorhynchus pallidactylus* (Mrsh.). *IOBC/wprs Bulletin* **23** (6), 131–4.

6

Parasitoids of Cabbage Seed Weevil

Ingrid H. Williams
Rothamsted Research, Harpenden, Hertfordshire, AL5 2JQ, UK

Introduction

Cabbage seed weevil (*Ceutorhynchus assimilis*) is attacked by parasitoids at all stages of its lifecycle, although most parasitism is directed towards the larval stage (Murchie & Williams, 1998a). All such parasitoids are hymenopterans (Table 6.1). The concealment of larvae inside the rape pod seems to provide particularly ideal conditions for ectoparasitism.

Parasitism of Adults

The braconid *Microctonus melanopus* is the most frequently cited parasitoid, attacking adult cabbage seed weevils in France, Germany and the Netherlands (Speyer, 1925b; Ankersmit, 1956; Bonnemaison, 1957; Jourdheuil, 1960; Harmon & McCaffrey, 1995). Miczulski (1967) also reports *Microctonus* cf. *deceptor* as a probable parasitoid of *Ceutorhynchus* spp. in oilseed rape in Poland; however, as only males were reared, identification to species was uncertain. Risbec (1953) found larvae of the subfamily Euphorinae parasitizing cabbage seed weevil but did not specify the species. In the UK, Murchie (1996) reared a specimen of *Microctonus* from cabbage seed weevil, but, again, since it was a male, identification to species was not possible. Variable levels of parasitism have been reported. Bonnemaison (1957), in France, found that *M. melanopus* was the most important parasitoid of cabbage seed weevil, achieving in some rape crops levels of 50–60% parasitism but only 1–3% parasitism on other rape crops nearby. Jourdheuil (1960), also in France, found 40–50% parasitism of cabbage seed weevil adults. Harmon & McCaffrey (1995), working in northern Idaho (USA), recorded levels of parasitism of 0–70%, and Murchie (1996), in southern England, up to 7% parasitism.

Parasitism of Eggs

Species of the family Mymaridae are cited as attacking the egg stage of cabbage seed weevil: *Patasson declinata* in France (Jourdheuil, 1960) and the Netherlands (Ankersmit,

Table 6.1 Systematic list of parasitoids of cabbage seed weevil. For further systematic details see Fitton *et al.* (1978) and Goulet & Huber (1993).

Order HYMENOPTERA

Superfamily ICHNEUMONOIDEA
Family ICHNEUMONIDAE (Subfamily TERSILOCHINAE)
Aneuclis Förster
– *melanarius* (Holmgren) (= *diversus* Szépligeti; = *petiolaris* Szépligeti)†
Tersilochus Holmgren
– sp.
Family BRACONIDAE (Subfamily DORYCTINAE)
Bracon Fabricius
– *fulvipes* Nees†
– sp.
– *variator* Nees (= *discoideus* Wesmael [Note 1]; = *maculiger* Wesmael)†
Family BRACONIDAE (Subfamily CHELONINAE)
Sigalphus Latreille
– *obscurellus* Nees†
Family BRACONIDAE (Subfamily HELCONINAE)
Diospilus Haliday
– *morosus* Reinhardt†
– *oleraceus* Haliday
Taphaeus Wesmael
– *affinis* Wesmael
Family BRACONIDAE (Subfamily EUPHORINAE)
Microctonus Wesmael
– *melanopus* Ruthe† [Note 2]
– cf. *deceptor* Wesmael

Superfamily CHALCIDOIDEA
Family CHALCIDIDAE
Spilochalcis Thomson
– *side* (Walker)‡
Family EURYTOMIDAE
Eurytoma Illiger
– *aciculata* (Ratzeburg) [hyperparasitoid]
– *curculionum* Mayr†
– sp.
Family PTEROMALIDAE (Subfamily PTEROMALINAE)
Anisopteromalus Ruschka
– *calandrae* (Howard)
Chlorocytus Graham
– *diversus* (Walker)
Habrocytus Thomson
– *dispar* (Curtis)
– *semotus* (Walker)
Mesopolobus Westwood (= **Amblymerus** Walker; = **Eutelus** Walker; = ***Xenocrepis*** Förster)
– *mayetiolae* (Gahan)‡
– *mediterraneus* (Mayr) [hyperparasitoid]
– *morys* (Walker) (= *pura* Mayr)†
Stenomalina Ghesquière
– *gracilis* (Walker)
– *muscarum* (Linnaeus)
Trichomalus Thomson
– *perfectus* (Walker) (= *decisus* (Walker); = *decorus* (Walker); = *laevinucha* (Thomson))†
– sp.

Table 6.1 *Cont'd*

Trimeromicrus Gahan
– *maculatus* Gahan‡
Zatropis Crawford
– sp.
Family EUPELMIDAE (Subfamily EUPELMINAE)
Eupelmus Dalman (= ***Eupelmella*** Masi; = ***Macroneura*** Walker)
– *vesicularis* (Retzius)†
Family EULOPHIDAE
Eulophus Müller
– *hegemon* Wek‡
– sp.
Necremnus Thomson
– *duplicatus* Gahan
– *tidius* (Walker)
Tetrastichus Haliday
–*galectobus* (Ratzeburg)
Family MYMARIDAE (Subfamily MYMARINAE)
Anaphes Haliday
– *fuscipennis* Haliday
Mymar Curtis
– *autumnalis* (Förster)
Patasson Walker
– *brachygaster* Debauche
– *declinata* (Soyka)

† Species included in the key to adults (Chapter 11).
‡ Species or genus not found in Europe.

Note 1. The status of *Bracon discoideus* Wesmael is unclear; some authorities treat it as a separate species.
Note 2. Also cited in the literature as *Perilitus melanopus* Ruthe.

1956); *Patasson brachygaster* in France (Risbec, 1953) and in the UK (Nasreldin, 1973); *Mymar autumnalis* in Germany (Godan, 1959). In Poland, Miczulski (1968) reared a species of *Patasson* and also recorded *Anaphes fuscipennis* as a probable parasitoid of cabbage seed weevil eggs.

In most studies, the percentage of eggs found parasitized was small: Jourdheuil (1960) recorded a maximum of 4% by *Patasson declinata*, Nasreldin (1973) recorded 2% by *P. brachygaster*, whereas Risbec (1953) states that parasitism by *P. brachygaster* was of negligible importance. In contrast, Godan (1959) found that *Mymar autumnalis* parasitized up to 90% of cabbage seed weevil eggs in 1952, although no parasitized eggs were found in subsequent years.

Parasitism of Larvae

Ectoparasitoids

The most commonly cited parasitoids of cabbage seed weevil are ectoparasitoids of the larval stage (Tables 6.2 and 6.3). Those involved belong either to the family Braconidae or to the superfamily Chalcidoidea, the most cited species being *Trichomalus perfectus*, *Mesopolobus morys* and *Stenomalina muscarum* (Table 6.2).

Table 6.2 Ectoparasitoids of the family Pteromalidae (Hymenoptera) that attack the larvae of cabbage seed weevil (*Ceutorhynchus assimilis*).

Species	**Country** (Citation)
Anisopteromalus calandrae (Howard)	**France** (Risbec, 1953)
Chlorocytus diversus (Walker)	**UK** (Nasreldin, 1973)
Chlorocytus sp.	**UK** (McKenna, 1972)
Habrocytus dispar (Curtis) [Note 1]	**Poland** (Dmoch & Sulgostowska, 1986)
Habrocytus semotus (Walker) [Note 1]	**Poland** (Dmoch & Sulgostowska, 1986); **UK** (Nasreldin, 1973)
Habrocytus sp.	**Poland** (Heymons, 1922); **Canada** (McLeod, 1953); **USA** (Hanson *et al.*, 1948; Walz, 1957)
Mesopolobus mayetiolae (Gahan) [Notes 2 & 3]	**USA** (Carlson *et al.*, 1951; McLeod, 1953 [Note 4])
Mesopolobus morys (Walker) [Note 5]	**France** (Risbec, 1953; Jourdheuil, 1960); **Poland** (Miczulski, 1968; Dmoch, 1975); **Sweden** (Herrström, 1964; Rosen, 1964); **Switzerland** (Günthart, 1949; Büchi, 1993); **UK** (McKenna, 1972; Nasreldin, 1973; Murchie, 1996); **Yugoslavia** (Maceljski *et al.*, 1980); **USA** (Doucette, 1948; Hanson *et al.*, 1948; Carlson *et al.*, 1951; McLeod, 1953; Walz, 1957 [Note 4])
Mesopolobus sp.	**UK** (Nasreldin, 1973); **USA** (Doucette, 1948; Hanson *et al.*, 1948)
Stenomalina gracilis (Walker)	**UK** (Nasreldin, 1973)
Stenomalina muscarum (Linnaeus)	**Poland** (Miczulski, 1968; Skrocki, 1972; Dmoch, 1975; Dmoch & Sulgostowska, 1986); **Sweden** (Herrström, 1964; Rosen, 1964); **Switzerland** (Büchi, 1993)
Trichomalus perfectus (Walker) [Note 6]	**France** (Risbec, 1953; Bonnemaison, 1957; Jourdheuil, 1960); **Germany** (Heymons, 1922; Speyer, 1925a, 1925b; Thiem, 1970; Laborius, 1972); **the Netherlands** (Ankersmit, 1956); **Poland** (Miczulski, 1968; Skrocki, 1972; Szczepanski, 1972; Dmoch, 1975; Dmoch & Sulgostowska, 1986); **Sweden** (Herrström, 1964; Rosen, 1964); **Switzerland** (Günthart, 1949; Büchi, 1993); **UK** (McKenna, 1972; Nasreldin, 1973; Murchie, 1996); **Yugoslavia** (Maceljski *et al.*, 1980); **Canada** (McLeod, 1953 [Note 4]); **USA** (Doucette, 1948; Hanson *et al.*, 1948; Carlson *et al.*, 1951; Crowell, 1952; McLeod, 1953; Walz, 1957)
Trichomalus sp.	**USA** (Doucette, 1948)
Trimeromicrus maculatus Gahan	**USA** (Carlson *et al.*, 1951; McLeod, 1953; Walz, 1957 [Note 4])
Zatropis sp.	**Canada** (McLeod, 1953); **USA** (Hanson *et al.*, 1948)

Note 1. Possibly hyperparasitic (Graham, 1969; Dmoch & Sulgostowska, 1986).
Note 2. Other names cited for this species: *Amblymerus mayetiolae* (Gahan); *Eutelus mayetiolae* (Gahan).
Note 3. Possibly hyperparasitic (McLeod, 1953).
Note 4. For further citations in North America see Peck (1963).
Note 5. Other names cited for this species: *Xenocrepis pura* Mayr; *Xenocrepis morys* (Walker); *Pteromalus morys* Walker.
Note 6. Other names cited for this species: *Trichomalus fasciatus* (Thomson); *Trichomalus herbidus* (Walker); *Trichomalus laevinucha* (Thomson); *Pteromalus decorus* (Walker); *Pteromalus decisus* (Walker).

Table 6.3 Ectoparasitoids other than pteromalids (Hymenoptera: Pteromalidae) that attack the larvae of cabbage seed weevil (*Ceutorhynchus assimilis*).

Family	Species	**Country** (Citation)
BRACONIDAE	*Bracon discoideus* Wesmael [Note 1]	**France** (Jourdheuil, 1960)
	Bracon fulvipes Nees	**France** (Hoffmann, 1951)
	Bracon sp.	**Canada** (McLeod, 1953)
	Bracon variator Nees [Note 2]	**France** (Hoffmann, 1951)
CHALCIDIDAE	*Spilochalcis side* (Walker)	**USA** (Carlson *et al.*, 1951; McLeod, 1953)
EULOPHIDAE	*Eulophus hegemon* Wek	**France** (Hoffmann, 1951)
	Necremnus duplicatus Gahan	**USA** (Gahan, 1941; Doucette, 1948; Hanson *et al.*, 1948; Carlson *et al.*, 1951; McLeod, 1953; Walz, 1957)
	Necremnus tidius (Walker)	**Poland** (Miczulski, 1968)
	Necremnus sp.	**Germany** (Laborius, 1972); **UK** (McKenna, 1972; Nasreldin, 1973)
	Tetrastichus galectobus (Ratzeburg) [Notes 3 & 4]	**UK** (Nasreldin, 1973)
EUPELMIDAE	*Macroneura vesicularis* (Retzius) [Notes 5 & 6]	**Poland** (Dmoch, 1975); **Canada** (McLeod, 1953); **USA** (Doucette, 1948; Hanson *et al.*, 1948)
EURYTOMIDAE	*Eurytoma curculionum* Mayr [Note 7]	**Poland** (Dmoch, 1975)
	Eurytoma sp.	**Canada** (McLeod, 1953); **USA** (Doucette, 1948; Hanson *et al.*, 1948)

Note 1. Regarded by some authorities as synonymous with *Bracon variator* Nees (see Table 6.1).
Note 2. Cited as *Bracon maculiger* Wesmael.
Note 3. '*galactopus* Ratzeburg' is probably a misspelling.
Note 4. Possibly hyperparasitic (Domenichini, 1966).
Note 5. Possibly hyperparasitic (Doucette, 1948).
Note 6. Other name cited for this species: *Eupelmella vesicularis* Retzius.
Note 7. Possibly hyperparasitic (Dmoch, 1975).

Study of the ectoparasitoids of cabbage seed weevil larvae is hampered by their synonymy and misidentification in the literature. Lerin (1987), for example, found that *Trichomalus perfectus* had been named by different authors as: *Pteromalus decorus*, *P. decisus*, *Trichomalus fasciatus*, *T. herbidus* and *T. laevinucha*. Further, the names *Pteromalus perfectus* and *Trichomalus fasciatus* are commonly used in older references for *T. perfectus* (see Peck, 1963). Likewise, *Mesopolobus morys* is cited in older papers as *Pteromalus morys*, *Xenocrepis pura* and *X. morys*; *Mesopolobus mayetiolae* as *Amblymerus mayetiolae* and *Eutelus mayetiolae*; and *Eupelmus vesicularis* as *Eupelmella vesicularis* (Peck, 1963). Also, it is sometimes unclear whether the species cited is a primary parasitoid or merely a hyperparasitoid. The only certain references to hyperparasitism of cabbage seed weevil are those of Nasreldin (1973) who recorded *Mesopolobus mediterraneus* as a parasitoid of *Chlorocytus diversus* (Walker) and *Habrocytus semotus* (Walker) (both hosts are pteromalids), and Dmoch (1975) who found *Eurytoma aciculata* as a parasitoid of the larval stage of *T. perfectus*.

Variable and often substantial rates of ectoparasitism of cabbage seed weevil larvae have been reported from several countries: in the UK, 31–73% (Murchie, 1996); in Switzerland, 7–9% (Günthart, 1949), <52% and 66% (Linz, 1991) and 3–70% (Büchi,

1993); in Sweden, 100% (Herrström, 1964); and in the USA, 13–91% (Hanson *et al.*, 1948), 11% (Carlson *et al.*, 1951) and 80–90% (McLeod, 1953).

Where the ectoparasitoids from parasitized larvae have been reared to determine species, *Trichomalus perfectus* has usually been the predominant species, with *Mesopolobus morys* and sometimes *Stenomalina muscarum* also reared in some numbers; other species of parasitoid are usually scarce. In the UK, Nasreldin (1973) found that *T. perfectus* was the most abundant ectoparasitoid, accounting for 32% and 52% of those reared; *M. morys* was second at 5% and 37%, in 1970 and 1971, respectively. More recently, Murchie (1996), also in the UK, found that 99% of the ectoparasitoids he reared (from cabbage seed weevil larvae) from winter rape were *T. perfectus* and <1% were *M. morys*, whereas from spring rape, almost half were *M. morys*, suggesting that *M. morys* may be relatively more important on spring than on winter rape. In Germany, Laborius (1972) found that 11–43% of ectoparasitoids were *T. perfectus*, and Ulber & Vidal (1998) reported that, in a two-year study, 84% and 95% of ectoparasitoids were *T. perfectus*, with few *M. morys* (<15%) and the occasional presence of *Eurytoma curculionum* and *Stenomalina gracilis.* In Poland, *T. perfectus* represented 90% and *M. morys* 6% of the ectoparasitoids reared (Dmoch, 1975). By contrast, in Switzerland, Büchi (1993) found *M. morys* to be the most abundant ectoparasitoid (86% of those reared), with *T. perfectus* (11%) and *S. muscarum* (5%) of lesser importance; in Sweden, Herrström (1964) found *M. morys* and *S. muscarum* predominating, and *T. perfectus* as scarce; finally, in the USA, Carlson *et al.* (1951) reported that *M. morys* was the most important and abundant parasitoid of cabbage seed weevil, occurring together with *Mesopolobus mayetiolae* and *Trimeromicrus maculatus.*

Endoparasitoids

The most comprehensive study on endoparasitism of cabbage seed weevil larvae is that of Jourdheuil (1960) in France, who recorded five species of hymenopteran parasitoid: two ichneumonids – *Aneuclis melanarius* and *Tersilochus* sp.; and three braconids – *Diospilus morosus*, *Diospilus oleraceus* and *Sigalphus obscurellus*. Börner *et al.* (1921) and Speyer (1925a), in Germany, and I. H. Williams, in the UK (*unpublished data*), also reared *D. oleraceus* from cabbage seed weevil larvae. Hoffmann (1951) refers to Speyer as citing *Perilitus melanopus* (= *Microctonus melanopus*) and to Reinhardt as citing *Diospilus oleraceus* as endoparasitoids of cabbage seed weevil larvae; in addition he reports, for example, that *Sigalphus obscurellus* and *Taphaeus affinis* (as *Diospilus affinis*) are endoparasitoids of cabbage seed weevil, but he gives no bibliographic references or further data on these species.

Unlike ectoparasitism, reported levels of endoparasitism are always low: Jourdheuil (1960) found that *Diospilus oleraceus* parasitized from 1% to 10% of cabbage seed weevil larvae; Herrström (1964) in Sweden and Laborius (1972) in Germany did not find any endoparasitoids.

The Pteromalid Wasp *Trichomalus perfectus*

The pteromalid *Trichomalus perfectus* (Fig. 6.1) is the most widespread, abundant and important parasitoid to attack cabbage seed weevil, occurring throughout most of Europe and North America. It is also the most studied parasitoid of this pest and the only one about which there is more substantial information than merely percentage incidence. A description of the small, greenish adult is given in Delucchi & Graham (1956).

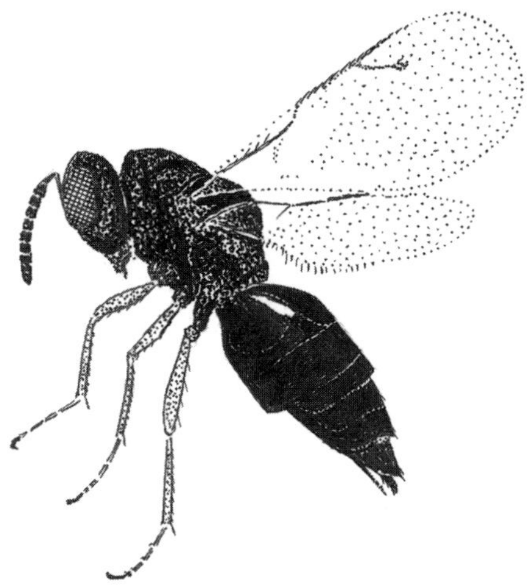

Fig. 6.1 Female *Trichomalus perfectus* – family PTEROMALIDAE (×20).

Status and importance

The high levels of parasitism of cabbage seed weevil by *Trichomalus perfectus* found in many studies suggests that this parasitoid plays a major role in reducing cabbage seed weevil densities, although the impact of *T. perfectus* on the population dynamics of its host has been little studied. To be an effective antagonist, a parasitoid should be efficient at locating its host and should have a positive density-dependent relationship with its host, so that parasitism rates increase with host density (Kidd & Jervis, 1996; Ulber & Vidal, 1998).

For the oilseed rape grower, the most important consequence of ectoparasitism by *T. perfectus*, as well as other species, is direct mortality of weevil larvae leading to reduction in adults for the following year. However, ectoparasitoids also reduce weevil damage to the standing crop; following parasitization, the immature larva is immobilized and thereby prevented from consuming its full complement of seeds.

Ectoparasitism of cabbage seed weevil larvae in a crop most often involves attack by several parasitoid species, but where the ectoparasitoid larvae have been reared to determine species, *T. perfectus* has usually been found to predominate. Thus, in the UK, Murchie (1996) found 37–73% ectoparasitism on winter rape, 99% of which were *T. perfectus*; on spring rape, he found that 31% were ectoparasitized but only 56% (of only nine specimens reared) were *T. perfectus*. In Germany, Thiem (1970) found 12–30%, Laborius (1972) found 11–43% and Ulber & Vidal (1998) found 84–95% of larvae parasitized by *T. perfectus*. In Canada, McLeod (1953) found 80–90% of larvae ectoparasitized, 86–94% of them by *T. perfectus*. By contrast, in Switzerland, Büchi (1993) found 3–70% ectoparasitism but only 11% attributable to *T. perfectus*.

Murchie (1996) found that cabbage seed weevil larvae parasitized by *T. perfectus* consumed fewer (an average of 3.2) seeds than unparasitized larvae that emerged successfully (an average of 5.2 seeds) – thus, parasitized larvae caused approximately 38% less damage than healthy ones. Dmoch (1975) reported a not dissimilar reduction (28%). Further, males of *T. perfectus* were reared from larvae that had consumed fewer seeds compared with those from which females were reared.

The relationship between host density, host distribution and parasitism by *T. perfectus* has been examined in field experiments (Ulber & Vidal, 1998), in which different host densities (16–57% pods infested), established in caged plots, were exposed to natural parasitization. The percentage of parasitized larvae was found to be almost constant (36–57%) over the whole range of larval densities, and no density-dependent relationship between parasitoid and host could be detected. The proportion of parasitized larvae did not differ with pod position on the plant, indicating good host location abilities by the parasitoid. This is also true on the field (crop) scale (Ferguson *et al.*, 2000; see below).

Phenology on the crop

In Europe, female *Trichomalus perfectus* begin their migration to rape crops in late April or early May but are not caught in substantial numbers until two to four weeks after peak migration of cabbage seed weevils. Thereafter, their phenology largely mirrors that of cabbage seed weevil, with numbers increasing throughout June and to mid-July but then declining (Laborius, 1972; Dmoch, 1975; Murchie *et al.*, 1997b; Murchie *et al.*, 1999) (Fig. 6.2). New-generation parasitoids mate on emergence and leave the rape crop shortly before harvest. Only females are thought to overwinter, possibly in evergreen foliage and in sheltered nooks and crannies (Graham, 1969; Szczepanski, 1972).

Thus, immigration is synchronized with the appearance of cabbage seed weevil larvae in pods (Dmoch, 1975). The dissociation between the immigration flights of the host and *T. perfectus* provides an opportunity for the temporal targeting of insecticide to kill cabbage seed weevil but not the parasitoid. This approach has been proposed as part of an integrated strategy for the management of cabbage seed weevil populations on winter rape (Alford *et al.*, 1996).

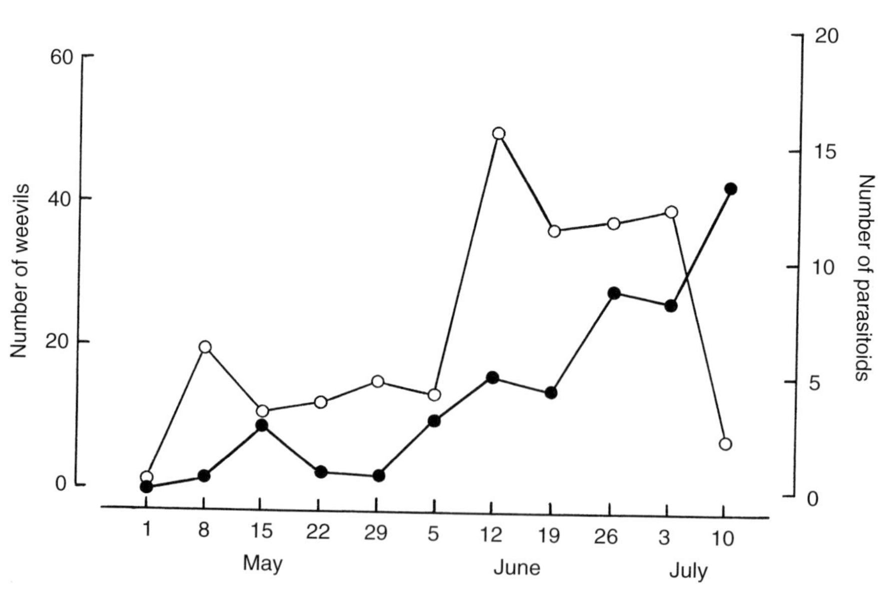

Fig. 6.2 Mean catches of cabbage seed weevil (*Ceutorhynchus assimilis*) (○) and *Trichomalus perfectus* (●) adults in yellow water traps in winter oilseed rape during 1992 (from Murchie *et al.*, 1999).

Number of generations

Trichomalus perfectus is probably univoltine (McKenna, 1972; Murchie, 1996). However, some authors have suggested that it might be multivoltine, with alternative hosts (McLeod, 1953; Rosen, 1964; Szczepanski, 1972); Nolte (1953), in Germany, reported that *T. perfectus* had three generations, the first parasitizing rape stem weevil (*Ceutorhynchus napi*), the second parasitizing cabbage seed weevil and the third parasitizing the cynipid *Phanacis* (= *Timaspis*) *papaveris* (Kieffer), but these possible alternative hosts are not present throughout Europe; for example, they do not occur in the UK.

Habitat and host finding

Habitat location by *Trichomalus perfectus* has not been studied and it is not known whether this species, like some other parasitoids of oilseed rape pests (Murchie *et al.*, 1997a) and the pests themselves (Bartlet, 1996), uses plant volatiles from the crop for habitat/host plant location.

Host-finding and host-acceptance studies (Dmoch & Rutkowska-Ostrowska, 1978) have shown that frass from the third (final) instar of cabbage seed weevil elicits oviposition by the parasitoid. To locate a host larva, the female parasitoid walks along the surface of the pod, performing 'radaring', 'touching', 'drumming' and 'stroking' movements with her antennae (Dmoch, 1998). If the pod does not contain a larva, these antennal movements are brief and the female soon moves to another pod. If a host larva is present, the intensity, duration and frequency of the 'drumming' increases until the larva is located and the female then commences to pierce the pod with her ovipositor (Dmoch, 1998). In laboratory studies, second-instar larvae were also parasitized, albeit in small numbers (Dmoch, 1975). Whether *T. perfectus* can distinguish qualitative and quantitative cues from host larvae of different sizes is not known.

Host feeding

Female *Trichomalus perfectus* probably feed from, as well as oviposit onto, the larvae of cabbage seed weevil. They are synovigenic, that is, they emerge from hibernation with undeveloped eggs. This type of parasitoid female typically feeds from its host to obtain the necessary nutrients to develop her eggs (Gauld & Bolton, 1988). Although host feeding has not been directly observed in *T. perfectus*, Doucette (1948) and Walz (1957) found many dead larvae with no evidence of parasitism, and Jourdheuil *et al.* (1974) speculated that larvae were used as food. Murchie (1996) also found many dead larvae, some with marks on their integument that were most likely made by parasitoids, perhaps during failed oviposition attempts or when host feeding. The mechanism of host feeding possibly used by *T. perfectus* is likely to be a feeding tube, as described by Fulton (1933) for the North American pteromalid *Habrocytus cerealellae* Ashmead.

Oviposition

Having located a suitable cabbage seed weevil larva within a pod, the female parasitoid raises her abdomen to a horizontal position, and protrudes her ovipositor ventrally from her vagina. She then turns her body to the side and penetrates the pod wall with her ovipositor and lays her egg onto the surface of the host larva. Her body is then raised, and she draws out the ovipositor and lets it return into the vagina (Dmoch & Rutkowska-Ostrowska, 1978). Egg laying lasts one to two minutes. Following one oviposition, the female usually rests for a short time, cleans her body and then starts to search for another host larva (Dmoch, 1998). *Trichomalus perfectus* is a solitary idiobiont, immobilizing its host at oviposition (Askew & Shaw, 1986). Following parasitization, the host larva discolours, at first around the site of penetration by the ovipositor of the parasitoid, and then, during the next two days, over the rest of the body until it has turned dark grey (Walz, 1957).

Stages of development

Dmoch & Klimek (1975) and Murchie (1996) have reared *Trichomalus perfectus* in the laboratory and have determined its development times. In essence, the parasitoid egg hatches in one to four days; the larva feeds externally from the host larva for seven to 10 days, eventually consuming it completely, except for its head capsule and skin, before pupating alongside the host remains without forming a cocoon (there are three larval instars); the pupal stage lasts eight to 15 days, and is on average slightly longer for the female than for the male. Total development time from egg to adult is, thus, 19–25 days. Female pupae are longer (2.1–3.2 mm) than male pupae (1.2–2.6 mm) (Murchie, 1996). On emergence from the pupa, the adult chews its way out through the pod wall, making a small, round exit hole; this hole is generally slightly smaller than that made by a healthy cabbage seed weevil larva for exiting the pod and has more jagged edges. Mating occurs soon after emergence.

Sex ratio

The sex ratio of *Trichomalus perfectus* is approximately equal but is plastic, being strongly influenced by the relative sizes of available larval hosts; if the host age stratum differs between crops the sex ratio will also differ. *T. perfectus*, as in the case of other hymenopteran parasitoids with their haplodiploid genetic system, can 'choose' the sex of their offspring by controlling the release of sperm from the spermatheca: unfertilized eggs produce males, whereas fertilized ones give rise to females. Murchie & Williams (1998b) found that parasitized larvae from which male *T. perfectus* were reared had eaten fewer seeds than those from which female *T. perfectus* were reared. This suggests that male parasitoids developed in smaller larvae than female parasitoids and that the mother parasitoid, selectively, chose to lay 'female' eggs in larger host larvae which have greater resources for her offspring, agreeing with the sex allocation theory of Charnov (1982).

Spatio-temporal distribution in field

The pattern of crop colonization by *Trichomalus perfectus*, like that of its host (Free & Williams, 1979), undoubtedly reflects the interplay of environmental factors, such as the location of overwintering sites and windbreaks relative to the position of the crop and the direction of the wind, with behavioural responses of the parasitoid, particularly those involved in crop location, selection of host plant and location of host larvae.

Recent studies have investigated the within-field spatio-temporal distributions of *T. perfectus*, relative to that of its host on winter rape, by sampling the insects using two-dimensional arrays of spatially referenced sampling points (Murchie *et al.*, 1999; Ferguson *et al.*, 2000); such information has implications for integrated crop management (ICM) strategies (see Chapter 3). Temporal and spatial precision targeting of insecticide treatments to crop area have potential for further reducing insecticide inputs with benefit to the parasitoid.

Murchie *et al.* (1999) used a randomization test to compare the observed counts of pest and parasitoid adults caught during May to July in water traps arranged in a grid on the crop. They found that the distributions of adults of *T. perfectus* and cabbage seed weevils were negatively correlated on the crop; the former were distributed at the crop edge only during their early phase of immigration into the crop in May, and not thereafter, and were always relatively less concentrated on the edge of the crop than were adults of their host.

Ferguson *et al.* (2000) mapped, analysed and compared the spatio-temporal distribution of cabbage seed weevils caught in flight traps in a winter oilseed rape crop with the distribution of their larvae and those of *T. perfectus*, using Spatial Analysis by Distance Indices (SADIE) (Perry, 1995); they found that, in the crop they studied, both host and parasitoid larvae were aggregated into regions 0–80 m from the crop edges and that they were strongly spatially associated (Fig. 6.3). The study showed that the parasitoid was highly effective in dispersing throughout its host range within the crop and in finding its host within that range.

Effects of insecticide

Many winter oilseed rape crops are treated with insecticides to kill cabbage seed weevils, either prophylactically or when threshold numbers are reached (Alford *et al.*, 1991). However, applied at the wrong time, the use of insecticides against this pest may be counter productive by removing the natural control exerted by *Trichomalus perfectus* (Murchie *et al.*, 1997b). In the UK, the widespread use of broad-spectrum insecticides, notably triazophos, towards or at the end of flowering of winter rape, is thought to have caused the rates of parasitism of cabbage seed weevil by *T. perfectus* to have declined from over 70% in the 1970s to much lower levels (<10%) in the 1980s and 1990s (Alford *et al.*, 1991; Cooper & Lane, 1991).

Murchie *et al.* (1997b) in the UK investigated the effects on *T. perfectus* of commercial applications of alpha-cypermethrin applied during flowering to kill cabbage seed weevil adults before oviposition, and of triazophos applied post-flowering to kill cabbage seed weevil larvae inside the pods. Triazophos had a detrimental effect on *T. perfectus*, substantially reducing the percentage of weevil larvae that were parasitized. The insecticide was applied at the time that parasitoids were searching for host larvae on which to oviposit, and it killed both adult *T. perfectus* and their host larvae. Alpha-cypermethrin was less harmful than triazophos, probably because it was applied before the main migration of *T. perfectus* females into the crop (Fig. 6.4). The recommended time for the application of pyrethroid insecticide to kill cabbage seed weevil adults and to conserve *T. perfectus* populations on winter rape in the UK is now during flowering, between 20 pod set and 80% petal fall on the main raceme, and then only if numbers of the pest have exceeded the treatment threshold (Alford *et al.*, 1996). The recent change to greater use of pyrethroids rather than of triazophos (Garthwaite *et al.*, 1995) may already have resulted in a national increase in the level of parasitism of cabbage seed weevils; the overall

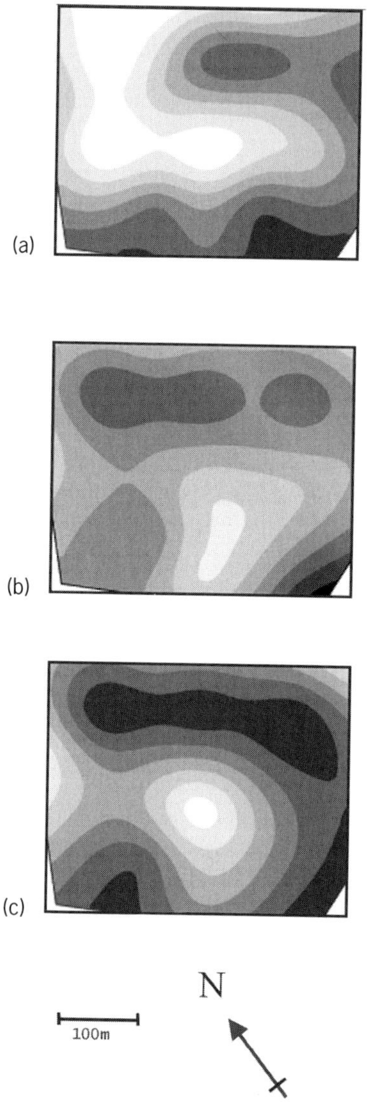

Fig. 6.3 Distributions of insects interpolated from samples taken at 19 locations in seven density classes with contours equally spaced on a logarithmic scale ($\log_{10} n + 1$); classes with larger values are represented by darker shades. Maximum and minimum classes, respectively, represent: (a) ≥15 and <0.6 adult female cabbage seed weevil (*Ceutorhynchus assimilis*) per trap (total for 20 April to 20 June); (b) ≥70 and ≤21 weevil larvae per 400 pods; (c) >27 and <8 *Trichomalus perfectus* larvae per 400 pods (from Ferguson *et al*., 2000).

mean parasitism in England and Wales increased from 8% in 1990 to 48% and 49% in 1993 and 1994, respectively (Alford *et al*., 1995). The effect of this kind of strategy on the parasitoids of other rape pests still needs to be determined.

Acknowledgements

Rothamsted Research receives grant-in-aid from the Biotechnology and Biological Sciences Research Council and this work was part-funded by the UK Ministry of Agriculture, Fisheries and Food and the European Commission Concerted Action: *Minimizing pesticide use and environmental impact by the development and promotion of bio-control strategies for oilseed rape pests, FAIR 96-1314.*

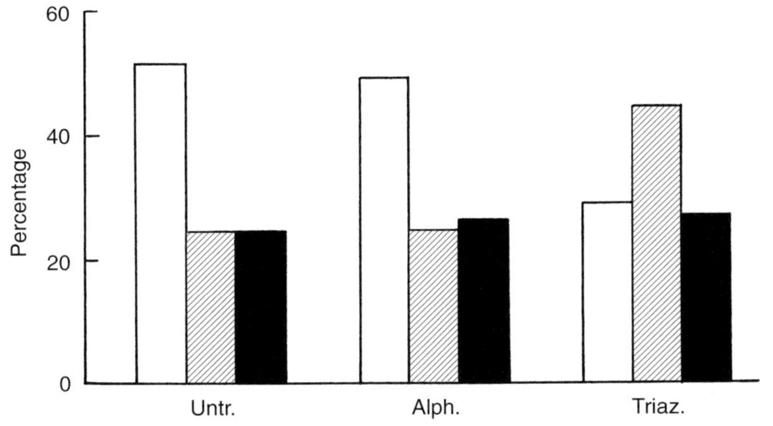

Fig. 6.4 Effect of applications of alpha-cypermethrin during flowering (Alph.), and triazophos post-flowering (Triaz.) or untreated (Untr.) on the percentage of parasitized (□) and dead (▨) cabbage seed weevil (*Ceutorhynchus assimilis*) larvae and weevil exit holes in pods (■) during 1994. Values are percentage of total weevil larvae and exit holes (n = 8) (from Murchie *et al.*, 1997b).

References

Alford, D. V., Cooper, D. A. & Williams, I. H. (1991). *Insect Pests of Oilseed Rape*. HGCA Oilseeds Research Review No. OS1. Home-Grown Cereals Authority, Research and Development: London, 130 pp.

Alford, D. V., Williams, I. H., Murchie, A. K. & Walters, K. F. A. (1995). *The Status and Potential of Parasitoids of Seed Weevil and Pod Midge on Winter Oilseed Rape*. HGCA Oilseeds Project Report No. OS14. Home-Grown Cereals Authority: London 89 pp.

Alford, D. V., Walters, K. F. A., Williams, I. H. & Murchie, A. K. (1996). A commercially viable low-cost strategy for the management of seed weevil populations on winter oilseed rape in the UK. *Proceedings of the Brighton Crop Protection Conference – Pests and Diseases* **2**, 609–14.

Ankersmit, G. W. (1956). De levencwijze en de bestrijding van de koolzad snuitkever (*Ceutorhynchus assimilis* Payk.) en de koolzadgalmug (*Dasyneura brassicae* Winn.). *Verslagen van landbouwkundige onderzoekingender Rijkslandbouw-proefstations* **62.9**, 1–59.

Askew, R. R. & Shaw, M. R. (1986). Parasitoid communities: their size, structure and development. *13th Symposium of the Royal Entomological Society of London*, pp. 225–64.

Bartlet, E. (1996). Chemical cues to host-plant selection by insect pests of oilseed rape. *Agricultural Zoology Reviews* **7**, 89–116.

Bonnemaison, L. (1957). Le charançon des siliques (*Ceuthorrhynchus assimilis* Payk.). *Annales des Épiphyties* **4**, 387–543.

Börner, C., Blunck, H., Speyer, W. & Dampf, A. (1921). Beiträge zur Kenntnis vom Massenwechsel (Graduation) schädlicher Insekten. *Arbeiten aus der Biologischen Reichsanstalt für Land- und Forstwirtschaft, Berlin* **10**, 405–66.

Büchi, R. (1993). Monitoring of parasitoids of the cabbage seed weevil, *Ceutorhynchus assimilis* during 1990 and 1991 in Switzerland. *IOBC/wprs Bulletin, IPM in Oilseed Crops* **16**, 145–9.

Carlson, E. C., Lange, W. H. & Sciaroni, R. H. (1951). Distribution and control of the cabbage seedpod weevil in California. *Journal of Economic Entomology* **44**, 958–65.

Charnov, E. L. (1982). *The Theory of Sex Allocation*. Princeton University Press: Princeton.

Cooper, D. A. & Lane, A. (1991). Monitoring of oilseed rape pests in England and Wales. *IOBC/wprs Bulletin, Integrated Control in Oilseed Crops* **14**, 5–13.

Crowell, H. H. (1952). Cabbage seedpod weevil control with parathion. *Journal of Economic Entomology* **45**, 545–6.

Delucchi, V. & Graham, M. W. R. de V. (1956). La révision du genre *Trichomalus* Thomson (1878) (Chalcoidea, Pteromalidae). *Beiträge zur Entomologie* **6**, 543–76.

Dmoch, J. (1975). Badania nad pasożytami chowacza podobnika (*Ceuthorrynchus assimilis* Payk.). I. Skład gatunkowy i znaczenie ektopasożytów larwalnych. *Roczniki nauk Rolniczych, Seria E* **5**, 99–112.

Dmoch, J. (1998). Kairomones and searching behaviour of *Trichomalus perfectus* Walker. *IOBC/wprs Bulletin, Integrated Control in Oilseed Crops* **21** (5), 171–6.

Dmoch, J. & Klimek, G. (1975). Badania nad pasożytami chowacza podobnika (*Ceuthorrynchus assimilis* Payk.). III. Obserwacje nad biologią *Trichomalus perfectus* (Walker). *Roczniki nauk Rolniczych, Seria E* **5**, 125–36.

Dmoch, J. & Rutkowska-Ostrowska, Z. (1978). Host-finding and host-acceptance mechanism in *Trichomalus perfectus* Walker (Hymenoptera, Pteromalidae). *Bulletin de l'Academie Polonaise des Sciences* **26**, 317–23.

Dmoch, J. & Sulgostowska, T. (1986). Parazytoidy chowacza podobnika (*Ceutorhynchus assimilis* PAYK.) (Coleoptera, Curculionidae) na uprawach kapusty nasiennej głowiastej. *Polskie Pismo Entomologiczne* **56**, 431–5.

Domenichini, G. (1966). Hymenoptera: Eulophidae. Palearctic Tetrastichinae. In: V. Delucchi & G. Remaudière (eds). *Index of Entomophagous Insects.* International Organisation of Biological Control: Paris.

Doucette, C. F. (1948). Field parasitization and larval mortality of the cabbage seedpod weevil. *Journal of Economic Entomology* **41**, 763–5.

Ferguson, A. W., Klukowski, Z., Walczak, B., Perry, J. N., Mugglestone, M. A., Clark, S. J. & Williams, I. H. (2000). The spatio-temporal distribution of adult *Ceutorhynchus assimilis* in a crop of winter oilseed rape in relation to the distribution of their larvae and that of the parasitoid *Trichomalus perfectus*. *Entomologia Experimentalis et Applicata* **95**, 161–71.

Fitton, M. G., Graham, M. W. R. de V., Bouček, Z. R. L., Fergusson, N. D. M., Huddleston, T., Quinlan, J. & Richards, O. W. (1978). A check list of British insects. Part 4: Hymenoptera. *Handbooks for the Identification of British Insects* **11** (4), 1–159.

Free, J. B. & Williams, I. H. (1979). The distribution of insect pests on crops of oilseed rape (*Brassica napus* L.) and the damage they cause. *Journal of Agricultural Science, Cambridge* **93**, 139–49.

Fulton, B. B. (1933). Notes on *Habrocytes cerealle*, a parasite of the angoumois grain moth. *Annals of the Entomological Society of America* **26**, 536–53.

Gahan, A. B. (1941). A revision of the parasitic wasps of the genus *Necremnus* Thomson (Eulophidae; Hymenoptera). *Journal of the Washington Academy of Sciences* **31**, 196–203.

Garthwaite, D. G., Thomas, M. R. & Hant, M. (1995). *Pesticide Usage Survey Report 127 – Arable Farm Crops 1994.* Ministry of Agriculture, Fisheries and Food: London.

Gauld, I. D. & Bolton, B. (eds) (1988). *The Hymenoptera.* Oxford University Press: Oxford.

Godan, D. (1959). *Mymar autumnalis* (Foerst.) [*Anaphes autumnalis* Foerst.] (*Chalcidoidea, Mymar.*), ein bisher unbekannter Eiparasit des Kohlschotenrüsslers. *Nachtrichtenblatt des Deutschen Pflanzenschutzdienstes* **11**, 161–4.

Goulet, H. & Huber, J. T. (eds) (1993). Hymenoptera of the world: an identification guide to families. *Agriculture Canada, Research Branch, Publication 1894/E*, 668 pp.

Graham, M. W. R. de V. (1969). The Pteromalidae of north-western Europe (Hymenoptera: Chalcidoidea). *Bulletin of the British Museum, Entomology Supplement* **16**, 1–908.

Günthart, E. (1949). Beiträge zur Lebensweise und Bekämpfung von *Ceutorhynchus quadridens* Panz. und *Ceutorhynchus napi* Gyll. *Mitteilungen der Schweizerischen Entomologischen Gesellschaft* **22**, 441–591.

Hanson, A. J., Carlson, E. C., Breakey, E. P. & Webster, R. L. (1948). Biology of the cabbage seedpod weevil in northwestern Washington. *State College of Washington Agricultural Experimental Station Bulletin* **498**, 1–15.

Harmon, B. L. & McCaffrey, J. P. (1995). Biology and impact of *Microctonus melanopus*, a parasite of *Ceutorhynchus assimilis* Paykull, in northern Idaho (USA). *Proceedings of the 9th International Rapeseed Congress, Cambridge, 4–7 July 1995*, pp. 1304–5.

Herrström, G. (1964). Untersuchungen über Parasiten von Ölfruchtschädlingen in Schweden. *Statens Växtskyddsanstalt Meddelanden* **12**, 437–48.

Heymons, R. (1922). Mitteilungen über den Rapsrüssler, *Ceutorrhynchus assimilis* Payk. und seinen Parasiten, *Trichomalus fasciatus* Thoms. *Zeitschrift für Angewandte Entomologie* **8**, 93–111.

Hoffmann, A. (1951). Contribution à l'étude de *Ceuthorrhynchus assimilis* Payk. (le Ceuthorrhynque des siliques). *Revue de Pathologie Végétale et d'Entomologie Agricole de France* **30**, 238–52.

Jourdheuil, P. (1960). Influence de quelques facteurs écologiques sur les fluctuations de population d'une biocénose parasitaire: étude relative à quelques Hyménoptères (Ophioninae, Diospilinae, Euphorinae) parasites de divers Coléoptères inféodés aux Crucifères. *Annales des Épiphyties* **11**, 445–660.

Jourdheuil, P., Coutin, R. & Lacote, J. P. (1974). Evolution du niveau des populations de *Ceuthorrhynchus assimilis* Payk. dans le Lauragais (Aude, France). *Proceedings of the 4th International Rape Congress, Giessen, 4–8 June 1974*, pp. 511–18.

Kidd, N. A. C. & Jervis, M. A. (1996). *Population dynamics.* In: M. Jervis & N. Kidd (eds). *Insect Natural Enemies – Practical Approaches to Their Study and Evaluation.* Chapman & Hall: London.

Laborius, A. (1972). Untersuchungen über die Parasitierung des Kohlschotenrüsslers (*Ceuthorrhynchus assimilis* Payk.) und der Kohlschotengallmücke (*Dasyneura brassicae* Winn.) in Schleswig-Holstein. *Zeitschrift für Angewandte Entomologie* **72**, 14–31.

Lerin, J. (1987). A short bibliographical review of *Trichomalus perfectus* Walker, a parasite of the seed pod weevil, *Ceuthorrhynchus assimilis* Payk. *IOBC/wprs Bulletin, Integrated Control in Oilseed Rape* **10**, 74–8.

Linz, B. (1991). Hymenopteren als Parasitoide des Rapsglanzkäfers (*Meligethes* spp., Nitidulidae) und des Kohlschotenrüsslers (*Ceutorhynchus assimilis*, Curculionidae) auf Raps und Rübsen. *Mitteilungen der Deutschen Gesellschaft für Allgemeine und Angewandte Entomologie* **8**, 90–2.

Maceljski, M., Balarin, I. & Danon, V. (1980). Rezultati višegodišnjih prou_avanja pojave i štetnosti insekata na uljanoj repici. *Zaštita Bilja* **31**, 317–24.

McKenna, D. (1972). *Study of the Effects of Insecticide Management on the Seedpod Weevil (*Ceuthorrynchus assimilis *Payk.) and its Parasites.* Unpublished MSc thesis, University of London, London.

McLeod, J. H. (1953). Notes on the cabbage seedpod weevil, *Ceutorhynchus assimilis* (Payk.) (Coleoptera: Curculionidae), and its parasites. *Proceedings of the Entomological Society of British Columbia* **49**, 11–18.

Miczulski, B. (1967). Błonkówki (Hymenoptera) w biocenozie upraw rzepaku. Częś III. Męczelkowate (Braconidae) i mszycarzowate (Aphidiidae). *Polskie Pismo Entomologiczne* **37**, 167–91.

Miczulski, B. (1968). Błonkówki (Hymenoptera) w biocenozie upraw rzepaku. Częś VI. Bleskotkowate (Chalcidoidea). *Polskie Pismo Entomologiczne* **38**, 341–85.

Murchie, A. K. (1996). *Parasitoids of Cabbage Seed Weevil and Brassica Pod Midge in Oilseed Rape.* Unpublished PhD thesis, University of Keele, UK.

Murchie, A. K. & Williams, I. H. (1998a). A bibliography of the parasitoids of the cabbage seed weevil (*Ceutorhynchus assimilis* Paykull). *IOBC/wprs Bulletin, Integrated Control in Oilseed Crops* **21** (5), 163–9.

Murchie, A. K. & Williams, I. H. (1998b). Effect of host size on the sex of *Trichomalus perfectus* (Walker) (Hymenoptera: Pteromalidae). *IOBC/wprs Bulletin, Integrated Control in Oilseed Crops* **21** (5), 177–81.

Murchie, A. K., Smart, L. E. & Williams, I. H. (1997a). Responses of *Dasineura brassicae* Winn. (Diptera: Cecidomyiidae) and its parasitoids *Platygaster subuliformis* Kieffer (Hymenoptera: Platygastridae) and *Omphale clypealis* Thomson (Hymenoptera: Eulophidae) to traps, baited with organic isothiocyanates, in the field. *Journal of Chemical Ecology* **23**, 917–26.

Murchie, A. K., Williams, I. H. & Alford, D. V. (1997b). Effects of commercial insecticide treatments to winter oilseed rape on parasitism of *Ceutorhynchus assimilis* Paykull (Coleoptera: Curculionidae) by *Trichomalus perfectus* (Walker) (Hymenoptera: Pteromalidae). *Crop Protection* **16**, 199–202.

Murchie, A. K., Williams, I. H. & Perry, J. N. (1999). Edge distributions of *Ceutorhynchus assimilis* and its parasitoid *Trichomalus perfectus* in a crop of winter oilseed rape (*Brassica napus*). *BioControl* **44**, 379–90.

Nasreldin, N. (1973). *A Study of the Weevil Complex on Mustard with Special Reference to the Seedpod Weevil.* Unpublished PhD thesis, University of London, London.

Nolte, H. W. (1953). *Trichomalus fasciatus* Först. als Parasit der Mohnstengelgallwespe (*Timaspis papaveris* Kieff.). *Mitteilungen Biologische Bundesanstalt für Land- und Forstwirtschäft, Berlin-Dahlem* **75**, 56–8.

Peck, O. (1963). A catalogue of the Nearctic Chalcoidea (Insecta: Hymenoptera). *The Canadian Entomologist Supplement* **30**, 1–1092.

Perry, J. N. (1995). Spatial analysis by distance indices. *Journal of Animal Ecology* **64**, 303–14.

Risbec, J. (1953). Contribution à l'étude des parasites du charançon du colza *Ceuthorrhynchus assimilis* Payk. *Revue de Pathologie Végétale et d'Entomologie Agricole de France* **32**, 37–43.

Rosen, H. von (1964). Untersuchungen über die Verbreitung und Biologie von zwei Pteromaliden in Rapsschoten *(Hym., Chalcidoidea)*. *Statens Växtskyddsanstalt – Meddelanden* **12**, 451–65.

Skrocki, C. (1972). Biologia chowacza podobnika (*Ceuthorrynchus assimilis* Payk.). *Roczniki nauk Rolniczych, Seria E* **2**, 7–20.

Speyer, W. (1925a). Kohlschotenrüssler (*Ceutorrhynchus assimilis* Payk.), Kohlschotenmücke (*Dasyneura brassicae* Winn.) und ihre Parasiten. *Arbeiten aus der Biologischen Reichsanstalt für Land- und Forstwirtschaft, Berlin* **12**, 79–108.

Speyer, W. (1925b). *Perilitus melanopus* Ruthe (Hym. Braconidae) als Imaginalparasit von *Ceutorrhynchus quadridens* Panz. zugleich eine kurze Zusammenfassung unserer bisherigen Kenntnisse von Schlupfwespen als Parasiten der Käferimagines. *Zeitschrift für Angewandte Entomologie* **11**, 132–46.

Szczepanski, H. (1972). Kosmacinek rzepakowy – *Trichomalus perfectus* (WALK.) (*Hymenoptera, Pteromalidae*) w biocenozie leśnej a problem biologicznej ochrony rzepaku. *Polskie Pismo Entomologiczne* **42**, 865–71.

Thiem, E. (1970). Die Bedeutung der Feldrandbehandlung für die Bekämpfung von *Dasyneura brassicae* Winn. (Dipt., Cecidomyiidae) und *Ceuthorrhynchus assimilis* Payk. (Col., Curculionidae) im blühenden Raps und der Einfluss dieser Insektizidmassnahme auf das Vorkommen von *Trichomalus perfectus* Walker (Hym., Chalcidoidea, Pteromalidae) an den Larven des Kohlschotenrüsslers. *Archiv Phytopathologie Pflanzenshutz* **6**, 83–98.

Ulber, B. & Vidal, S. (1998). Influence of host density and host distribution on parasitism of *Ceutorhynchus assimilis* by *Trichomalus perfectus*. *IOBC/wprs Bulletin, Integrated Control in Oilseed Crops* **21**, 185–95.

Walz, A. J. (1957). Observations on the biologies of some hymenopterous parasites of the cabbage seedpod weevil in northern Idaho. *Annals of the Entomological Society of America* **50**, 219–20.

7

Parasitoids of Brassica Pod Midge

Ingrid H. Williams
Rothamsted Research, Harpenden, Hertfordshire, AL5 2JQ, UK

Introduction

In total, 28 species of parasitoids (some cited only to genus) from five different families are reported to attack brassica pod midge (*Dasineura brassicae*) (Williams & Walton, 1990; Tables 7.1–7.4), but study of the literature is hampered by synonymy and misidentification. Also, knowledge of the parasitoids attacking brassica pod midge is relatively poor and their biology scantily known. The families Platygastridae (Table 7.2), Eulophidae (Table 7.3) and Ceraphronidae (Table 7.4) are of particular importance; they all attack the egg or larval stages of the host. Two species of Eurytomidae – *Eurytoma aciculata* (Czajkowska, 1978a) and *E. dentata* (Miczulski, 1968) – are reported to be endoparasitic on brassica pod midge larvae in Poland (Table 7.3). The former species, however, has also been reported to hyperparasitize the larvae of the pteromalid *Trichomalus perfectus* (Dmoch, 1975), which is an ectoparasitoid of the larvae of cabbage seed weevil (*Ceutorhynchus assimilis*) (see Chapter 6). As brassica pod midge and cabbage seed weevil larvae often co-habit in the same pods, misidentification seems possible. Several species are reported to be ectoparasitoids; these include *Sigmophora brevicornis* (Speyer, 1925), *Aphanogmus abdominalis*, *Ceraphron longipennis*, *C. xanthosoma* (Coutin & Riom, 1971; Laborius, 1972) and *Conostigmus rufescens* (Laborius, 1972). However, the life strategies and biology of most of these parasitoids are poorly known and would benefit from further study; the ceraphronids, for example, have been reared as endoparasitoids from Cecidomyiidae (Goulet & Huber, 1993). Each larva of *Pseudotorymus napi* (Table 7.4) feeds externally on an average of ten brassica pod midge larvae within the rape pod (Czajkowska, 1978a); therefore, it may be termed an 'ectoparasitic predator', as it attacks more than one host rather than the more normal situation of a parasitoid larva attacking a single host.

The most commonly cited parasitoids of brassica pod midge are:

- *Platygaster* spp. (family Platygastridae)
- *Omphale clypealis* (family Eulophidae)
- *Pseudotorymus napi* (family Torymidae)
- *Aprostocetus epicharmus* (family Eulophidae)
- *Aphanogmus abdominalis* (family Ceraphronidae)
- *Ceraphon xanthosoma* (family Ceraphronidae).

Table 7.1 Systematic list of parasitoids of brassica pod midge. For further systematic details see Fitton *et al.* (1978) and Goulet & Huber (1993).

Order HYMENOPTERA

Superfamily CHALCIDOIDEA
Family EURYTOMIDAE
Eurytoma Illiger
– *aciculata* Ratzeburg
– *dentata* Mayr†
Family TORYMIDAE
Pseudotorymus Masi
– *napi* Amerling & Kirchner (= *brassicae* Ruschka)†
Family EULOPHIDAE
Aprostocetus Westwood
– *epicharmus* (Walker) (= *variegatus* Szelényi)†
Necremnus Thomson
– *leucarthros* (Nees)†
Neochrysocharis Kurdjumov
– sp.
Omphale Haliday (= **Secodes** Förster)
– *clypealis* (Thomson)†
– *coilus* (Walker)
Sigmophora Rondani
– *brevicornis* (Panzer)†

Superfamily PLATYGASTROIDEA
Family PLATYGASTRIDAE
Amblyaspis Förster
– sp.
Inostemma Haliday
– *boscii* (Jurine)†
– nr *reticulatum* (Szelényi)‡
– *walkeri* Kieffer†
Isocybus Förster
– *thomsoni* Kieffer†
Piestopleura Förster
– sp.
Platygaster Latreille (= **Prosactogaster** Kieffer)
– *boscii* Nees
– *gladiator* Zetterstedt†
– *iolas* Walker†
– *munita* Walker
– *niger* Nees
– *nitida* (Thomson)
– *oebalus* Walker†
– *subuliformis* (Kieffer)†
– *tisias* Walker†
Synopeas Förster
– nr *lugubris* Thomson#
– sp.

Superfamily CERAPHRONOIDEA
Family CERAPHRONIDAE
Aphanogmus Thomson
– *abdominalis* (Thomson)†
– *tenuicornis* Thomson†
Ceraphron Jurine = **Calliceras** Nees
– *longipennis* (Kieffer) (= *insularis* (Kieffer))†
– *pallipes* Thomson
– *serraticornis* Kieffer
– *xanthosoma* (Kieffer)†
Family MEGASPILIDAE
Conostigmus Dahlborn
– *rufescens* Kieffer†

† Species included in the key to adults (Chapter 11).
‡ Keyed in Chapter 11 to *Inostemma reticulatum* (Szelényi).
# Keyed in Chapter 11 to *Synopeas lugubris* Thomson.

All parasitize the eggs or larvae within the pods, except for the two ceraphronids which attack the larvae within their cocoons in the soil. No parasitoids are known to attack brassica pod midge adults.

The taxonomy of the parasitoids of brassica pod midge is in need of revision. Identification to species of the Platygastridae and of the Ceraphronidae is particularly difficult and Murchie *et al.* (1999) advise that all records to species should be treated with extreme caution.

Impact of Parasitoids

Assessing the impact of parasitoids on brassica pod midge populations is complex and difficult, for a variety of reasons: (a) the host and some of its parasitoids have up to three generations each year; (b) there are temporal variations in the abundance of the para-

Table 7.2 Endoparasitoids of the family Platygastridae (Hymenoptera) that attack the eggs or larvae of brassica pod midge (*Dasineura brassicae*).

Species	**Country** (Citation)
Amblyaspsis sp.	**Denmark** (J. Axelsen, *pers. comm.*)
Inostemma boscii (Jurine)	**France** (de Gaulle, 1908); **Germany** (Buhl, 1960)
Inostemma nr *reticulatum* (Szelényi)	**France** (Coutin & Coulon, 1966; Coutin & Riom, **Germany** Laborius, 1972)
Inostemma sp.	**France** (Coutin, 1961); **Germany** (Laborius, 1972); **Sweden** (C. Nilsson, *pers. comm.*); **Switzerland** (Büchi & Keller, 1994)
Inostemma walkeri (Kieffer)	**Germany** (Laborius, 1972); **Poland** (Czajkowska, 1978a)
Isocybus thomsoni Kieffer [Note 1]	**Denmark** (Axelsen, 1992); **France** (Coutin & Coulon, 1966; Coutin & Riom, 1970); **Germany** (Laborius, 1972)
Piestopleura sp.	**Denmark** (J. Axelsen, *pers. comm.*)
Platygaster boscii Nees	**Switzerland** (Forel, 1866)
Platygaster gladiator Zetterstedt	**Poland** (Czajkowska, 1978a)
Platygaster iolas (Walker)	**Poland** (Czajkowska, 1978a)
Platygaster niger Nees [Note 2]	**Switzerland** (Forel, 1866)
Platygaster nitida (Thomson) [Note 3]	**Poland** (Czajkowska, 1978a, 1978b)
Platygaster oebalus Walker [Note 4]	**Denmark** (Axelsen, 1992, 1994); **France** (Coutin & Coulon, 1966; Coutin & Riom, 1970); **Germany** (Speyer, 1925; Laborius, 1972); **Poland** (Czajkowska, 1978a, 1978b)
Platygaster sp. [Note 5]	**France** (Coutin, 1961); **Germany** (Speyer, 1925; Buhl, 1960); **Poland** (Miczulski, 1965, 1968; Czajkowska, 1978a); **Sweden** (Sylvén, 1949; Herrström, 1964)
Platygaster subuliformis (Kieffer)	**UK** (Murchie, 1996; Murchie *et al.*, 1999)
Platygaster tisias Walker	**France** (de Gaulle, 1908); **Switzerland** (Büchi & Keller, 1994)
Synopeas nr *lugubris* (Thomson)	**France** (Coutin & Coulon, 1966; Coutin & Riom, 1970)
Synopeas sp.	**Poland** (Miczulski, 1965; Czajkowska, 1978a); **Sweden** (Sylvén, 1949; Herrström 1964); **Switzerland** (Büchi & Keller, 1994)

Note 1. Other names cited for this species: *Piestopleura thomsoni* Kieffer; *Synopeas thomsoni* (Kieffer) (trivial name often also spelt *thomsonii*).
Note 2. Cited by Forel as *Platygaster noir* (= *niger*?).
Note 3. Other name cited for this species: *Prosactogaster nitida* (Thomson).
Note 4. Other name cited for this species: *Prosactogaster oebalus* (Walker).
Note 5. Other name cited for this species: *Prosactogaster* sp.

sitoid species; (c) the proportions of hosts and parasitoids that emerge pre- and post-diapause can vary with season; and (d) the host can remain in diapause for more than one year. However, the impact of parasitization on host populations is greatest with species that attack the earliest generations in the greatest numbers.

Table 7.3 Parasitoids of the superfamily Chalcidoidea (Hymenoptera) that attack the larvae of brassica pod midge (*Dasineura brassicae*).

Family	Species	**Country** (Citation)
EULOPHIDAE	*Necremnus leucarthros* (Nees)	**Poland** (Czajkowska, 1978a)
	Neochrysocharis sp.	**Germany** (Buhl, 1960); **Poland** (Czajkowska, 1978a)
	Omphale clypealis (Thomson) [Note 1]	**Denmark** (Axelsen, 1992); **France** (Coutin & Coulon, 1966; Coutin & Riom, 1970); **Germany** (Laborius, 1972); **Poland** (Miczulski 1968); **Sweden** (Sylvén, 1949; Herrström, 1964); **Switzerland** (Büchi & Keller, 1994); **UK** (Murchie, 1996)
	Omphale coilus (Walker)	**Poland** (Czajkowska, 1978a)
	Sigmophora brevicornis (Panzer) [Note 2]	**Germany** (Speyer, 1925); **Poland** (Miczulski, 1968)
	Tetrastichus epicharmus (Walker) [Note 3]	**Czechoslovakia** (Delucchi & Remaudière, 1966); **France** (Coutin & Coulon, 1966; Delucchi & Remaudière, 1966; Domenichini, 1966; Coutin & Riom, 1970); **Hungary** (Delucchi & Remaudière, 1966); **Italy** (Delucchi & Remaudière, 1966); **Poland** (Czajkowska, 1978a, 1978b); **UK** (Delucchi & Remaudière, 1966; Domenichini, 1966)
	Tetrastichus sp. [Note 4]	**Poland** (Miczulski, 1968; Czajkowska, 1978a); **Sweden** (Sylvén, 1949)
EURYTOMIDAE	*Eurytoma aciculata* Ratzeburg	**Poland** (Czajkowska, 1978a)
	Eurytoma dentata Mayr	**Poland** (Miczulski, 1968)
TORYMIDAE	*Pseudotorymus napi* Amerling & Kirchner [Note 5]	**Germany** (Ruschka, 1923; Speyer, 1925); **Poland** (Miczulski, 1968; Czajkowska, 1978a, 1978b); **Switzerland** (Büchi & Keller, 1994)

Note 1. Other name cited for this species: *Secodes clypealis* Thomson.
Note 2. Other names cited for this species: *Geniocerus brevicornis* (Panzer); *Tetrastichus brevicornis* (Panzer).
Note 3. Other names cited for this species: *Aprostocetus epicharmus* (Walker); *Cirrospilus epicharmus* Walker; *Tetrastichus variegatus* Szelényi.
Note 4. Other name cited for this species: *Syntomosphyrum* sp.
Note 5. Cited as: *Pseudotorymus brassicae* Ruschka.

Parasitization rates

Reports of the incidence of parasitism are variable and are difficult to compare – some authors give parasitism by a single species, some give cumulative parasitism by several species, and others give data on different generations or on pre- and post-diapause populations. There have been few recent investigations on the incidence of parasitoids of brassica pod midge.

The most extensive investigation is that of Czajkowska (1978a) in Poland. She reared 10 species of parasitoid from field-collected brassica pod midge larvae, and found an average parasitization rate of 2–11% pre-diapause. Her references to *Mesopolobus morys* and *Trichomalus perfectus* are probably errors, as these are well-known parasitoids of cabbage seed weevil (see Chapter 6), although *M. morys* is also reported by Rosen (1960).

Table 7.4 Parasitoids of the superfamily Ceraphronoidea (Hymenoptera) that attack the larvae of brassica pod midge (*Dasineura brassicae*).

Family	Species	**Country** (Citation)
CERAPHRONIDAE	*Aphanogmus abdominalis* (Thomson) [Note 1]	**Denmark** (Axelsen, 1992); **France** (Dessart, 1964; Coutin & Coulon, 1966; Coutin & Riom, 1970); **Germany** (Laborius, 1972); **Poland** (Czajkowska, 1978a)
	Aphanogmus tenuicornis (Thomson) [Note 2]	**Sweden** (Sylvén, 1949; Herrström, 1964)
	Ceraphron longipennis (Kieffer) [Note 3]	**France** (Dessart, 1964; Coutin & Coulon, 1966; Coutin & Riom, 1970)
	Ceraphron pallipes (Thomson)	**Germany** (Laborius, 1972)
	Ceraphron xanthosoma (Kieffer) [Note 4]	**France** (Dessart, 1964; Coutin & Coulon, 1966; Coutin & Riom, 1970); **Germany** (Laborius, 1972)
MEGASPILIDAE	*Conostigmus rufescens* Kieffer	**Germany** (Laborius, 1972)

Note 1. Other names cited for this species: *Calliceras abdominalis* Thomson; *Ceraphron abdominalis* (Thomson).
Note 2. Other names cited for this species: *Calliceras tenuicornis* Thomson; *Ceraphron tenuicornis* (Thomson).
Note 3. Other names cited for this species: *Calliceras insularis* (Kieffer); *Ceraphron insularis* (Kieffer).
Note 4. Other names cited for this species: *Calliceras xanthosoma* (Kieffer); *Ceraphron serraticornis* Kieffer.

Three species – *Platygaster oebalus*, *Aprostocetus epicharmus* and *Pseudotorymus napi* – accounted for more than 96% of the individuals reared. On average, only 1.8%, 1.5% and 0.7% of larvae were parasitized by these species, respectively, although maximal site parasitization rates were 13%, 29% and 16%, respectively. Some (18%) brassica pod midge larvae in overwintering cocoons were also parasitized, the most numerous species reared from them post-diapause being *Synopeas* sp. (39%), *Platygaster iolas* (28%), *Aphanogmus abdominalis* (15%) and *Platygaster oebalus* (13%), with *Platygaster nitida*, *Omphale coilus*, *Tetrastichus* sp. and *Neochrysocharis* sp. together accounting for only 6% of those reared.

A similar investigation (Murchie, 1996), the first in the UK, revealed a less diverse parasitoid complex attacking brassica pod midge, with only two species (*Platygaster subuliformis* and *Omphale clypealis*) reared pre-diapause from larval samples. *P. subuliformis* was the more abundant on winter rape (96% of those reared from first-generation larvae), whereas *O. clypealis* was the more abundant on spring rape (68% of those reared). *Platygaster subuliformis* emerged pre-diapause from 3% to 13% and from 0% to 18% of the larvae collected at weekly intervals (for four weeks) from two crops of winter rape and from 4% to 67% and from 27% to 74% of the larvae post-diapause. Brassica pod midge cocoons from a spring rape crop produced a further two species of parasitoid post-diapause (namely, a species of *Ceraphron* and one of *Aphanogmus*), in addition to *O. clypealis* (64% of parasitized individuals) and *P. subuliformis* (26% of parasitized individuals), although numbers were small (<1% of parasitized individuals); in total, 10% of cocoons were parasitized.

In France, Coutin (1961) reported a species of *Platygaster* (cited as *Prosactogaster*) to be the most important parasitoid of brassica pod midge, parasitizing 29% of overwintering larvae, followed by a species of *Inostemma* (16%); Coutin & Riom (1971) found <50% parasitized by *Aphanogmus abdominalis*, *Ceraphron longipennis* and *C. xanthosoma*.

In Germany, Laborius (1972) reared 10 parasitoid species from cocoons collected over three years from four different sites and found 1–14% parasitized, the dominant species being *Platygaster oebalus*, *Aphanogmus abdominalis* and *Ceraphron serraticornis*. Also in Germany, Buhl (1960) found a species of *Platygaster* (cited as *Prosactogaster*) to be the main egg parasitoid, with an average of 18% of the first and 7% of the second generation parasitized – a species of *Neochrysocharis* (2%) and *Inostemma boscii* (<1%) were also found; Speyer (1925) found 7–68% of larvae parasitized by the two species *Pseudotorymus napi* and *Sigmophora brevicornis*.

In Sweden, Herrström (1964) reported 0–36% of larvae parasitized by a species of *Platygaster* (cited as *Prosactogaster*) and, in Denmark, Axelsen (1992) reported 3% parasitism, the most common parasitoids reared from overwintered cocoons being *Platygaster oebalus* followed by *Aphanogmus abdominalis*. In Switzerland, Büchi & Keller (1994) reported the following rates of parasitism: *Platygaster tisias* 5–17%, *Omphale clypealis* 1–7%, *Inostemma* spp. 0–5%.

Biology of the Parasitoids

The biology of the majority of the parasitoids attacking brassica pod midge is scantily known; published information on the two most important and most studied (*Platygaster* spp. and *Omphale clypealis*) is summarized below.

Hymenoptera: Platygastridae – *Platygaster* spp.

Nomenclature

Platygaster spp. appear to be the most widespread and frequently reported parasitoids of brassica pod midge in Europe (Table 7.2). However, because the taxonomy of this genus is in need of revision, identification to species is difficult and records to species must be treated with caution (Murchie *et al.*, 1999). This problem is highlighted by the recent finding by Murchie (1996) that *P. subuliformis* (Fig. 7.1), the most abundant parasitoid attacking brassica pod midge in his study, is a new species record for Britain. In their description of *P. subuliformis*, Murchie *et al.* (1999) point out that the species is close to (and can easily be confused with) *Platygaster gladiator*, *P. munita*, *P. oebalus* and *P. tisias*, and they suggest that species of *Platygaster* mentioned in the literature (under either *Platygaster* or the frequently cited synonym *Prosactogaster*) (Table 7.2) may be misidentifications and may in fact refer to *P. subuliformis*.

Number of generations

The number of generations of *Platygaster* spp. attacking brassica pod midge in oilseed rape remains uncertain. Most authors consider them to be multivoltine, with three genera-

Fig. 7.1 Female *Platygaster subuliformis* – family PLATYGASTRIDAE (×30).

tions per year – two on winter rape and a third on spring rape, in common with their host (Buhl, 1960; Coutin, 1961; Coutin & Riom, 1971; Czajkowska, 1978b). However, others have suggested that the apparent separate generations may result from the staggered emergence of overwintered parasitoids (Herrström, 1964; Axelsen, 1994). Murchie (1996), however, found no evidence of staggered emergence, although insects collected earlier tended to emerge earlier post-diapause and he suggested that, as *P. subuliformis* took more than three times as long as brassica pod midges to emerge pre-diapause, it may have fewer generations per year than its host.

Phenology on the crop

Investigations of the phenology of *Platygaster* spp. on oilseed rape crops in different parts of Europe have given variable patterns of seasonal abundance but mostly with a long period of presence on the crop and with peak abundance co-incident with peak abundance of brassica pod midge. In Sweden, spring emergence of a species of *Platygaster* was two-peaked (i.e. bimodal), coinciding with the first and second generations of its host, and the parasitoid was present from early May to early July with peak abundance in mid-May (Herrström, 1964); these two emergence peaks and the part of the offspring emerging within the same season can give three population peaks in Germany (Buhl, 1960), in France (Coutin & Riom, 1971) and in Poland where Czajkowska (1978b) caught *P. oebalus* from early May to late July, the three peaks co-incident with the three host peaks in late May, late June and mid-July. In Denmark, according to Axelsen (1994), there is little parasitization by *P. oebalus* of first-generation brassica pod midge larvae, and emergence coincides with the second and third host generations, whereas in France, Coutin & Coulon (1966) caught *P. oebalus* in the field almost continuously from late March to late May, with a peak in early April co-incident with the time of first emergence of host adults. In the UK, spring emergence of *P. subuliformis* also appears to be well synchronized with that of its host. Murchie (1996) caught the first females of both parasitoid and host during the first week of May. Phenology of both was bimodal. Brassica pod midge had peaks in mid-May and late June. The parasitoid increased through May and early June, decreased during mid-June and increased again in late June and early July.

Developmental stages

Species of *Platygaster* are egg or larval endoparasitoids, attacking brassica pod midge eggs within the pods of oilseed rape. Each female parasitizes several eggs and lays a single egg in each, possibly exerting her ovipositor through the same opening in the pod wall as that utilized by the host midge. Several females can parasitize the same egg cluster. The parasitoid is a koinobiont, as its egg develops only after the host is nearly full-grown and at the prepupal or pupal stage within a cocoon in the soil. On maturity, the parasitoid pupates within the host's larval skin (Buhl, 1960; Coutin, 1961; Czajkowska, 1978b; Axelsen, 1994). A part of the population emerges the same season, whereas the remainder enters diapause. Diapause is spent in the soil inside brassica pod midge cocoons. Mating occurs soon after emergence and the mated females then migrate to rape crops.

Murchie (1996) reared *Platygaster subuliformis* from parasitized brassica pod midge larvae collected from winter rape at weekly intervals during June. At 18°C, brassica pod midges emerged *c.* 14 days after their larvae had burrowed into sand to pupate, whereas adults of *P. subuliformis* emerged after 50 days. The proportions of brassica pod midge and *P. subuliformis* larvae from the weekly samples that entered diapause increased with later collection of the midge larvae: for example, from one crop, midge larvae entering diapause increased from 5% to 11% to 16% to 46% and, similarly, *P. subuliformis* larvae increased from 52% to 63% to 75% to 93%. This concurs with the findings of Buhl (1960) for *Platygaster* sp. The greater proportion of *P. subuliformis* entering diapause on any given date is probably the result of the longer development time for the parasitoid than for the host, and explains why the ratio of *P. subuliformis* to brassica pod midges emerging is greater post-diapause than pre-diapause. As brassica pod midge larval development in the pod is about 14 days (Sylvén, 1949), egg to adult development for the parasitoid is about 64 days. Emergence of *P. subuliformis* from spring rape pre-diapause took *c.* 41 days.

Post-diapause, brassica pod midges and female *P. subuliformis* emerged at the same time, 21 days after being transferred from 5°C to 18°C, whereas male *P. subuliformis* emerged about eight days earlier. Therefore, the parasitoid is one generation out of synchrony with its host pre-diapause but regains sychrony post-diapause.

The times to emergence of both *P. subuliformis* and its host are, thus, influenced both by crop and by date of larval drop. The parasitoid took longer to emerge pre-diapause than its host, probably because it must develop from an egg or early larva whereas the brassica pod midge larva is already mature when it drops to the soil. Post-diapause, the time to emergence was equal to or less than that for the brassica pod midge, suggesting that the parasitoid is at a similar or more advanced developmental stage than the host when it diapauses. Buhl (1960) found that *Platygaster* sp. overwintered in both larval and adult forms.

Sex ratio

The sex ratio for both *Platygaster oebalus* (Laborius, 1972; Axelsen, 1994) and *P. subuliformis* (Murchie, 1996; Murchie *et al.*, 1999) is unity.

Habitat and host finding

Murchie *et al.* (1997) found that both male and female *Platygaster subuliformis* were attracted to flight traps baited with 2-phenylethyl isothiocyanate, a volatile compound released when the oilseed rape plant is damaged. Presumably, the females use this chemical to locate their host habitats, thereby increasing the chances of males finding females for mating and females locating host eggs or larvae in which to oviposit. Traps baited with attractant isothiocyanates have potential for development as monitoring tools for the parasitoid.

Spatio-temporal distribution in field

The within-field spatio-temporal distributions of adult *Platygaster subuliformis* and adults of its host have been investigated by Murchie (1996), using yellow water traps arranged on a grid system in and around a crop of winter rape. Catches of both were spatially aggregated, with strong positive correlations between those of pest and parasitoid. During May and early June, parasitoid catches were sporadic and showed little spatial pattern. However, from mid-June onwards the parasitoid was more abundant in the centre of the crop. Correlations between dates suggested that the spatial distribution of first-generation parasitoids determined that of the second. These patterns of distribution indicate that *P. subuliformis* is efficient in dispersing throughout the area occupied by its host.

Hymenoptera: Eulophidae – *Omphale clypealis*

Omphale clypealis (Fig. 7.2) is a larval endoparasitoid of brassica pod midge, but, as with *Platygaster* spp. and other parasitoids of this pest, its biology is poorly known. The number of generations on oilseed rape crops is uncertain, but it is probably multivoltine as is its host. It overwinters within midge larval cocoons in the soil and emerges in the spring.

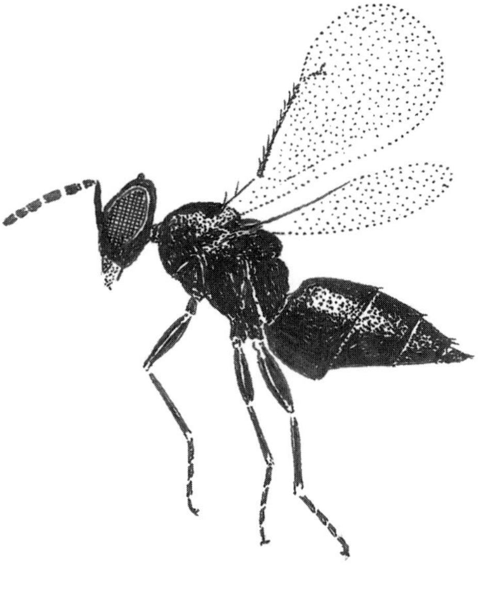

Fig. 7.2 Female *Omphale clypealis* – family EULOPHIDAE (×30).

In Sweden, Herrström (1964) caught females in winter rape from late May, the time of peak emergence of first-generation adults of its host brassica pod midge, until late June, the time of peak occurrence of second-generation host adults. The females oviposit into mature brassica pod midge larvae through the wall of the oilseed rape pod. The parasitoid larva feeds within the host during its larval and pupal stages. Murchie (1996) found that, at 18°C, some adults emerged 31 days after host larvae from spring rape had burrowed into the soil, whereas others emerged post-diapause, appearing 42 days after their transfer from 5°C (at which temperature they were overwintered) to 18°C. Pre-diapause emergence from winter rape samples took 28 days, whereas post-diapause emergence took 37 days. Sex ratio is strongly female biased (deuterotokous); 97% of parasitoids reared by Murchie (1996) from brassica pod midge larvae from spring rape and all those reared from winter rape were female. The parasitoid probably uses plant volatiles as cues to habitat and host location; Murchie *et al.* (1997) found that flight traps baited with allyl isothiocyanate, a chemical produced by brassicaceous plants on being damaged, caught female *O. clypealis* as well as both male and female brassica pod midges. Such traps have potential for development as monitoring tools for both the pest and its parasitoid.

Acknowledgements

Rothamsted Research receives grant-in-aid from the Biotechnology and Biological Sciences Research Council and this work was part-funded by the UK Ministry of Agriculture, Fisheries and Food and the European Commission Concerted Action: *Minimizing pesticide use and environmental impact by the development and promotion of bio-control strategies for oilseed rape pests, FAIR 96-1314.*

References

Axelsen, J. (1992). The population dynamics and mortalities of the pod gall midge (*Dasyneura brassicae* Winn.) (Dipt., Cecidomyiidae) in winter rape and spring rape (*Brassica napus* L.) in Denmark. *Journal of Applied Entomology* **114**, 463–71.

Axelsen, J. (1994). Host–parasitoid interactions in an agricultural ecosystem: a computer simulation. *Ecological Modelling* **73**, 189–203.

Büchi, R. & Keller, S. (1994). Die Parasitierung der Kohlschotengallmücke durch Nützlinge. *Agrarforschung* **1** (9), 400–2.

Buhl, C. (1960). Beobachtungen über vermehrtes Schadauftreten der Kohlschotenmücke (*Dasyneura brassicae* Winn.) an Raps und Rübsen in Schleswig-Holstein. *Nachrichtenblatt des Deutschen Pflantzenschutzdienstes (Braunschweig)* **12** (1), 1–6.

Coutin, R. (1961). Recherchés récentes sur la Cecidomyie des siliques de Colza (*Dasyneura brassicae* (Winn.)). Méthodes de lutte envisagées. *Compte Rendus des Journées d'Information Technique sur les Oléagineux des Climats Tempérés*. CETIOM: Paris, pp. 46–58.

Coutin, R. & Coulon, J. (1966). Toxicités comparées de plusiers insecticides vis-à-vis de *Dasyneura brassicae* Winn. et de son parasite principal *Prosactogaster oebalus* Walk. *Phytiatrie – Phytopharmacie* **15**, 105–12.

Coutin, R. & Riom, J. (1971). Biologie des populations de *Dasyneura brassicae* Winn. *Journées Intern sur le Colza, Paris, 26–30 May 1970*, pp. 250–71.

Czajkowska, M. (1978a). Badania nad prszczarkiem kapustnikiem (*Dasyneura brassicae* Winn.). IV. Sklad gatunkowy I znaczenie pasozytów. *Roczniki Nauk Rolniczych, Seria E* **8** (2), 159–75.

Czajkowska, M. (1978b). Badania nad pryszczarkiem kapustnikiem (*Dasyneura brassicae* Winn.). V. Biologia i ekologia pasozytów. *Roczniki Nauk Rolniczych, Seria E* **8** (2), 177–88.

Delucchi, V. & Remaudière, G. (eds) (1966). *Index of Entomophagous Insects. 1. Hymenoptera: Eulophidae. Palearctic Tetrastichinae.* International Organisation of Biological Control: Paris, pp. 1–101.

Dessart, P. (1964). Contribution à l'étude des Hyménoptères Proctotrupoidea (IV). Trois ceraphronidae parasites de la Cecidomyie du Colza: *Dasyneura brassicae* (Winnertz), en France. *Bulletin et Annales de la Societe Royale d'Entomologie de Belgique* **100**, 109–30.

Dmoch, J. (1975). Badania nad pasożytami chowacza podobnika (*Ceuthorrynchus assimilis* Payk.). I. Skład gatunkowy i znaczenie ektopasożytów larwalnych. *Roczniki nauk Rolniczych, Seria E* **5**, 99–112.

Domenichini, G. (1966). *Hymenoptera: Eulophidae. Palearctic Tetrastichinae.* In: V. Delucchi & G. Remaudière (eds). *Index of Entomophagous Insects.* International Organisation of Biological Control: Paris.

Fitton, M. G., Graham, M. W. R. de V., Bouček, Z. R. L., Fergusson, N. D. M., Huddleston, T., Quinlan, J. & Richards, O. W. (1978). A check list of British insects. Part 4: Hymenoptera. *Handbooks for the Identification of British Insects* **11** (4), 1–159.

Forel, A. (1866). Notes sur quelques insectes nuisibles au colza dans le canton de Vaudoise. *Bulletin de la Sociète Vaudoise. Sciences Naturelles* **9**, 72–84.

Gaulle, J. de (1908). *Catologue Systematique et Biologique des Hymenoptères de France.* P. Klincksieck: Paris.

Goulet, H. & Huber, J. T. (eds) (1993). Hymenoptera of the world: an identification guide to families. *Agriculture Canada, Research Branch, Publication 1894/E*, 668 pp.

Herrström, G. (1964). Untersuchungen über Parasiten von Ölfruchtschädlingen in Schweden. *Statens Växtskyddsanstalt Meddelanden* **12**, 433–48.

Laborius, A. (1972). Untersuchungen über die Parasitierung des Kohlschotenrüsslers (*Ceuthorrhynchus assimilis* Payk.) und der Kohlschotengallmücke (*Dasyneura brassicae* Winn.) in Schleswig-Holstein. *Zeitschrift für Angewandte Entomologie* **72**, 14–31.

Miczulski, B. (1965). Blonkówski z nadrodziny Proctrotrupoidea w biocenozie upraw rzepaku. *Annales Universitatis Mariae Curie-Sktodowska C 20*, 133–47.

Miczulski, B. (1968). Blonkowski (Hymenoptera) w biocenozie upraw rzepaku. Czesc VI. Bleskotkowate (Chalcidoidea). *Polskie Pismo Entomologiczne* **38** (2), 341–83.

Murchie, A. K. (1996). *Parasitoids of Cabbage Seed Weevil and Brassica Pod Midge in Oilseed Rape.* Unpublished PhD thesis, University of Keele, UK.

Murchie, A. K., Smart, L. E. & Williams, I. H. (1997). Responses of *Dasineura brassicae* and its parasitoids *Platygaster subuliformis* and *Omphale clypealis* to field traps baited with organic isothiocyanates. *Journal of Chemical Ecology* **23**, 917–26.

Murchie, A. K., Polaszek, A. & Williams, I. H. (1999). *Platygaster subuliformis* (Kieffer) (Hym., Platygastridae) new to Britain, an egg-larval parasitoid of the brassica pod midge *Dasineura brassicae* Winnertz (Dipt., Cecidomyidae). *Entomologist's Monthly Magazine* **135**, 217–22.

Rosen, H. von (1960). Zur Kenntnis des Pteromaliden – genus *Mesopolobus* Westwood 1833 (Hym., Chalcidoidea) V. *Opuscula Entomologica* **25**, 1–15.

Ruschka, F. (1923). Die europäischen Arten der mit *Monodontomerus* Westw. verwandten Gattungen. (Chalcidenstudien IV Teil). *Zeitschrift für Angewandte Entomologie* **9**, 395–406.

Speyer, W. (1925). Kohlschotenrüssler (*Ceutorrhynchus assimilis* Payk.), Kohlschotenmücke (*Dasyneura brassicae* Winn.) und ihre Parasiten. *Arbeiten aus der Biologischen Reichsanstalt für Land- und Forstwirtschaft* **12**, 79–108.

Sylvén, E. (1949). Skidgallmygan, *Dasyneura brassicae* Winn. *Statens Växtskyddsanstalt Meddelande* **54**, 1–120.

Williams, I. H. & Walton, M. (1990). A bibliography of the parasitoids of the brassica pod midge (*Dasineura brassicae* Winn.). *IOBC/wprs Bulletin, Integrated Control in Oilseed Rape* **13** (4), 46–52.

8

Parasitoids of Flea Beetles

Bernd Ulber
Institut für Pflanzenpathologie und Pflanzenschutz, Universität Göttingen, Grisebachstraße 6, D-37077 Göttingen, Germany

Ingrid H. Williams
Rothamsted Research, Harpenden, Hertfordshire, AL5 2JQ, UK

Introduction

Flea beetles from two genera are associated with oilseed rape: *Psylliodes* and *Phyllotreta*, the former represented by one species (cabbage stem flea beetle) and the latter by several (collectively known as cabbage flea beetles). See Chapter 2 for further details.

Cabbage stem flea beetle (*Psylliodes chrysocephala*) is attacked by parasitoids in both its larval and adult stages: the larvae by four species of ichneumonid wasps, three species of braconid and one species of pteromalid; the adults by a species of braconid (Table 8.1).

Parasitism of cabbage flea beetles (*Phyllotreta* spp.) has been little studied. The most extensive data have been presented by Jourdheuil (1960), Sommer (1981) and Wylie (1980, 1981, 1982). Parasitism of the larvae has been reported only from leaf-mining larvae of *P. nemorum*, but parasitism of the adult beetles has been shown for a wide range of *Phyllotreta* species. Several species of *Phyllotreta* are afforded vernacular English names (e.g. crown flea beetle = *P. diademata*; large striped flea beetle = *P. nemorum*; small striped flea beetle = *P. undulata*) (for further details see Seymour, 1989). However, vernacular names for species of *Phyllotreta* are not used in this account.

Parasitoids Attacking Larvae of Cabbage Stem Flea Beetle

Hymenoptera: Ichneumonidae – *Tersilochus tripartitus*

Distribution

Tersilochus tripartitus is the most widely distributed species attacking the larvae of cabbage stem flea beetle; it has been found in Austria, Denmark, Ireland and Sweden (Horstmann, 1981), in France (Aubert & Jourdheuil, 1958), in the Czech Republic (Šedivý, 1983) and in Germany (Dosse, 1961; Lehmann, 1965; Horstmann, 1971).

Phenology

Tersilochus tripartitus is a solitary, univoltine, ichneumonid endoparasitoid which attacks cabbage stem flea beetle larvae in the spring (Jourdheuil, 1960; Dosse, 1961). Adults

Table 8.1 Systematic list of parasitoids of cabbage stem flea beetle (*Psylliodes chrysocephala*).

Order HYMENOPTERA

Superfamily ICHNEUMONOIDEA
Family ICHNEUMONIDAE
Aneuclis Förster
– *melanarius* (Holmgren) (= *diversus* Szépligeti; = *petiolaris* Szépligeti)† [Note 1]
Cremastus Gravenhorst
– *carinifer* Thomson [Note 2]
Tersilochus Holmgren (= ***Thersilochus*** Holmgren)
– *microgaster* (Szépligeti)† [Note 3]
– *tripartitus* Brischke (= *melanogaster* Thomson; = *nigricans* Szépligeti)†
Family BRACONIDAE
Diospilus Haliday
– *morosus* Reinhardt†
– *oleraceus* Haliday
Microctonus Wesmael
– *melanopus* Ruthe† [Note 4]

Superfamily CHALCIDOIDEA
Family PTEROMALIDAE
Trichomalus Thomson
– *lucidus* (Walker)†
– nr *lucidus* (Walker)

† Species included in the key to adults (Chapter 11).

Note 1. Also cited in the literature under the genera *Isurgus*, *Perilitus* and *Thersilochus*.
Note 2. Also cited in the literature as *Temelucha carinifera* (Thomson).
Note 3. Also cited in the literature as *Isurgus microgaster* Szépligeti.
Note 4. Also cited in the literature as *Perilitus melanopus* Ruthe.

of *T. tripartitus* can be found in the rape crop from the early bud stage to flowering (Miczulski, 1966; Šedivý, 1983). In Germany, peak numbers were caught by yellow water traps and sweep-netting just before flowering in late April–early May (Lehmann, 1965). The emergence of adult *T. tripartitus* in the previous rape crop began when the soil temperature rose to 9–10°C at 5 cm depth or at air temperature sums of 200–220 day degrees (DD) (starting on 1 January; >0°C) (Jourdheuil, 1960). When reared with water at 18°C in the laboratory, the longevity of females and males was three to four days and two days, respectively.

Status and importance

The level of parasitism has been assessed in only two countries; it ranged from 30% to 61% in 1954 to 1956 in France (Jourdheuil, 1960) and from 3% to 27% in 1959 and 1960 in Germany (Dosse, 1961). Schulz (1983) estimated the spring parasitism of cabbage stem flea beetle in north-eastern Germany by merely recording the percentage of larvae that contained encapsulated, dark-pigmented parasitoid eggs and larvae in their bodies. In 1981 and 1982, he found 2% and 8% of parasitized larvae, respectively; however, by using this method the level of parasitism was underestimated and the parasitoid species could not be identified exactly. In the Czech Republic, *T. tripartitus* was particularly abundant in 1955 and 1956 when populations of cabbage stem flea beetle exceeded the economic damage threshold, but it occurred only rarely in subsequent years when host densities were low (Šedivý, 1983); however, there is no information on per cent parasitism in this publication. Superparasitism by *T. tripartitus* has been reported to be common, by both Jourdheuil (1960) and Dosse (1961), with superparasitized host larvae showing enhanced levels of encapsulation of parasitoid eggs and young larvae.

Life history

The biology of *Tersilochus tripartitus* has been studied extensively by Jourdheuil (1960) and Dosse (1961). The host-seeking female forages on the petioles and stems of the rape plant, inspecting the surface with her antennae. When she detects a feeding gallery of cabbage stem flea beetle she probes by inserting the ovipositor into the plant tissue repeatedly, moving along the gallery. On locating a host larva, she stops walking and lays a single egg into the host; this takes only 10 seconds. The *T. tripartitus* egg hatches after three days (at 18–20°C) and the larva feeds on the host haemolymph for three to four days; after this time, food ingestion is interrupted until the full-grown host larva has emigrated to the ground. Then, the parasitoid larva completes its development within the prepupal stage of the host. At 15°C, the development of the parasitoid larva, prepupa and pupa takes 56–63 days. The pupae develop into adults in summer, but the wasps remain in diapause within their pupal cocoons until the following spring.

Hymenoptera: Ichneumonidae – *Tersilochus microgaster*

Tersilochus microgaster is another solitary, univoltine, ichneumonid endoparasitoid that attacks cabbage stem flea beetle larvae in the spring. In Germany, specimens of this parasitoid were reared from cabbage stem flea beetle larvae sampled from rape crops in early June by Klingenberg & Ulber (1994); the parasitoids were identified to species, using the key of Horstmann (1971) [and the determinations later confirmed by K. Horstmann (*pers. comm.*)], thereby being designated for the first time to a specific host. Peak emergence of overwintered adults of *T. microgaster* from fields in which oilseed rape had been grown in the previous year occurred in mid-March or April (Klingenberg & Ulber, 1994). Female parasitoids were caught in the new rape crop by yellow water traps and Malaise traps in March and April (Nitzsche, 1998). In total, 72 cabbage stem flea beetle adults and 23 adults of *T. microgaster* were reared from 196 cabbage stem flea beetle larvae in 1991 by Klingenberg & Ulber (1994).

Hymenoptera: Ichneumonidae – *Cremastus carinifer*

The solitary, ichneumonid endoparasitoid *Cremastus carinifer* has been reported to parasitize cabbage stem flea beetle larvae in France (Bonnemaison & Jourdheuil, 1954) and in northern Germany (Meuche, 1940). However, the identification by Bonnemaison & Jourdheuil (1954) was later revised by Jourdheuil (1960) to *Aneuclis melanarius* (see below).

Hymenoptera: Ichneumonidae – *Aneuclis melanarius*

Aneuclis melanarius is different from the ichneumonids presented above because it is a multivoltine larval endoparasitoid of various coleopteran species (Jourdheuil, 1960;

Horstmann, 1981), with three generations per year (Aubert & Jourdheuil, 1958). *Aneuclis melanarius* attacks the larvae of cabbage stem flea beetle and turnip gall weevil (*Ceutorhynchus pleurostigma*) in the autumn and overwinters within these hosts as a first-instar larva (Horstmann, 1971; Šedivý, 1983). The adults of *A. melanarius* emerge during May or June and switch to parasitize the larvae of cabbage seed weevil (*Ceutorhynchus assimilis*) (and probably other summer hosts), which were found to be parasitized in June and July (Jourdheuil, 1960). However, even at high population densities of cabbage seed weevil the parasitism of cabbage stem flea beetle larvae by *A. melanarius* was generally very low. In France, only 1.5%, 0.5% and 0.2% of the larvae were parasitized in 1953, 1954 and 1955, respectively (Jourdheuil, 1960). *Aneuclis melanarius* has been recorded from almost all European countries, ranging from Ireland in the west to Turkey in the east (Horstmann, 1971, 1981).

Hymenoptera: Braconidae – *Diospilus morosus*

The solitary, multivoltine, braconid larval endoparasitoid *Diospilus morosus* has four generations per year (Jourdheuil, 1960). In France and Germany, it has been reported to parasitize the larvae of cabbage stem flea beetle and, additionally, the larvae of turnip gall weevil and *Ceutorhynchus leprieuri*, over the winter period (Godan, 1950; Jourdheuil, 1960), while the larvae of *Phyllotreta* spp., cabbage seed weevil and other coleopteran species are parasitized by three generations of *D. morosus* during the summer. However, in many years, the autumnal synchrony of the fourth generation of *D. morosus* with that of cabbage stem flea beetle is insufficient to allow significant levels of parasitsm: only parasitoid females that survive until late September can succesfully parasitize the very first larvae of this winter host (Jourdheuil, 1960). Accordingly, the level of parasitism is usually very low; Godan (1950) found only one cocoon of *D. morosus* in 13,000 examined plants. The biology and phenology of *D. morosus* is presented in greater detail under the parasitism of *Phyllotreta* larvae (see below).

Hymenoptera: Braconidae – *Diospilus oleraceus*

The braconid endoparasitoid *Diospilus oleraceus* parasitizes the larvae of cabbage stem flea beetle during the autumn in France and Germany (see Jourdheuil, 1960); this species has a wide host range, including *Ceutorhynchus* spp. (see Chapter 5; see also footnote in Table 4.1). It appears to have only a minor impact on cabbage stem flea beetle.

Hymenoptera: Pteromalidae – *Trichomalus lucidus*

The solitary, multivoltine, pteromalid ectoparasitoid *Trichomalus lucidus* has been reared from cabbage stem flea beetle larvae in northern Germany (Nissen, 1997), and a species of *Trichomalus* (determined by G. H. L. Nixon as near *T. lucidus*) was found parasitizing cabbage stem flea beetle larvae in Cambridgeshire, England (D. V. Alford, *pers. comm.*). *Trichomalus lucidus* has also been found to parasitize the larvae of cabbage stem weevil (*Ceutorhynchus pallidactylus*) in rape stems in Germany. Its occurrence on either species appears to be rare (Nissen, 1997).

Parasitoids Attacking Adults of Cabbage Stem Flea Beetle

Hymenoptera: Braconidae – *Microctonus melanopus*

The braconid *Microctonus melanopus* has been reared from cabbage stem flea beetle adults in France and in the UK (Jourdheuil, 1960; I. H. Williams, *unpublished*). This solitary multivoltine endoparasitoid is known as an abundant parasitoid of cabbage seed weevil adults in June and July (Jourdheuil, 1960) (see Chapter 6). The third and fourth generations occur towards the end of the summer and these parasitize the adults of various species of *Ceutorhynchus* and also those of cabbage stem flea beetle, the latter then becoming the overwintering hosts. There are no quantitative data in the literature on levels of parasitism of cabbage stem flea beetle, and Jourdheuil (1960) reported only low incidences of *M. melanopus* in France.

Parasitoids Attacking Larvae of *Phyllotreta nemorum*

The larvae of *Phyllotreta nemorum* are attacked by several hymenopterous parasitoids. These include the braconid endoparasitoid *Diospilus morosus*, an as yet unidentified braconid, the ichneumonid endoparasitoid *Aneuclis brevicauda* and two chalcid ectoparasitoids: *Eulophus* sp. and *Pnigalio soemius* (Jourdheuil, 1960; Sommer, 1981) (Table 8.2).

Table 8.2 Systematic list of parasitoids of cabbage flea beetles (*Phyllotreta* spp.).

Order HYMENOPTERA

Superfamily ICHNEUMONOIDEA
Family ICHNEUMONIDAE
Aneuclis Förster
– *brevicauda* Thomson
Mesochorus Gravenhorst
– *minutus* Szépligeti (= *phyllotretae* Jourdheuil; = *curvulus* Thomson)‡
– *plumosus* Dasch‡#
Family BRACONIDAE
Diospilus Haliday
– *morosus* Reinhardt†
Microctonus Wesmael
– *areolatus* Thomson [Note 1]
– *vittatae* Muesebeck#
Townesilitus Haeselbarth & Loan
– *bicolor* (Wesmael) [Note 2]

Superfamily CHALCIDOIDEA
Family EULOPHIDAE
Eulophus Olivier
– sp.
Pnigalio Schrank
– *soemius* Walker

† Species included in the key to adults (Chapter 11).
‡ Hyperparasitoid.
# A non-European species.

Note 1. Also cited in the literature as *Perilitus areolatus* Thomson.
Note 2. Also cited in the literature as *Microctonus bicolor* Wesmael and *Perilitus bicolor* Nees.

Hymenoptera: Braconidae – *Diospilus morosus*

Distribution

The solitary, multivoltine, braconid endoparasitoid *Diospilus morosus* is the most common parasitoid of *Phyllotreta nemorum* reported from France (Jourdheuil, 1960), Germany, Switzerland (Sommer, 1981) and the UK (Newton, 1931). Besides *P. nemorum*, this parasitoid has also been shown to parasitize the larvae of cabbage seed weevil during the summer and the larvae of cabbage stem flea beetle over the winter.

Phenology

The phenology of *Diospilus morosus* has been studied in France by Jourdheuil (1960) and in south-western Germany/western Switzerland by Sommer (1981). The adults of the first generation begin to emerge and reproduce in mid-May; the parasitoid larvae can be found within the host larvae from the end of May onwards. Adults of the second and third generations of *D. morosus*, the phenologies of which are not separated clearly, emerge at the end of June or early July and in August, respectively. The second generation is usually more abundant than the first. As the abundance of host larvae becomes reduced in August, the density of the fourth generation of the parasitoid is also often small in many years. The adults of this generation, emerging at the beginning of September, switch to the larvae of cabbage stem flea beetle. However, the presence of this winter host in good time is crucial and in many years will be too late, which results in a further strong decline in parasitoid density. As *Phyllotreta nemorum* is usually more abundant on mustard (*Sinapis arvensis*) and radish (*Raphanus raphanistrum*) growing in semi-natural habitats outside the rape crop, the change of *D. morosus* to the winter host (namely, cabbage stem flea beetle larvae developing in rape crops) can also be biased by spatial non-coincidence.

Status and importance

Because of this poor synchrony, the abundance of the first generation of *Diospilus morosus* is usually very small, resulting in low levels of parasitism of *Phyllotreta nemorum* in May to June. Sommer (1981), who studied the parasitism of *P. nemorum* at 44 sites in south-western Germany and western Switzerland, found levels of parasitism in May and June below 20% at around three-quarters of the sites; only 9% of the sites showed levels of parasitism higher than 40%. In the second and third generation, in July and August, at over half (57%) of the sites parasitism increased to over 40% and reached a maximum of 90%. In France, from 17% to 27% of *P. nemorum* was parasitized by *D. morosus* during 1954–1956 (Jourdheuil, 1960).

Life history

The life history of *Diospilus morosus* has been studied in detail by Jourdheuil (1960) and Sommer (1981). The females oviposit into the host larvae while these mine within leaf tissue. Superparasitism occurs commonly, and up to 13 eggs per host have been observed; however, in such cases, only one parasitoid larva develops to the second instar. As long as the host is feeding within the leaves, the parasitoid larva persists in its first instar. However, larval development is completed very quickly (in three days at 21°C) when the host has emigrated to the ground for pupation. The parasitoid (a koinobiont parasitoid) kills its host in the prepupal stage. The fifth larval instar of *D. morosus* pupates outside the host but within the host's earthen pupal chamber. Under laboratory conditions (at 21°C), pupal development is 11 days; longevity of females and males was five and four days, respectively (Sommer, 1981).

Hymenoptera: Ichneumonidae – *Aneuclis brevicauda*

The solitary, multivoltine ichneumonid endoparasitoid *Aneuclis brevicauda* was related for the first time to a specific host, *Phyllotreta nemorum*, by Sommer (1981). The biology of this koinobiont is very similar to that of *Aneuclis melanarius*. The larvae of *P. nemorum* were found to be parasitized from the end of May to the end of July. The adults of the first parasitoid generation were caught in May, those of the second and third generations in July and in August to September, respectively (Sommer, 1981). The parasitoid needs to change to other host species for overwintering, but these hosts have not yet been identified. This requirement is considered to be the most limiting factor for populations of *A. brevicauda* (Sommer, 1981). Compared with *Diospilus morosus*, *A. brevicauda* is thought, therefore, to be less effective at regulating the density of *P. nemorum*. *Aneuclis brevicauda* has been reported from the southern Rhine valley (Germany) and from western Switzerland (Sommer, 1981); the level of parasitism was only 0.1–5% at half of the sampling sites, although maximum parasitism at one site was 22% at the end of May.

Hymenoptera: Eulophidae – *Pnigalio soemius*

The solitary, multivoltine, eulophid ectoparasitoid *Pnigalio soemius* has a wide host range and attacks plant-mining larvae of various insect orders (Coleoptera, Lepidoptera, Diptera, Hymenoptera). This parasitoid was obtained by Sommer (1981) from 15 out of 44 sampling sites in the southern Rhine valley (Germany) and in western Switzerland. In this study, it was first related to the host *Phyllotreta nemorum*, parasitism of larvae being observed from June to August. As the parasitoid develops very quickly, three to four generations can occur within this period on this host. The female parasitoids prefer to oviposit onto the second or third larval instars of *P. nemorum*, the parasitoid larvae eventually pupating within the leaf gallery formed by the host. In the autumn, *P. soemius* must switch to other host species, which have not yet been identified; Sommer (1981) speculated that leaf-mining dipteran larvae might serve as winter hosts. In the period 1975–1977, the level of parasitism was generally low, ranging below 5% at over three-quarters of the sampling sites; the maximum parasitism was 7%. In each year, there was a slight increase of parasitism from June to August.

Hymenoptera: Eulophidae – *Eulophus* sp.

Another ectoparasitoid of *Phyllotreta nemorum* larvae, an undetermined species of *Eulophus*, was reported by Jourdheuil (1960) from France.

Parasitoids Attacking Adults of *Phyllotreta* spp.

Several species of braconid wasp have been reported to parasitize the adults of various species of *Phyllotreta*. The most widely distributed of these is the braconid *Townesilitus bicolor* (Speyer, 1925; Jourdheuil, 1960; Sommer, 1981; Haeselbarth & Loan, 1983; Ekbom, 1990). In addition, two unidentified species of the genus *Microctonus* have been reared from *Phyllotreta* spp. in low numbers (Sommer, 1981). One of these was later found to be identical to *Microctonus vittatae* (Wylie, 1988). It was reared from adults of *P. atra, P. cruciferae* and *P. vittata* and some other uniformly dark-pigmented species of *Phyllotreta*. The other species of *Microctonus* occurred exclusively in *P. nemorum* (a yellow-striped species of flea beetle) (see Fig. 2.6). Neither species was found in the adults of *P. undulata* (another yellow-striped species). *Microctonus areolatus* has been reported in the UK to parasitize *Phyllotreta* spp.; this species emerged in July in a culture of adult beetles of various species: namely, *P. atra*, *P. consobrina*, *P. cruciferae*, *P. diademata*, *P. nigripes* and *P. undulata* (Newton, 1931).

Hymenoptera: Braconidae – *Townesilitus bicolor*

Distribution

Townesilitus bicolor is a solitary, multivoltine endoparasitoid. It has been observed in Austria, France, Germany, Poland and Russia (Jourdheuil, 1960, Sommer, 1981). Moreover, this parasitoid has been introduced from Europe to Canada for biological control of the imported flea beetles *Phyllotreta cruciferae* and *P. striolata* (Wylie, 1988). In total, 1,936 adult *T. bicolor* and an estimated number of 2,639 parasitized flea beetle adults (*P. cruciferae* and *P. striolata*) were released at two sites in Manitoba from 1978 to 1983. However, up to 1985, no evidence of successful establishment of *T. bicolor* was found. This 'failure' was attributed both to the small number of parasitoids released and to the dispersal of the host population between oilseed rape crops and deciduous woodlands twice annually.

Phenology

The phenology of *Townesilitus bicolor* has been investigated by Jourdheuil (1960) and Sommer (1981). The first-instar larva overwinters within the body of adult *Phyllotreta* host. Larval development is resumed in spring, coinciding with the beginning of the reproduction of the host beetle. The first adult generation of *T. bicolor* emerges at the end of May or beginning of June and parasitizes, again, the overwintered beetle generation. The second generation of *T. bicolor*, which attacks the young flea beetle generation, was observed in France by Jourdheuil (1960) in June. In contrast, in Germany and western Switzerland, Sommer (1981) caught adults of the second generation of *T. bicolor* in July and August. The number of generations per year was estimated to be two or three.

Host range and preference

Although *Townesilitus bicolor* has a wide host range within the genus *Phyllotreta*, a distinct host preference has been found in the field – 73% of adult *T. bicolor* sampled in various locations and years were reared from *P. undulata*, 24% from *P. atra*, *P. cruciferae* or other dark-pigmented *Phyllotreta* species and only 1% from each of *P. nemorum* and *P. vittata* (Sommer, 1981). Similarly, when given a choice in oviposition experiments in the laboratory, *T. bicolor* preferred *P. undulata* over other species of *Phyllotreta*: the level of parasitism of *P. undulata*, *P. atra*, *P. vittata* and *P. nemorum* was 62%, 24%, 13% and 1%, respectively (Sommer, 1981). In Canada, oviposition tests showed that *T. bicolor* oviposited readily in *P. striolata*, occasionally in *P. cruciferae* but not at all in the flea beetle *Psylliodes punctulata*; immature parasitoids developed and emerged from both *P. cruciferae* and *P. striolata* (Wylie, 1983a).

Status and importance

The reported level of parasitism of *Phyllotreta* spp. by *Townesilitus bicolor* varies from country to country. In France, regularly high levels of up to 100% were observed by Jourdheuil (1960). In contrast, at 61 sampling sites in Austria, southern Germany and western Switzerland, only low levels were found by Sommer (1981), i.e. at 80% of the sites from 0% to 10% parasitism and at 20% of the sites from 10% to 50% parasitism. Superparasitism, with up to four eggs/host, was detected in 15% of flea beetles examined. In Sweden, the level of parasitism of male and female *P. undulata* caught from plots of rape, turnip and white mustard was only 16% and 15%, respectively (Ekbom, 1990), and only few adults of *T. bicolor* emerged from the beetles. The parasitism increased during the years 1987 to 1989, which was attributed to favourable weather conditions and to the fact that no pesticides had been applied for the previous five years (Ekbom, 1991).

Effect on host

The parasitism of *Phyllotreta* spp. by *Townesilitus bicolor* (and by *Microctonus* spp.) has been found to significantly affect both the reproduction and the longevity of hosts (Sommer, 1981). The proportion of female *Phyllotreta undulata* sampled in the field that were mature was 67% when they were not parasitized but only 51% when parasitized by first-instar larvae of *T. bicolor*. When *P. undulata* females were parasitized by later instars of *T. bicolor* (instars three to five), maturation of the ovaries and eggs was completly inhibited. This was also confirmed by Ekbom (1990) in Sweden, who found no development of ovaries or eggs in parasitized females of *P. undulata*. The effect of parasitism on reproduction has also been observed in laboratory experiments, using *P. undulata* females collected from fields in spring (Sommer, 1981). Unparasitized females laid an average of 28 eggs, whereas females parasitized by *T. bicolor* laid only an average of four eggs over the same period. In addition, as the host dies a few hours after the full-grown parasitoid larva has departed and entered the soil to pupate (Sommer, 1981), the longevity of adult *Phyllotreta* spp. is reduced by parasitism; this applies in the case of both *T. bicolor* and *Microctonus* spp.

Life history

The oviposition behaviour and biology of *Townesilitus bicolor* have been studied in great detail by Sommer (1981). Having located a suitable host, the female pursues the flea beetle at a short distance and then very quickly climbs up onto the host's abdomen from the back. It clings to the host with its legs before it bends down its abdomen under the host's abdomen to insert the ovipositor through an intersegmental membrane and to lay a single egg into the host's body.

The host reacts to the encounter with various defence mechanisms (jumping, turning around, waggling and preening), and this eventually results in the repulsion of the parasitoid female. Encapsulation of the egg or young larva of *T. bicolor* within the host's haemolymph has been found in only very few hosts (<1%). Under laboratory conditions, at 23°C, the developmental time of the egg, larval and pupal stages of *T. bicolor* was 5, 18 and 15 days, respectively. Supplied with water, adult females and males lived for 8.7 and 6.7 days, respectively. The average number of eggs laid per female was 87, and the average number of eggs produced within the ovaries was 123.

Hyperparasitism

The larvae of *Townesilitus bicolor* and various species of *Microctonus* parasitizing *Phyllotreta atra*, *P. nemorum*, *P. nigripes*, *P. undulata*, *P. vittuala*, *P. zimmermani* and a species of *Aphthona* (also a flea beetle) can be hyperparasitized by the secondary, solitary, multivoltine, ichneumonid endoparasitoid *Mesochorus minutus* (Jourdheuil, 1960; Sommer, 1981).

Mesochorus minutus has been reared from *Microctonus* sp. in Switzerland, the Rhine valley (Germany) and in different regions of Austria (Sommer, 1981). However, in these countries only 44% of the sampled *Microctonus* populations showed any hyperparasitism by *M. minutus*. The level was relatively high during June to September, ranging from 20% to 50% in two-thirds of the parasitized *Microctonus* populations (Sommer, 1981). In France, hyperparasitism of *Microctonus* sp. by *M. minutus* increased with increasing densities of the host; for example, at levels of parasitism by *Microctonus* below 50%, hyperparasitism did not exceed 20%. However, when the parasitism of *Microctonus* spp. was greater than 75–80%, hyperparasitism increased to 90–100% (Jourdheuil, 1960). Thus, *M. minutus* may have a significant effect on the population dynamics of the primary parasitoid in France. *Mesochorus minutus* is present also in some parts of Canada (Sommer, 1981).

The phenology of *M. minutus* is similar to that of *Microctonus* spp. (Sommer, 1981). The hyperparasitoid overwinters within the first-instar host larva. The adults emerge synchronously with adults of *Microctonus* in May. The longevity of females and males under laboratory conditions was 21 days and 19 days, respectively. The adults of subsequent generations occur from June to August. In France, adults of *M. minutus* have been observed up to October (Jourdheuil, 1960).

Hymenoptera: Braconidae – *Microctonus vittatae*

Distribution

Microctonus vittatae parasitizes the flea beetles *Phyllotreta cruciferae* and *P. striolata* in Canada (Wylie, 1980). This parasitoid is not recorded in Europe.

Phenology

Microctonus vittatae is a multivoltine parasitoid with four generations per year. It overwinters as a first-instar larva in its diapausing host adult. Parasitoid larvae resume their development in May after their hosts have left their overwintering sites in woodland leaf litter and soon moult to become second-instar larvae. Fully grown parasitoid larvae emerge from their hosts in mid-May and spin cocoons from which the adults emerge in late May or early June. Further generations of adult parasitoids emerge in July, August and September to October although successive generations are not discrete. Most of the larvae of the first three generations are non-diapausing, whereas most larvae of the fourth generation develop briefly and then enter diapause, whilst still in their first instar, in September to October and then overwinter. Wylie (1980) showed that *M. vittatae* produces a greater proportion of non-diapausing progeny when reared under a long-day (16h light:8h dark) cycle at a temperature of 20–25°C than under a shorter-day cycle and a lower temperature.

Status and importance

In Canada, *Microctonus vittatae* is the only parasitoid reported to attack adults of *Phyllotreta cruciferae* and *P. striolata*. However, the level of parasitism is low. Parasitism of *P. cruciferae* and *P. striolata* adults collected from leaf litter at overwintering sites and from crucifers was usually <5% (Wylie, 1981). However, a higher level of parasitism (15% in *P. striolata* and 53% in *P. cruciferae*) was observed in the first adults of both species to leave overwintering sites in spring (Wylie, 1982).

Effect on host

Parasitization by *Microctonus vittatae* affects the behaviour of the host in two ways. Firstly, it affects the time of host emergence from overwintering – adults of *Phyllotreta cruciferae* and *P. striolata* parasitized by *M. vittatae* in Canada emerged earlier from overwintering sites than unparasitized adults of the same species (Wylie, 1982); early emergence of parasitized beetles ensures early emergence of parasite adults, probably increasing the impact of the parasitoid on host populations. Secondly, adults of *P. cruciferae* parasitized by *M. vittatae* do not reproduce, as the eggs of the host cease to develop soon after the parasitoid egg hatches (Smith, 1952). This change in host behaviour is reflected in trap catches. Wylie (1981) found that, in spring and autumn, when the host was not reproducing, similar percentages of parasitized adult and female *P. cruciferae* were caught in traps baited with allyl isothiocyanate and in a D-Vac insect suction sampler. However, during the host's reproductive period in May–July, a lower percentage of parasitized adults and a greater percentage of females were caught in the baited traps than in the D-Vac sampler.

Life history

Female *Microctonus vittatae* pursue walking adults of *Phyllotreta cruciferae* and *P. striolata* and insert the ovipositor soon after the beetles stop walking (Wylie & Loan, 1984). The females stand beside their hosts while piercing, lay a single egg into the haemolymph and usually withdraw the ovipositor within five seconds. When exposed singly with just one specimen of *P. striolata*, parasitoid females made 71% of insertions into the head, usually through the mouth, 18% into the thorax and 11% into the abdomen. Some oviposition attempts were unsuccessful because the host beetle turned away, jumped or flew as the parasitoid began its ovipositor insertion. Two females enclosed with a large number of host beetles oviposited when <48 hours old and continued ovipositing up to one to two days, before dying 12 and 14 days later. Each female oviposited in 114 hosts, and they laid 166 and 146 eggs. Female parasitoids laid eggs in beetles of both sexes and did not discriminate between parasitized and unparasitized adults of *P. striolata*, so that some beetles received up to 13 parasitoid eggs.

At 20°C, 70–80% RH and a 16 h light : 8 h dark cycle, the eggs hatch about five to nine days after oviposition (Wylie & Loan, 1984). First-instar larvae were found in the head, thorax and abdomen, but after 10 days only in the abdomen. Later-instar larvae were found only in the abdomen. Parasitoid eggs hatch within the host's head or thorax, and the first-instar larvae eventually move into the abdomen, where later-instar larvae remain throughout their development (Wylie, 1985). The parasitoid larvae appear not to cause any feeding damage to host organs, except possibly the testes; however, adipose (fat-body) tissue in parasitized hosts was progressively reduced. Smith (1952) believed that the larvae fed on teratocytes in the blood and, thereby, caused shrinkage of the adipose tissue and the eggs of female beetles.

Only one parasitoid larva survives in each host (Wylie & Loan, 1984). It emerges from the posterior end of the abdomen 16–19 days after oviposition by pushing out the membranous area between the pygidium of the beetle and its parallel ventral sclerite. The host usually dies within one hour of parasitoid emergence, and the larva spins a cocoon within five to six hours of emerging. The parasitoid adult emerges 11–13 days later. The mean time from oviposition to adult emergence for 35 female *M. vittatae* was 29 days (range 27–32 days). Superparasitism of *P. cruciferae* by *M. vittatae* prolongs development of the single-surviving parasitoid in each beetle (Wylie, 1983b).

In the field, few male *M. vittatae* are found and most or all reproduction is probably by deuterotoky (Wylie & Loan, 1984).

Hyperparasitism

The ichneumonids *Mesochorus minutus* and *M. plumosus* have been reared from *Microctonus vittatae* in the USA (Krombein *et al.*, 1979). Both species of *Mesochorus* oviposit in larvae of *M. vittatae* inside the flea beetle adults (Jourdheuil, 1960; Krombein *et al.*, 1979).

Acknowledgements

Rothamsted Research receives grant-in-aid from the Biotechnology and Biological Sciences Research Council. This work was part-funded by the UK Ministry of Agriculture, Fisheries and Food and the European Commission Concerted Action: *Minimizing pesticide use and environmental impact by the development and promotion of bio-control for oilseed rape pests, FAIR 96-1314.*

References

Aubert, J.-F. & Jourdheuil, P. (1958). Nouvelle description et biologie de quelques Ichneumonides appartenant aux genres *Aneuclis* Först., *Isurgus* Först. et *Thersilochus* Holm. *Revue de Pathologie Végétale et d'Entomologie Agricole de France* **37**, 175–93.

Bonnemaison, L. & Jourdheuil, P. (1954). L'altise d'hiver du Colza (*Psylliodes chrysocephala* L.). *Annales des Épiphyties* **4**, 345–524.

Dosse, G. (1961). *Thersilochus melanogaster* Thoms. als Larvenparasit des Rapserdflohs *Psylliodes chrysocephala* L. *Zeitschrift für Pflanzenkrankheiten und Pflanzenschutz* **68**, 575–80.

Ekbom, B. (1990). Flea beetles (*Phyllotreta* spp.) in spring oilseed rape in Sweden. *IOBC/wprs Bulletin* **13** (4), 57–61.

Ekbom, B. (1991). The effects of abiotic and biotic factors on flea beetle populations in spring rape in Sweden. *IOBC/wprs Bulletin* **14** (6), 21–7.

Godan, D. (1950). Parasitierung von Rapserdflohlarven. *Anzeiger für Schädlingskunde* **23**, 150.

Haeselbarth, E. & Loan, C. (1983). *Townesilitus*, a new genus for a species group in *Microctonus* (Hymenoptera: Braconidae, Euphorinae). *Contributions of the American Entomological Institute* **20**, 384–7.

Horstmann, K. (1971). Revision der europäischen Tersilochinen (Hym.: Ichneumonidae). Teil I. *Veröffentlichungen der Zoologischen Staatssammlung München* **15**, 47–138.

Horstmann, K. (1981). Revision der europäischen Tersilochinae II (Hymenoptera, Ichneumonidae). *Spixiana, Supplementum* **4**, 1–76.

Jourdheuil, P. (1960). Influence de quelques facteurs écologiques sur les fluctuations de population d'une biocénose parasitaire: étude relative à quelques Hyménoptères (Ophioninae, Diospilinae, Euphorinae) parasites de divers Coléoptères inféodés aux Crucifères. *Annales des Épiphyties* **11**, 445–658.

Klingenberg, A. & Ulber, B. (1994). Untersuchungen zum Auftreten der Tersilochinae (Hym., Ichneumonidae) als Larvalparasitoide einiger Rapsschädlinge im Raum Göttingen 1990 und 1991 und zu deren Schlupfabundanz nach unterschiedlicher Bodenbearbeitung. *Journal of Applied Entomology* **117**, 287–99.

Krombein, K. V., Hurd, P. D. & Smith, D. R. (1979). *Catalog of Hymenoptera in America North of Mexico*. Smithsonian Institution Press: Washington, DC.

Lehmann, W. (1965). Einfluss chemischer Bekämpfungsmassnahmen auf einige Rapsschädlinge und ihre Parasiten. II. Knospen- und Stengelschädlinge. *Archiv für Pflanzenschutz* **1**, 209–19.

Meuche, A. (1940). Untersuchungen am Rapserdfloh (*Psylliodes chrysocephala* L.) in Ostholstein. *Zeitschrift für Angewandte Entomologie* **27**, 464–95.

Miczulski, B. (1966). Community studies on Hymenoptera found on *Brassica napus* L. Part II. Ichneumonids (Ichneumonidae). *Polskie Pismo Entomologiczne* **20**, 263–90. [In Polish.]

Newton, H. C. F. (1931). Notes on some parasites reared from flea beetles of the genus *Phyllotreta* (Chrysomelidae). *Entomologist's Monthly Magazine* **67**, 82–4.

Nissen, U. (1997). *Ökologische Studien zum Auftreten von Schadinsekten und ihren Parasitoiden an Winterraps norddeutscher Anbaugebiete*. PhD thesis, Christian-Albrechts-University of Kiel, Germany.

Nitzsche, O. (1998). *Auftreten und Effizienz von Parasitoiden als natürliche Gegenspieler von Schadinsekten im Winterraps unter besonderer Berücksichtigung unterschiedlicher Bodenbearbeitungsmaßnahmen nach Winterraps*. PhD thesis, Georg-August-University of Göttingen, Germany.

Schulz, R.-R. (1983). *Zur Biologie, Ökologie und Schadwirkung des Rapserdflohs* (Psylliodes chrysocephala *L.) am Winterraps* (Brassica napus *L. var.* oleifera *Metzg.*). PhD thesis, Wilhelm-Pieck-University of Rostock, Germany.

Šedivý, J. (1983). Tersilochinae as parasitoids of insect pests of winter rape (Hymenoptera: Ichneumonidae). *Contributions of the American Entomological Institute* **20**, 266–76.

Seymour, P. R. (1989). *Invertebrates of Economic Importance in Britain. Common and Scientific Names. MAFF Technical Bulletin 6.* 4th edn. HMSO: London.

Smith, O. J. (1952). Biology and behaviour of *Microctonus vittatae* Muesebeck (Braconidae) with descriptions of its immature stages. *University of California Publications in Entomology* **9**, 315–44.

Sommer, G. (1981). *Biologie und Parasitenkomplex der Halticinen Gattung* Phyllotreta. PhD thesis, Albert-Ludwigs-Universität zu Freiburg, Germany.

Speyer, W. (1925). *Perilitus melanopus* Ruthe (Hym. Braconidae) als Imaginalparasit von *Ceutorrhynchus quadridens* Panz. *Zeitschrift für Angewandte Entomologie* **11**, 132–46.

Wylie, H. G. (1980). Factors affecting facultative diapause of *Microctonus vittatae* (Hymenoptera: Braconidae). *Canadian Entomologist* **112**, 747–9.

Wylie, H. G. (1981). Effects of collection method on estimates of parasitism and sex ratio of flea beetles (Coleoptera: Chrysomelidae) that infest rape crops in Manitoba. *Canadian Entomologist* **113**, 665–71.

Wylie, H. G. (1982). An effect of parasitism by *Microctonus vittatae* (Hymenoptera: Braconidae) on emergence of *Phyllotreta cruciferae* and *Phyllotreta striolata* (Coleoptera: Chrysomelidae) from overwintering sites. *Canadian Entomologist* **114**, 727–32.

Wylie, H. G. (1983a). Oviposition and survival of the European parasite *Microctonus bicolor* (Hymenoptera: Braconidae) in crucifer-infesting flea beetles (Coleoptera: Chrysomelidae) in Manitoba. *Canadian Entomologist* **115**, 55–8.

Wylie, H. G. (1983b). Delayed development of *Microctonus vittatae* (Hymenoptera: Braconidae) in superparasitized adults of *Pyllotreta cruciferae* (Coleoptera: Chrysomelidae). *Canadian Entomologist* **115**, 441–2.

Wylie, H. G. (1985). Posterior dispersal of eggs and larvae of *Microctonus vittatae* (Hymenoptera: Braconidae) in crucifer-infesting flea beetles (Coleoptera: Chrysomelidae). *Canadian Entomologist* **117**, 541–5.

Wylie, H. G. (1988). Release in Manitoba, Canada, of *Townesilitus bicolor* (Hym.: Braconidae), a European parasite of *Phyllotreta* spp. (Col.: Chrysomelidae). *Entomophaga* **33**, 25–32.

Wylie, H. G. & Loan, C. (1984). Five Nearctic and one introduced Euphorine species (Hymenoptera: Braconidae) that parasitize adults of crucifer-infesting flea beetles (Coleoptera: Chrysomelidae). *Canadian Entomologist* **116**, 235–46.

9

Parasitoids of Miscellaneous Pests

David V. Alford
Orchard House, Oakington Road, Dry Drayton, Cambridge CB3 8DD, UK

Yannick Ballanger
CETIOM, Centre de Grignon, BP 4, 78850 Thiverval-Grignon, France

Bernhard Kromp
Ludwig Boltzmann-Institute for Biological Agriculture and Applied Ecology, Rinnböckstraße 15, A-1110 Wien, Austria

Introduction

In this chapter brief mention is made of parasitoids associated with several minor pests of oilseed rape not covered elsewhere, namely: aphids, *Baris* weevils, cabbage root fly (*Delia radicum*) and turnip sawfly (*Athalia rosae*). Although several other minor or incidental pests occur on oilseed rape (see Chapter 2), information on their natural enemies is either wanting or cited only in relation to the occurrence of the pests on crops other than oilseed rape or in non-cropping situations.

In common with parasitoids associated with the major pests of oilseed rape (Chapters 4–8), those affecting the minor pests considered here are also mainly members of the order Hymenoptera. However, additionally, in the case of turnip sawfly (see below), parasitoids of another major insect group (tachinid flies – Diptera: Tachinidae) are important. Tachinids are well-known parasitoids of a wide range of crop pests, but their hosts are mainly caterpillars of butterflies, moths and sawflies (Clausen, 1940; Askew, 1971; Herting, 1984; Smith, 1989); they are not recorded as enemies of the coleopterous and dipterous pests (i.e. beetles, weevils and midges) that form the bulk of the pest complex on oilseed rape crops. Brassicaceous pests such as cabbage moth (*Mamestra brassicae*), large white and small white butterflies (*Pieris brassicae* and *P. rapae*) (see Chapter 2), however, are recorded as hosts of tachinids (see, for example, Bisset, 1938; Smith, 1989). Tachnid flies may be identified using the keys of Benshaw (1993).

Members of the usual range of oilseed rape pests are not known to be attacked by egg parasitoids (e.g. Hymenoptera: Trichogrammatidae), although incidental lepidopterous pests of oilseed rape (such as *Pieris* spp.) are known to be hosts of, for example, *Trichogramma evanescens* and (along with cabbage moth) *T. brassicae* (Li, 1994).

Parasitoids of Aphids

Aphids are subject to attack by a wide range of natural enemies, including various parasitoid wasps in the family Aphidiidae. Such parasitoids in oilseed rape crops are likely to be present in greatest numbers during the summer, in colonies of cabbage aphid (*Brevicoryne brassicae*), but they have received little attention on this crop.

The most frequently reported parasitoid of cabbage aphid (on brassica crops, including oilseed rape) is *Diaeretiella rapae* (M'Intosh) (= *Aphidius brassicae* Marshall) (Hymenoptera: Aphidiidae) (e.g. Kelm, 1988), a species which pupates in pale-beige mummies formed by the remains of outer skin of the host. It is unlikely, however, that cabbage aphid populations are greatly depleted by such parasitoids. *Diaeretiella rapae* is widely known as the 'cabbage aphid parasite' (Seymour, 1989), i.e. the 'cabbage aphid parasitoid'; it also attacks peach/potato aphid (*Myzus persicae*) (George, 1957). Other parasitoids known to attack field populations of peach/potato aphid (at least on potato crops) include *Aphidius picipes* (Nees) (= *A. avenae* Haliday) and *Praon volucre* (Haliday) (both family Aphidiidae) (Dunn, 1949); their impact on peach/potato aphid populations in oilseed rape crops (along with that of other aphidophagous parasitoids), however, is likely to be minimal.

Parasitoids of *Baris* Weevils

Six species of hymenopterous parasitoid are reported to parasitize weevils of the genus *Baris* (see Tables 9.1 and 9.2). However, no data and no records relate to *B. coerulescens*, the main species associated with oilseed rape (Koubaiti, 1992) (see Chapter 2). All publications on the natural enemies of *Baris* weevils either deal with other species of *Baris* on vegetable brassica crops (such as cabbage) (e.g. Faure, 1923a, 1923b, 1924) or refer to entomopathogenic nematodes – see Chapter 17.

Faure (1924) considered that the parasitoid *Bracon glaphyrus* was well adapted to parasitize *Baris laticollis*, the larvae of which (as in the case of *B. coerulescens*) live in the roots; only occasionally would this species parasitize *B. chlorizans* (when larvae were feeding at the collar of the plant, at the junction of the stem and root) or other species whose larvae live within the stems. *Bracon variator*, however, is adapted for attacking stem-inhabiting *Baris* larvae (Faure, 1924).

Table 9.1 Hymenopterous parasitoids of *Baris* weevils (from Faure, 1923b, 1924). [Notes 1 and 2]

Parasitoid	*Baris chlorizans*	*Baris cuprirostris*	*Baris laticollis*
Bracon baridii	–	–	(+) [Note 3]
Bracon glaphyrus	(+) [Note 3]	–	+
Bracon variator	+	+	–
Entedon pharnus	–	–	(+)
Eurytoma curculionum	+	+	–
Pteromalus sp.	+	+	–

– Not recorded.
+ Present.
(+) Present (according to other authors cited by Faure).

Note 1. Includes references cited from older literature.
Note 2. In addition, Hoffmann (1954) cites *Asecodes pharnus* and *Bracon baridii* as parasitoids of *Baris laticollis*, and *Entedon confinis* as a parasitoid of *Baris chlorizans*.
Note 3. Recorded from plants infested by more than one species of *Baris*.

Table 9.2 List of hymenopterous parasitoids associated with *Baris* spp. (after Faure, 1924). [Note 1]

Family BRACONIDAE ***Bracon*** Fabricius – *baridii* Marshall – *glaphyrus* Marshall – *variator* Nees Family EURYTOMIDAE ***Eurytoma*** Illiger – *curculionum* Mayr	Family PTEROMALIDAE ***Pteromalus*** Swederus – sp. Family EULOPHIDAE ***Entedon*** Dalman – *confinis* (Ratzeburg) [Note 2] – *pharnus* Walker [Note 2]

Note 1. Species determined by Dr. Ferrière of the Natural History Museum, Berne, Switzerland.
Note 2. Additional records cited in Hoffmann (1954).

Although in recent years many oilseed rape plants have been dissected for eggs, larvae and pupae of *Baris* weevils, and in many cases plants have been caged for monitoring the emergence of *Baris* adults, there have been no reports of the presence of hymenopterous ectoparasitoids in these samples and no adult parasitoids have ever appeared in the emergence cages, either in the autumn or in the following spring.

Parasitoids of Cabbage Root Fly

The most well-known parasitoid of cabbage root fly is the rove beetle *Aleochara bilineata*, whose larvae feed as endoparasitoids in the larvae of the pest (Finch & Collier, 2000). This rove beetle is also a predator of cabbage root fly eggs and, perhaps, young larvae (Langlet & Brunel, 1996). The biology of *A. bilineata* was described by Wadsworth (1915), but suggestions by this and other authors that the beetle is also an effective predator of the eggs of cabbage root fly have been refuted (see Finch *et al.*, 1999). Biocontrol of cabbage root fly larvae with *A. bilineata* involves inundative release of the beetles (at rates of many thousands per hectare), and this requires prior mass-rearing of the parasitoid in the laboratory (e.g. Bromand, 1980; Hertveldt *et al.*, 1984; Whistlecraft *et al.*, 1985).

A second species of *Aleochara* (*A. bipustulata*) and the hymenopterous parasitoid *Trybliographa rapae* (Hymenoptera: Chalcidae) (sometimes cited in the literature as *Idiomorpha rapae*) are also reported as parasitoids of cabbage root fly (Hughes *et al.*, 1959); in their studies these authors reported up to 65% parasitism of cabbage root fly by *Aleochara* spp. and *T. rapae*. At least in southern England, *T. rapae* has two generations per year, but adults are sufficiently long-lived that they are able to attack all three generations of the host (Kacem *et al.*, 1996). A further hymenopterous parasitoid (*Phygadeuon* sp.) (family Ichmeumonidae) of cabbage root fly larvae is reported (Langlet & Brunel, 1996), which is capable in oilseed rape crops of causing significant mortality, e.g. 31 dead larvae in a sample of 80, and (after overwintering) the surviving 49 pupae giving rise to 32 cabbage root flies and 14 adult parasitoids (Y. Ballanger, *unpublished*).

Biocontrol of cabbage root fly is aimed at infestations on vegetable brassicas (e.g. cauliflower crops), and further studies on such crops are currently underway in France and the UK (Y. Ballanger, *unpublished*; R. H. Collier, *pers. comm.*).

Parasitoids of Turnip Sawfly

There are no recent studies on parasitoids of turnip sawfly and very few published references, the most recent being that of Reich (1962). Prior to that, the most substantive work was that by Riggert (1939).

According to the literature, tachinid flies (Diptera: Tachinidae) are important natural enemies of turnip sawfly larvae. Cited genera include *Exorista*, *Blondelia* and *Meigenia* (e.g. Cameron, 1882; Reich, 1962), and high incidences of parasitism have been reported for some species, for example:

- 80% of larvae parasitized in Finland by *Meigenia mutabilis* (Kanervo & Talvitie, 1946);
- 60% of second-generation larvae parasitized in Germany by *Exorista rustica* (Reich, 1962).

Cameron (1882) lists numerous hymenopterous wasps as parasitioids of turnip sawfly larvae, including various chalcids and ichneumonids. However, these are probably of

Table 9.3 Examples of parasitoids of turnip sawfly (*Athalia rosae*).

Order DIPTERA

  Family TACHINIDAE
    Blondelia Robineau-Desvoidy
    – *nigripes* (Fallén)
    Exorista Meigen (= ***Tachina*** auctt., nec Meigen)
    – *rustica* (Fallén)
    Meigenia Robineau-Desvoidy
    – *mutabilis* (Fallén) (= *bisignata* (Meigen))†

Order HYMENOPTERA

  Superfamily ICHNEUMONOIDEA
    Family ICHNEUMONIDAE
      Arbelus Townes
      – *athaliaeperdus* Curtis† [Note 1]
      Campolestis Förster (= ***Anilasta*** Thomson)
      Mesoleius Holmgren
      – *armillatorius* (Gravenhorst)†
      – *ciliatus* Holmgren†
      Mesochorus Gravenhorst
      – *areolaris* Ratzeburg
      Monoblastus Hartig (= ***Coeloconus*** Förster)
      – *brachyacanthus* (Gmelin in Linnaeus)†
      Perilissus Holmgren
      – *lutescens* Holmgren†
      Tryphon Fallén
      – *succinctus* Gravenhorst†
      – *marginellus* Gravenhorst†
  Superfamily CHALCIDOIDEA
    Family PERILAMPIDAE
      Perilampus Latreille
      – *aeneus* (Rosssius) (= *splendidus* Dalman)†
      – *ruficornis* (Fabricius) (= *violaceus* (Fabricius))†

†Cited in Cameron (1882).

Note 1. Also cited as *Bassus athaliaeperda*.

lesser importance than tachinids (Reich, 1962). Frequently reported genera of parasitoids within the family Ichneumonidae include: *Campolestis, Mesoleius, Mesochorus, Monoblastus, Perilissus* and *Tryphon*; chalcid parasitoids cited by Reich (1962) are all from the genus *Perilampus* (family Perilampidae). Rusinov (1977) recorded high levels of parasitism by *Perlissus lutescens* in Kasachstan: 70% in the first generation and 12% in the second.

For further details see Table 9.3.

Acknowledgements

This work was part-funded by the UK Ministry of Agriculture, Fisheries and Food and the European Commission Concerted Action: *Minimizing pesticide use and environmental impact by the development and promotion of bio-control strategies for oilseed rape pests, FAIR 96-1314.*

References

Askew, R. R. (1971). *Parasitic Insects.* Heinemann Educational: London.

Benshaw, R. (1993). Tachinid flies. *Handbooks for the Identification of British Insects* **10** (4ai), 1–169.

Bisset, G. A. (1938). Larvae and pupae of tachinids parasitizing *Pieris rapae* L. and *P. brassicae* L. *Parasitology* **31**, 111–22.

Bromand, B. (1980). Investigations on the biological control of the cabbage root fly (*Hylemya brassicae*) with *Aleochara bilineata*. *Bulletin OILB/SROP* **3** (1), 49–62.

Cameron, P. (1882). *A Monograph of the British Phytophagous Hymenoptera. Vol. I.* Ray Society: London.

Clausen, C. P. (1940). *Entomophagous Insects.* McGraw-Hill: New York.

Dunn, J. A. (1949). The parasites and predators of potato aphids. *Bulletin of Entomological Research* **40**, 97–122.

Faure J. C. (1923a). Les Baridiés, Charançons nuisibles aux choux. Causes naturelles de destruction des Baridiés. *Revue de Zoologie Agricole Appliquée* **22**, 84–93.

Faure, J. C. (1923b). Observations sur les *Baris* et leurs parasites. *Annales des Épiphyties* IX **2**, 109–20.

Faure, J. C. (1924). Observations biologiques sur *Bracon glaphyrus* Marsh. *Revue de Pathologie Végétale et d'Entomologie Agricole* **11**, 70–2.

Finch, S. & Collier, R. H. (2000). Integrated pest management in field vegetable crops in northern Europe – with focus on two key pests. *Crop Protection* **19**, 817–24.

Finch, S., Elliott, M. S. & Torrance, M. T. (1999). Is the parasitoid staphylinid beetle *Aleochara bilineata* an effective predator of the egg stage of its natural host, the cabbage root fly? *Integrated Control in Field Vegetable Crops. IOBC Bulletin* **22** (5), 109–12.

George, K. S. (1957). Preliminary investigations on the biology and ecology of the parasites and predators of *Brevicoryne brassicae* (L.). *Bulletin of Entomological Research* **48**, 619–29.

Herting, B. (1984). Catalogue of Palearctic Tachinidae (Diptera). *Stuttgart Beiträge zur Naturkunde (A)* **369**, 1–228.

Hertveldt, L., Keymeulen, M. van & Pelerents, C. (1984). Large scale rearing of the entomophagous rove beetle *Aleochara bilineata* (Coleoptera: Staphylinidae). *Mitteilungen aus der Biologischen Bundesanstalt für Land- und Forstwirtschaft* **218**, 70–5.

Hoffmann, A. (1954). Coléoptères curculionides (Tribu des Barini). *Faune de France* **59** (2), 1053–84.

Hughes, R. D., Salter, D. D. & Mitchell, B. (1959). The natural control of cabbage root fly at Wellesbourne. *Report of the National Vegetable Research Station for 1958*, p. 31.

Kacem, N., Neveun, N. & Nénon, J. P. (1996). Development of *Trybliographa rapae* a larval parasitoid of the cabbage rootfly *Delia radicum. IOBC/wprs Bulletin* **19** (11), 156–61.

Kanervo, V. & Talvitie, Y. K. K. (1946). Tutkimuksia sinappikuriaiskärpäsestä *Meigenia mutabilis. Annales Zoologici Societatis Zoologico-Botanicae Fennicae* **11**, 1–45.

Kelm, M. (1988). The effectiveness of *Diaeretiella rapae* (McIntosh) in the control of the cabbage aphid *Brevicoryne brassicae* L. on oilseed rape in Great Britain. In: E. Niemczyk & A. F. G. Dixon (eds). *Ecology and Effectiveness of Aphidophaga.* SPB Academic Publishers: The Hague.

Koubaiti, K. (1992). *Reproduction, Développement et Nuisibilité de* Baris coerulescens *Scop. (Coléoptère, Curculionidae) au Contact du Colza (*Brassica napus *var.* oleifera*).* Thesis, Tours University, France.

Langlet, X. & Brunel, E. (1996). Preliminary results on predation by *Aleochara bilineata* Gyll. (Coleoptera: Staphylinidae). *IOBC/wprs Bulletin* **19** (11), 173–8.

Li, Li-Yang (1994). Worldwide use of *Trichogramma* for biological control on different crops: a survey. In: E. Wajnberg & S. A. Hassan (eds). *Biological Control with Egg Parasitoids.* CAB International: Ascot.

Reich, R. (1962). Natürliche Feinde der Rübsenblattwespe (*Athalia rosae* L.). *Nachrichtenblatt Deutscher Pflanzenschutzdienst (Berlin) N.F.* **16**, 33–7.

Riggert, E. (1939). Untersuchungen über die Rübsenblattwespe *Athalia colibri* Christ (*A. spinarum* F.). *Zeitschrift für Angewandte Entomologie* **26**, 462–516.

Rusinov, V. I. (1977). [Parasites of mustard pests.] *Zashchita Rastenii* **2**, 27. [In Russian.]

Seymour, P. R. (1989). *Invertebrates of Economic Importance in Britain. Common and Scientific Names.* 4th edn. HMSO: London.

Smith, K. G. V. (1989). An introduction to the immature stages of British flies. Diptera larvae, with notes on eggs, puparia and pupae. *Handbooks for the Identification of British Insects* **10** (14), 1–280.

Wadsworth, J. T. (1915). On the life-history of *Aleochara bilineata* Gyll., a staphylinid parasite of *Chortophila brassicae*, Bouché. *Journal of Economic Biology* **10**, 1–27.

Whistlecraft, J. W., Harris, C. R., Tolman, J. H. & Tomlin, A. D. (1985). Mass-rearing technique for *Aleochara bilineata* (Coleoptera: Staphylinidae). *Journal of Economic Entomology* **78**, 995–7.

10

Sampling, Trapping and Rearing Oilseed Rape Pests and Their Parasitoids

Ingrid H. Williams
Rothamsted Research, Harpenden, Hertfordshire, AL5 9TX, UK

Rudolf Büchi
Swiss Federal Research Station for Agroecology and Agriculture, CH-8046 Zürich, Reckenholzstrasse, Switzerland

Bernd Ulber
Institut für Pflanzenpathologie und Pflanzenschutz, Universität Göttingen, Grisebachstrasse 6, D-37077 Göttingen, Germany

Introduction

Various methods are used to sample the pests of oilseed rape, and many of them can also be used to sample their parasitoids. Species of parasitoid attack specific life stages of their host pest, and various methods are used to sample these different stages, either in flight, from the plant or from the soil. Adult pest and parasitoid may need to be sampled simultaneously to determine their relative phenologies. Pest eggs, larvae, cocoons or adults may be sampled so that parasitoids can be dissected-out from them to determine the percentage parasitized, or so that adult parasitoids can be reared and, where possible, identified. In this chapter, the main sampling, trapping and rearing methods used for oilseed rape pests and their parasitoids are reviewed. Sampling, trapping and rearing methods for predators are reviewed in Chapter 14. Further information on general insect sampling techniques, their respective advantages and disadvantages and the factors influencing their effectiveness can be found in Muirhead-Thomson (1991), Powell *et al.* (1996), Dent & Walton (1997) and Southwood & Henderson (2000).

Sampling Adults

Pest and parasitoid adults can be sampled as they emerge from pupation in the soil (by emergence traps), as they infest the crop plant (by direct counting, sweep-netting, beating, suction sampling) or as they fly to, from or within the crop canopy (flight interception traps: water traps, sticky traps, omni-directional flights traps, malaise traps, cone traps).

Emergence traps

Emergence traps or ground photo-eclectors (see Behre, 1989) have been used to sample both overwintered and new-generation adults of the coleopterous and dipterous pests of rape and some of their hymenopterous parasitoids as they emerge from pupation in the soil:

- brassica pod midge (*Dasineura brassicae*) and its parasitoids (Williams *et al.*, 1987a, 1987b; Axelsen, 1992);
- cabbage stem flea beetle (*Psylliodes chrysocephala*) and rape winter stem weevil (*Ceutorhynchus picitarsis*) (Büchi, 1986);
- cabbage stem weevil (*Ceutorhynchus pallidactylus*) and rape stem weevil (*Ceutorhynchus napi*) and their tersilochine parasitoids (Klingenberg & Ulber, 1994; Nissen, 1997; Ulber, 2000);
- pollen beetle (*Meligethes aeneus*) and its parasitoids (Nilsson & Andreasson, 1987; Nitzsche, 1998; Büchi, 2002).

Emergence traps vary in design detail but usually comprise a box, cone or pyramid of a dark material with an open base which is placed on the soil surface; typically, the trap includes a transparent, removable container, towards which positively phototactic insects emerging from the soil will move, and in which they are collected. R. Büchi (*unpublished*) used an inverted plastic funnel (42 cm diameter; 33 cm high), the inner walls of which were covered with a black sheet. The top opening of the funnel (4 cm diameter) was surmounted by a horn-shaped glass tube which led into a plastic collecting vial (100-ml volume). The vial contained 50% ethanol, to kill the insects that gained access. The surface of the lower inner wall of the horn-shaped tube was roughened, to enable the insects coming from the funnel to climb along it into the vial. The traps used by Williams *et al.* (1987a, 1987b) comprised a wooden box (60 × 60 cm; 32 cm high) supporting a pyramidal timber frame with dark mesh walls. An inverted glass jar at the top of the pyramid (58 cm above ground) collected the insects, which were prevented from escaping by an internal, inverted plastic funnel. Büchi (1986) used a cube-shaped emergence box (50 × 50 × 20 cm) made from iron sheeting, with its top covered with a fine mesh gauze. On one side of the box, there was a hole (4 cm diameter), onto which was glued the metal lid of a glass jar. The cover also had a matching hole (4 cm diameter). A glass jar screwed onto the metal lid on the side of each trap was used to catch the insects emerging from the soil. A funnel made from copper mesh prevented the insects from flying back into the box. Each box was covered with an aluminium frame, to avoid damage to the traps from foxes or other animals; the frame was perforated with square holes (8 × 8 mm) to allow light to enter the box. For collecting adult pollen beetles, Nilsson & Andreasson (1987) used wooden emergence boxes (0.11 m^2) with a screened glass tube inserted in one side; they painted their boxes white, to avoid overheating within the box and moved them once, just after first emergence of the beetles, to minimize the impact of the box on the beetle population.

Direct sampling of plants

Some pests, such as pollen beetle (Nilsson & Andreasson, 1987; Cook *et al.*, 1999) and cabbage seed weevil (*Ceutorhynchus assimilis*), and also cabbage aphid (*Brevicoryne brassicae*) colonies, can be counted on, and sampled directly into collecting vials from, the oilseed rape plants they infest, particularly when these are at an early growth stage (green or yellow bud and early flowering), when they are most easily seen; some parasitoids can also be removed directly from the plants. Some insects, particularly Coleoptera, fall from the plants on disturbance, so care must be taken to minimize disturbance during sampling.

Sweep netting

Sweep netting has been used extensively for sampling most kinds of rape pest (Free & Williams, 1979a, 1979b; Nilsson, 1988a), and their parasitoids (Winfield, 1963; Lehmann, 1965; Miczulski, 1968; Šedivý, 1983; Nilsson & Andreasson, 1987), present in the crop canopy. Typically, at each sampling point in the crop, the collector makes ten sweeps with the net (450 mm diameter) while advancing ten paces, before emptying the contents of the net into a container (Free & Williams, 1979a, 1979b); Nilsson (1988a) used a 320-mm net, making five sweeps of it at each point. Sweep netting is most efficient when the crop canopy is dry and when there is little or no wind.

Beating onto a tray

Pollen beetles and cabbage seed weevils can more readily be counted if they are first dislodged from the plants they infest, or disturbed sufficiently to evoke their natural response of dropping from the plant into or onto a receptacle. The beating method used by Cooper & Lane (1991) involved 'beating' the main flowering racemes of plants over a white tray, on which the dislodged insects could then be counted.

Suction sampling

Suction samplers for insects are based on the principles of domestic vacuum cleaners. A mobile D-vac suction sampler (Riverside, California, USA) has been used in the autumn to sample winter rape for cabbage stem flea beetles (Alford, 1979) and a similar device (Dietrick, 1961) has been used in the spring to capture parasitoids of rape pests both at field margins and in the crop (R. Büchi, *unpublished*); the insects were caught live in a nylon gauze bag with a fine mesh. Ulber (1995) used an intensive D-vac technique in the spring to estimate densities of cabbage stem weevils and rape stem weevils. Shoots enclosed within cylinders (0.25 m^2 area) were shaken and excised at ground level, then the area was vacuumed for three minutes using the D-vac with a nozzle 20 cm in diameter. However, a limitation to the use of suction samplers is that the habitat to be sampled must be dry.

Suction traps are used to sample air-borne populations of insects. The diel periodicity of flight of brassica pod midges has been monitored in the field by using a stationary Johnson–Taylor suction trap (Johnson & Taylor, 1955 – Burkard Manufacturing Company, Rickmansworth, UK). This has a 22-cm horizontal Ventaxia fan with an inlet 1 m above the ground, leading into an inverted metal mesh cone with a cylindrical brass tube at its base; the tube houses a disc-dropping mechanism which segregates the catch temporally into 5- to 60-minute samples (I. H. Williams, *unpublished*). A similar trap has been used to determine the diel periodicity of the parasitoids of pollen beetle and cabbage stem weevil (Šedivý, 1983).

The insect pooter (see Theron, 1985) is a useful device for sucking individual pests and parasitoids from the crop plant into a container.

Water traps

Water traps have been used to sample all the coleopterous and dipterous pests of oilseed rape (Free & Williams, 1978, 1979b; Williams *et al.*, 1987a, 1987b; Axelsen, 1992; Ulber, 1995; Walczak *et al.*, 1998), turnip sawfly (*Athalia rosae*) (B. Kromp, *pers. comm.*) and some of their parasitoids, e.g. *Trichomalus perfectus* (Murchie *et al.*, 1999b), *Phradis* spp. and *Tersilochus* spp. (Nitzsche, 1998), *Omphale clypealis* and *Platygaster subuliformis* (I. H. Williams, *unpublished*). Typically, a water trap is a plastic bowl (*c.* 210 mm in diameter and 90 mm deep), containing about 1.5 litres of water to which is often added a few drops of detergent (e.g. Teepol®), to decrease surface tension so that captured insects sink, and some formaldehyde or sodium benzoate to preserve the insects until the trap is emptied; omission of the detergent reduces the catch. Yellow has long been recognized as one of the most effective colours for trapping insects on the rape crop. Trap height also affects catch. To sample insects that fly within the crop canopy, the trap is usually supported on a post at the top of the canopy, and raised as the crop grows. Axelsen (1992) placed his water traps at half crop height to capture brassica pod midge and its parasitoids. Yellow water traps installed at the top of the canopy were more effective for trapping parasitoids of pollen beetles (*Meligethes* spp.), e.g. *Tersilochus heterocerus*, whereas traps placed on the ground level were most effective for catching parasitoids of cabbage stem weevil and rape stem weevil (Nitzsche, 1998). For pests that move at or near the ground, e.g. cabbage stem flea beetles, water traps have been placed on the ground around field perimeters to monitor their immigration to crops (Walters & Lane, 1994), and D. V. Alford (*pers. comm.*) used natural puddles along farm tracks for a similar purpose; later, when cabbage stem flea beetles walk or jump rather than fly, they are best monitored using water trays dug into the soil so that the upper edges are level with the soil surface (Büchs, 1993).

The attractiveness of yellow water traps can be enhanced for some pests by baiting them with host plant volatiles (e.g. isothiocyanates and methyl salicylate) released from dispensers hung above the trap: pollen beetles and cabbage seed weevils (Free & Williams, 1978; Smart *et al.*, 1995, 1996), cabbage stem weevils and rape stem weevils (Walczak *et al.*, 1998). Isothiocyanates also attract the parasitoids *Omphale clypealis* and *Platygaster subuliformis* (Murchie *et al.*, 1997; I. H. Williams, *unpublished*). Similarly, the capture efficiency of water traps for brassica pod midges and rape stem weevils can be increased by adding an extract of rape seed with a high glucosinolate content to the trapping fluid (Erichsen & Daebeler, 1987).

Sticky traps

Traps coated with a long-lasting glue to make them sticky to flying insects that impact with them have been extensively used to monitor several different species of pest on oilseed rape. They are not often used for their parasitoids, possibly because of the difficulties of removing small fragile insects intact from their sticky surfaces for identification, although this can be done using a solvent such as white spirit.

Different designs of sticky trap have been used, varying in size, shape, colour, orientation and height in the crop; some have been baited with chemical lures. R. Büchi (*unpublished*) used yellow sticky traps (Rebell®, Swiss Federal Research Station for Fruit-Growing, Viticulture and Horticulture, Wädenswil, Switzerland), originally developed for European cherry fruit fly (*Rhagoletis cerasi*), to monitor the immigration into rape crops of rape stem weevil in the spring and of rape winter stem weevil in autumn. He placed two sticky plates (20.5 × 17.5 cm), mounted vertically on wooden posts, 30–40 cm above the ground, 2 m from the border on each of the long sides of a crop. Sticky traps (*c.* 25 × 15 cm), coated with Boltac grease, placed flat on the ground have been used to monitor cabbage stem flea beetles (Alford, 1979).

Sticky traps baited with a chemical bait and inclined at 45° have been used to monitor pollen beetles and cabbage seed weevils (Smart *et al.*, 1996). The angled sticky-card trap (20 × 10 cm) consisted of an oblong piece of grey plastic sheet [acrylonitrile butadiene-styrene (ABS), 0.3 cm thick], glued at an angle of 45° to a 10-cm length of grey plastic (ABS) tubing. The tubing fitted over the end of a rigid pole so that the trap was fixed in position at, or just above, crop height. A yellow sticky card, coated with Oecotak A5 (Oecos Limited, Kimpton, Hertfordshire, UK), was clipped to the upper side of the angled face. The chemical bait, the plant volatile 2-phenylethyl isothiocyanate, was released at *c.* 2.5 mg/day from a sealed polythene bag, suspended just above the trap from a wire loop.

Transparent cylindrical sticky traps have been used to sample brassica pod midges within rape canopies (Williams *et al.*, 1987a, 1987b). Here, each trap consisted of three cylinders of clear acetate sheet, held vertically, one above the other, around an aluminium pole (1.5 m high), inserted into the ground. An area (20 × 20 cm) of the outer surface of each cylinder was coated with adhesive (again, Oecotak A5). The sticky surface of the lowest cylinder was 10–30 cm above ground level, that of the middle one 50–70 cm above ground level and that of the uppermost one 90–110 cm above ground level.

Cardboard Delta traps (Oecos Limited, Kimpton, Hertfordshire, UK) with sticky base plates, some baited with virgin females or extracts of them, have been used to monitor male brassica pod midges in rape crops (Williams, 1989, 1990). Each trap had a sticky plate (19.0 × 7.5 cm) held horizontally 10 cm above the ground, beneath a canopy that was triangular in cross section. Those baited with virgin females contained one to three females; these were housed in a cylindrical glass tube (5 cm long × 3 cm diameter) with muslin ends, suspended below the canopy ridge. The glass tube contained drinking water in a small glass vial plugged with cotton wool. Traps baited with hexane washes of brassica pod midge had test solutions on polyethylene blocks (10 mm long) suspended below the canopy ridge of the trap.

Flight traps

Flying Coleoptera and other insects that fall on hitting an obstacle during flight have been sampled on the rape crop using window traps. A typical window trap is a large sheet of transparent glass or Plexiglass held vertically, with a trough (containing water with a wetting agent and a little preservative) below it, into which the insects fall.

Omni-directional flight traps have been used to sample brassica pod midge and its parasitoids *Omphale clypealis* and *Platygaster subuliformis* (Murchie *et al.*, 1997). Each trap has three main parts: the fins, the funnel unit and the collecting jar. The fins consist of four sheets of polyvinyl chloride (PVC) hot-air welded onto PVC tubing and painted yellow or grey. The funnel unit was attached to the fins by two self-tapping screws. The collecting jar was partially filled with 70% methanol and connected to the funnel unit by screwing the universal tube into its cap. The trap was mounted on steel tubing attached to a wooden post, and its height adjusted (to keep the top of the fins level with the top of the crop canopy) by sliding it up or down the tubing to which it was secured by tightening a nylon screw. Some traps were baited with plant volatiles (allyl isothiocyanate or 2-phenylethyl isothiocyanate), dispensed from a cellulose sponge sealed in a polyethylene bag and mounted inside the funnel unit slightly above the fins. In the omnidirectional flight traps, the insects move upwards on impacting the fins of the trap.

Malaise traps

A malaise trap is a large net-like tent, open on one or two sides, with a collecting vessel containing alcohol at its highest point, into which the insects climb. Nitzsche (1998) used malaise traps for monitoring the activity and phenology of various parasitoids, both at overwintering sites and in oilseed rape crops. The traps proved to be very efficient, particularly for trapping adults of *Phradis* spp., *Tersilochus fulvipes* and *T. obscurator.* To sample parasitoids alive, the collecting vessel can be replaced by a large net (bag) containing some rape inflorescences. Murchie *et al.* (2001) developed a mechanism for time-sorting the insect catches into two-hourly samples; it used a 24-h rotation quartz clock motor powered by a single 1.5-volt battery to rotate a funnel directing positively phototactic insects from the malaise trap into consecutive sample tubes; they used a malaise trap fitted with this mechanism to determine the diel periodicity of flights of brassica pod midges.

Cone traps

Modified boll weevil traps (Glinwood *et al.*, 1993) have been used to monitor adult cabbage stem weevils and rape stem weevils in winter rape (Walczak *et al.*, 1998). Each trap consisted of a cylindrical fluorescent-green plastic base unit, with a wire mesh cone and a clear collecting unit fitted above it (overall dimensions 220 × 110 mm diameter). Four semi-circular holes, cut in the base of each cone, allowed the weevils to walk up inside. A semiochemical lure, consisting of a polythene bag dispenser that released 2-phenylethyl isothiocyanate at *c.* 5 mg/day was suspended inside each trap. However, the cone trap was not as effective as a yellow water trap (235 mm diameter) or a yellow sticky trap (100 × 200 cm, Oecos Limited, Kimpton, Hertfordshire, UK) in catching these two pests.

Sampling Eggs and Larvae

Plant dissection

Eggs and larvae of stem-infesting pests can be sampled by taking rape plants from the field and dissecting the relevant part of the plants in the laboratory: petioles, stem bases and root crowns for cabbage stem flea beetle and rape winter stem weevil; stems for rape stem weevil; and petioles and stems for cabbage stem weevil (Büchi & Keller, 1994; Ulber, 1995; Nitzsche, 1998). Pollen beetle larvae can be collected directly from buds and flowers of rape plants (Büchi & Keller, 1994; Cook *et al.*, 1999). The timing of sampling is crucial; to assess the impact of parasitism completely, full-grown host larvae should be collected from plant samples just before the start of their migration to the soil.

Cabbage seed weevil larvae and their ectoparasitoids can be collected from the field by opening maturing pods of rape (at BBCH growth-stage code 81 – Lancashire *et al.*, 1991: '10% of pods ripe, seeds black and hard'; equivalent to GS 6.4 of Sylvester-Bradley & Makepeace, 1984: 'most seeds green–brown mottled'), finding those infested by cabbage seed weevil larvae and examining the host larvae for the attached eggs or larvae of the ectoparasitoids. Ectoparasitized larvae can also be obtained by allowing rape plants in the greenhouse to become infested by host larvae and then introducing the adult ectoparasitoid to parasitize them. Murchie (1996) enclosed field-collected cabbage seed weevil adults on greenhouse-grown rape plants (between BBCH growth-stage codes 71 and 73), using muslin bags (1.5 × 0.5 m), for 48 h at 18°C. Three weeks later, he collected the pods from the plants. Groups of 25 pods were presented to ten field-collected female *Trichomalus perfectus* in a 140-mm-diameter Petri dish. After 24 h, the pods were opened, the host larvae with parasitoid eggs attached were removed and reared to adults as described below.

Plastic funnels below the crop canopy

Plastic or polyethylene funnels have been placed below the flowering rape crop canopy to collect full-grown pollen beetle larvae (Büchi, 2002; R. Büchi, *unpublished*) and brassica pod midge larvae (Axelsen, 1992) as they drop to the ground to pupate. Those used for pollen beetle larvae were 31 cm in diameter and those for brassica pod midge larvae 18.8 cm in diameter. R. Büchi (*unpublished*) placed each funnel in a plastic pot (500 ml) which had a flexible opening to grasp the funnel. The pot contained 50 ml of 50% ethanol. Small drainage holes (1 mm diameter) in the upper part of the pot allowed rain water to drain away. Full-grown larvae dropping from flowers into the funnel accumulated in the pot, to be killed and preserved by the ethanol. Collected larvae could be deep frozen, and afterwards dissected under the microscope for the presence of endoparasitoids. To estimate the density of pollen beetle larval populations in a rape crop, plants were bent together above the funnels and the number of flowering racemes above each counted.

Trays below the crop canopy

Trays placed on the soil below the crop canopy have been used to sample full-grown larvae of pollen beetle, cabbage seed weevil and brassica pod midge as they drop from the canopy to pupate in the soil (Williams *et al.*, 1987a, 1987b; Nilsson & Andreasson, 1987; Nilsson, 1988b; Klingenberg & Ulber, 1994; Warner *et al.*, 2000). To trap larvae of rape stem weevil and cabbage stem weevil which migrate from stems, the trays must be located as close to the stem base as possible (Nitzsche & Ulber, 2002). The trays may be left empty to collect live larvae, can be part-filled with water or a weak solution of detergent to drown the larvae, or can be part-filled with soil to allow the larvae to pupate for future emergence. A similar system can be used in the laboratory or insectary, using cut plants; phenology of brassica pod midge larval emergence from pods was determined by cutting infested rape plants from the field, placing them with their stems in water and their pods over an empty tray in an outdoor insectary. Daily counts were made of the larvae that emerged from the pods and fell into the tray (I. H. Williams, *unpublished*). Trays containing soil, placed below the canopy in the field or below infested field-collected winter rape plants cut from the field, can be used to collect mature brassica pod midge larvae and allow them to pupate in the soil for later recovery of cocoons (Isidoro *et al.*, 1992).

Brassica pod midge larvae can be obtained by allowing them to emerge on 'maturity' from infested rape pods into a container; infested pods can be identified by their swollen and yellowed appearance. Murchie (1996), in the UK, collected infested rape plants from winter rape crops during June and from spring rape crops during August. He bundled them into groups of 20 or 40 and placed each bundle inside a large black polythene bag. These he kept in an outdoor insectary. After 48 h, brassica pod midge larvae that had emerged from the pods were retrieved from the bags. R. Büchi (*unpublished*) has used a similar method in Switzerland. In winter rape crops, he collected pods infested with brassica pod midge larvae from mid-May to mid-June and placed them into plastic boxes (9.7 × 200 × 8 cm) lined with a layer of moist peat. The boxes were held in a climate chamber at 25°C and 70% RH. After two weeks, the midge larvae had emerged from the pods and burrowed into the peat.

Sampling Cocoons

Soil cores

Soil cores (5 cm diameter, 10 cm deep) have been used to collect cocoons of brassica pod midge, *Tersilochus* spp. and *Phradis* spp. from the soil beneath the rape crop canopy and/or from overwintering sites (Williams *et al.*, 1987a, 1987b; Nissen, 1997; Nitzsche, 1998), using a steel soil corer. This is pushed into the ground vertically and then extracted together with the soil containing the cocoons. The cocoons are then recovered by water-washing the soil through fine sieves. Live cocoons as well as vegetative matter usually float, whereas mineral matter usually sinks. The Ladell system for soil washing and flotation (Ladell, 1936), using magnesium sulphate, has been used successfully to extract brassica pod midge cocoons from soil (I. H. Williams, *unpublished*). The cocoon contents can be determined after opening under a microscope or following emergence of the adult.

Rearing Pests and Parasitoids

The continuous rearing of parasitoids necessitates the continuous culture of suitable hosts; there have been few attempts to develop such culturing methods, except for a few of the pests and their parasitoids, e.g. cabbage aphid. The potential for the use of alternative hosts or artificial diets for the parasitoids of rape pests has not been explored. However, methods have been developed to rear some of the pests and their parasitoids through various parts of their lifecycles and these are outlined below.

Rearing pests

Many studies of the pests of oilseed rape have involved their collection from the field and their subsequent maintenance in the laboratory until the end of their adult life or until the end of the season, rather than their culture through several generations. Many of the pests are univoltine, with a prolonged period of winter or summer diapause which is obligatory before reproduction; the provision of conditions for diapause that result in the emergence of a high percentage of good-quality, behaviourally normal insects appears to be a factor currently limiting the continuous culture of some pests, e.g. cabbage seed weevil (I. H. Williams, *unpublished*).

Overwintered pollen beetle adults, collected from rape crops in the spring, can be kept alive in the laboratory for several months if enclosed on flowering racemes of oilseed rape at 18–25°C; also, the eggs they lay in the buds can be reared to adults (Bromand, 1983; I. H. Williams, *unpublished*). Bromand (1983) dissected the buds, removed the eggs and placed them on dry paper within a Petri dish; this contained moist filter paper to provide a humid atmosphere. Following egg hatch, he transferred the neonate larvae to the buds of cut racemes of rape and, after a further three to four days, to fresh rape flowers. When the larvae were fully grown and left the flowers, he placed them in glass vials of sterile soil for subsequent pupation and adult emergence. Full-grown larvae will also pupate and develop into adults if kept in Petri dishes lined with moist filter paper (I. H. Williams, *unpublished*). Bromand (1983) found that overwintered females, field-collected in June and kept at 20°C, 75% RH and 16h light, continued to lay fertile eggs until the end of October, whereas new generation females, collected in mid-September, produced only sterile eggs after 31 days, presumably because they had not mated. By contrast, new-generation females that emerged in the laboratory from field-collected larvae did lay fertile eggs, albeit few (17 per female) and not until 66–95 days after emergence; of these, 90% hatched to larvae, but only 6% of the larvae produced adults. Pollen beetle adults that had been overwintered in a refrigerator started egg-laying after 15 days, a maturation time similar to that observed under field conditions. Continuous rearing of pollen beetle in the laboratory, thus, appears feasible if rearing conditions can be optimized.

Overwintered cabbage seed weevil adults, collected from rape crops in the spring, can (in common with pollen beetles) be kept alive in the laboratory for several months, enclosed on flowering oilseed rape racemes at 18–25°C; they will also lay eggs readily in young pods (I. H. Williams, *unpublished*). When maintained on intact rape plants over trays of moist soil, the larvae will 'mature' in the pods, drop to the soil for pupation and subsequently give rise to new-generation adults. After two to three weeks feeding on rape, these adults can be induced to enter diapause by placing them at 5°C. After four months at this temperature, their return to 18–25°C will result in adult emergence (I. H. Williams, *unpublished*).

Brassica pod midges that emerged from soil containing their cocoons have been reared continuously for three to five generations, kept at 23°C, on flowering oilseed rape plants with young pods, into which holes have been punctured with a pin to provide oviposition sites; the plants were placed over a tray of soil for pupation of full-grown larvae that dropped from the plants. The adult population of the culture gradually declined as increasing numbers of larvae entered diapause (I. H. Williams, *unpublished*).

Cabbage stem flea beetle adults, collected from the field in autumn, have been maintained in the laboratory until late spring, at 10°C, in ventilated plastic boxes lined with moist filter paper and supplied with fresh leaves of oilseed rape. The females laid eggs in the boxes; these were either removed and stored at 5°C for several months until required, or kept at 10°C until they hatched; the neonate larvae were then transferred to the stems of rape plants, which they entered to begin feeding. On becoming fully grown, the larvae vacated the plants to pupate in soil; eventually, adult beetles emerged (I. H. Williams, *unpublished*).

Larvae of cabbage stem weevil and rape stem weevil, field-collected from infested rape stems, have been reared in stem sections at 20–23°C in the laboratory (B. Ulber, *unpublished*). On reaching 'maturity', the larvae migrated from the stem sections to pupate in sieved, moistened field soil. New-generation weevils of cabbage stem weevil, which emerged after 14 days, were allowed to feed on rape leaves for three weeks before being cooled to 10°C for two weeks and kept at 2°C for 12 weeks for diapause. They were then brought back to the laboratory and, after two to three weeks of feeding on rape leaves and maturation of ovaries, egg-laying began. It seems feasible that this method could be used to establish a continuous rearing system, producing three to four generations per year. Adults of rape stem weevil do not emerge from their earthen pupal cell before diapause; for this they are exposed to cold conditions within the soil.

Aphids are easily reared in their parthenogenetic phase, given an appropriate host plant, suitable cage, environmental conditions that inhibit production of sexual morphs and a management procedure that limits their numbers; general methods are reviewed by Blackman (1988). Cabbage aphid can be reared continuously on oilseed rape plants (I. H. Williams, *unpublished*) and on fresh cabbage leaves (Dimetry & Marei, 1992) and peach/potato aphid (*Myzus persicae*) can be reared on bean plants (Blackman, 1974) and on potato (Blackman, 1988).

Rearing parasitoids

The parasitoids that attack the pests of oilseed rape can, mostly (except for a few larval endoparasitoids), be identified to species only by examining the adult stage (and often then only if female). To determine the extent of parasitization of the various life stages of the host it is often necessary, therefore, to rear-out the parasitoids from the egg, larval, pupal or adult stages of the host. Rearing methods are now well established in some European research laboratories, but few of them have been published. These methods are reviewed below. More general advice on rearing parasitic Hymenoptera can be found in Shaw (1997).

Egg and larval endoparasitoids

The eggs and larvae of rape pests, obtained by the sampling and trapping methods outlined above, need to be kept in suitable environmental conditions to allow any endoparasitoids within them to develop further and emerge as adults. Rearing success is generally greater from host larvae that have completed their development and are ready to pupate; hence, collection of full-grown larvae as they drop from the plants to pupate in the soil is preferable to their removal, at uncertain age, from the parts of rape plants they infest. With the latter, rearing of larvae in their final instar usually ensures better survival than does removal of younger ones. Nilsson (1988b) compared emergence of pollen beetle adults from larvae of different ages. He placed groups of first-, second- and third-instar larvae in pots of moist soil. The pots were covered with black plastic into which a glass tube, with its open end covered with netting, was inserted. Pots were kept at 17°C or at 20–25°C. No pollen beetle adults emerged from first-instar larvae nor from larvae of any age kept at 20–25°C. At 17°C, 35% of early-second-instar and 72% of late-second-instar larvae produced adults. A few larval endoparasitoids can be identified to species within the body of the larva, because they are pigmented and visible through the larval cuticle

– e.g. the egg of *Tersilochus heterocerus* has a black chorion and the first-instar larva of *Diospilus capito* has a black head capsule, visible through the cuticle of their host pollen beetle larvae.

Büchi & Keller (1994) reared endoparasitoids from the larvae of brassica pod midges, cabbage stem weevils, pollen beetles, rape stem weevils and rape winter stem weevils, using the emergence containers shown in Fig. 10.1. These were modified screw-capped, black-plastic bottles (12.5 cm high × 6.7 cm diameter) containing damp peat to a depth of 5 cm. The bottles were upturned into a tray part-filled with water (depth 1 cm). There was a gauze-covered hole in the screw cap of each bottle, allowing water to enter and moisten the retained peat. Each bottle was surmounted by a plastic collecting tube, with a lid to allow removal of emerged insects; the tube contained an inverted funnel at its base to prevent the return of emerged insects into the bottle. Host larvae introduced onto the surface of the peat burrowed down in order to pupate. The bottles were kept at 25°C and 70% RH for three to four weeks to allow for emergence of unparasitized adult pests and, later, adult parasitoids. To obtain further emergence of adults of any pests and parasitoids that had entered diapause, the bottles were then kept at 4°C for four months before being returned once more to 25°C.

Williams & Martin (1986) reared brassica pod midges from cocoons washed from soil in the autumn. They placed groups of 50 cocoons in glass vials on moist sand (20 mm deep) and covered them with a thin (2-mm) layer of moist sand. Diapause was maintained at 5°C for 150 days and the vials then kept at 23°C on a light/dark (LD) 18:6 h cycle until emergence of adults 11–31 days later; many endoparasitoids also emerged from these cocoons (I. H. Williams, *unpublished*). Murchie (1996) and Murchie *et al.* (1999a) reared endoparasitoids from brassica pod midge larvae using similar methods; they placed groups of 100–400 full-grown larvae onto moist silver sand, either in 90-mm triple-vented Petri dishes, or in glass vials (72 mm high × 25 mm diameter) stoppered by cotton wool; sand depth in dishes was 5 mm, that in vials was 25 mm. The cultures were

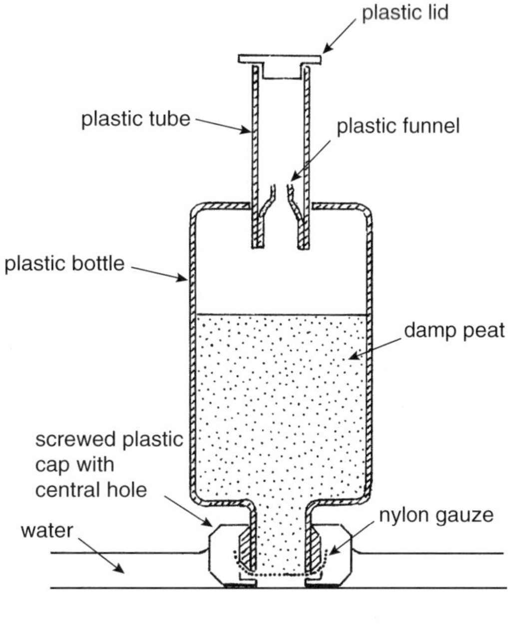

Fig. 10.1 Plastic bottle containing 5 cm of damp peat, to rear adults of oilseed rape pests and endoparasitoids from fully grown host larvae.

kept at 18°C until emergence of adult midges and endoparasitoids was complete. To rear-out diapausing parasitoids and hosts from the sand, the containers were placed from mid-September until late January (140 days) at 5°C for overwintering and then returned to 18°C for emergence.

In summary, pests of oilseed rape and their endoparasitoids have emerged from host larvae, both pre-diapause and post-diapause, when kept on a moist substrate (sand or peat) at 17–25°C; for diapause 120–150 days at 4–5°C appears to be sufficient to allow emergence of insects on their return to 17–25°C.

Larval ectoparasitoids

Ectoparasitoids of cabbage seed weevil larvae, whether field-collected or laboratory cultured (as outlined above), can be successfully reared to the adult stage if provided with suitable conditions of temperature and humidity. Since the parasitized host larva does not feed, it can be removed from the pod it infests and the attached ectoparasitoid egg or larva reared to adulthood in the absence of plant material. This greatly reduces the chances of mould developing in cultures.

Murchie & Williams (1998) reared *Trichomalus perfectus* eggs and larvae to the adult stage by removing them from the pods, still attached to their host larvae, and placing them in a vented Petri dish (50 mm diameter) kept at 18°C and lined with moist filter paper. The filter paper was kept moist during the development of the parasitoid eggs and larvae until the pupal stage was reached, when it was allowed to dry. At 18°C, the development times for the ectoparasitoid *T. perfectus* were: egg, one to three days; larvae, 7–10 days; and pupa, 10–15 days. R. Büchi (*unpublished*) reared ectoparasitized larvae of cabbage seed weevil at 25°C. Walz (1956) used a smilar method but transferred pupae to individual vials capped with cotton wool. B. Ulber (*unpublished*) reared the larval ectoparasitoids of cabbage seed weevil to adulthood inside gelatin capsules normally used to contain medicines (Capsulae Operculatae, volume 0.5 ml; G. Pohl-Boskamp GmbH & Co., Hohenlockenstedt, Germany) and kept them at room temperature. After emergence, he stored the parasitoids within the capsule at –20°C for later identification.

Aphid endoparasitoids

The endoparasitoids of cabbage aphid are readily reared from the host. R. Büchi (*unpublished*) collected oilseed rape plants infested with cabbage aphid colonies in May. After removing any pods infested with brassica pod midge or cabbage seed weevil larvae, he placed their stems in water and enclosed the flowers and pods in a bag of fine mesh gauze. He kept the cultures at 25°C and 70% RH and, within one to two weeks, aphid mummies were formed; subsequently, adult parasitoids emerged. If necessary, adult parasitoids can be reared from individual mummies which have been transferred to gelatin capsules.

Acknowledgements

Rothamsted Research receives grant-in-aid from the Biotechnology and Biological Sciences Research Council and this work was part-funded by the UK Ministry of Agriculture, Fisheries and Food and the European Commission Concerted Action: *Minimizing pesticide use and environmental impact by the development and promotion of bio-control strategies for oilseed rape pests, FAIR 96-1314.*

References

Alford, D. V. (1979). Observations on the cabbage stem flea beetle, *Psylliodes chrysocephala*, on winter oil-seed rape in Cambridgeshire. *Annals of Applied Biology* **93**, 117–23.

Axelsen, J. (1992). The population dynamics and mortalities of the pod gall midge (*Dasyneura brassicae* Winn.) (Dipt., Cecidomyiidae) in winter rape and spring rape (*Brassica napus* L.) in Denmark. *Journal of Applied Entomology* **114**, 463–71.

Behre, G. F. (1989). Freilandökologische Methoden zur Erfassung der Entomofauna (Weiter- und Neuentwicklung von Geräten). *Jahresbericht des Naturwissenschaftlichen Vereins Wuppertal* **42**, 238–42.

Blackman, R. (1974). *Aphids.* Ginn & Co.: London.

Blackman, R. L. (1988). Rearing and handling aphids. In: A. K. Minks & P. Harrewijn (eds). *Aphids, Their Biology, Natural Enemies and Control. Vol. B.* Elsevier: Oxford.

Bromand, B. (1983). Possibility of continuous rearing of *Meligethes aeneus* Fabr. (Col.). *Journal of Applied Entomology* **96**, 419–22.

Büchi, R. (1986). Biologie und Bekämpfung des Schwarzen Triebrüsslers, *Ceutorhynchus picitarsis* Gyll. (Col. Curculionidae). *Anzeiger für Schädlingskunde, Pflanzenschutz und Umweltschutz* **59**, 51–6.

Büchi, R. (2002). Mortality of pollen beetle (*Meligethes* spp.) larvae due to parasitoids and predators in rape fields and the effect of conservation strips. *Agriculture, Ecosystems and Environment* **90**, 255–68.

Büchi, R. & Keller, S. (1994). Parasitierung der Kohlschotengallmücke durch Nützlinge. *Agrarforschung* **9**, 400–2.

Büchs, W. (1993). Investigations on the occurrence of pest insects in oil seed rape as a basis for the development of action thresholds, concepts for prognosis and strategies for the reduction of the input of insecticides. *IOBC/wprs Bulletin* **16**, 216–34.

Cook, S., Murray, D. A. & Williams, I. H. (1999). Pollen beetle, *Meligethes aeneus* Fabricius, incidence in the composite hybrid winter oilseed rape, Synergy. *Proceedings 10th International Rapeseed Congress, Canberra, Australia, 26–29 September 1999.*

Cooper, D. A. & Lane, A. (1991). Monitoring of oilseed rape pests in England and Wales. *IOBC/wprs Bulletin* **14**, 5–13.

Dent, D. R. & Walton, M. P. (1997). *Methods in Ecological and Agricultural Entomology.* CAB International: Wallingford.

Dietrick, E. J. (1961). An improved backpack motor fan for suction sampling of insect populations. *Journal of Economic Entomology* **54**, 394–5.

Dimetry, N. Z. & Marei, S. S. (1992). Laboratory evaluation of some pesticides on the cabbage aphid, *Brevicoryne brassicae* L. and their side effects on some important natural enemies. *Anzeiger für Schädlingskunde, Pflanzenschutz und Umweltschutz* **65**, 16–9.

Erichsen, E. & Daebeler, F. (1987). Zur Überwachung der Kohlschotenmücke (*Dasyneura brassicae* Winn.) im Winterraps. *Nachrichtenblatt für den Pflanzenschutzdienst der DDR* **41**, 33–4.

Free, J. B. & Williams, I. H. (1978). The responses of the pollen beetle *Meligethes aeneus* and the seed weevil *Ceuthorhynchus assimilis* to oil-seed rape *Brassica napus*, and other plants. *Journal of Applied Ecology* **15**, 761–74.

Free, J. B. & Williams, I. H. (1979a). The distribution of insect pests on crops of oilseed rape (*Brassica napus* L.) and the damage they cause. *Journal of Agricultural Science, Cambridge* **93**, 139–49.

Free, J. B. & Williams, I. H. (1979b). The infestation of oilseed rape (*Brassica napus* L.) by insect pests. *Journal of Agricultural Science, Cambridge* **92**, 203–18.

Glinwood, R. T., Blight, M. M. & Smart, L. E. (1993). Development of a monitoring trap for the pea and bean weevil, using the aggregation pheromone. *IOBC/wprs Bulletin* **16**, 347–50.

Isidoro, N., Williams, I. H., Salinas, M. & Martin, A. (1992). Mating behaviour and identification of the female sex pheromone gland in the brassica pod midge (*Dasineura brassicae* Winn.; *Cecidomyidae*, Diptera). *Bolletino Instituto Entomologia "G. Grandi" Universitas Bologna* **47**, 27–48.

Johnson, C. G. & Taylor, L. R. (1955). The development of large suction traps for airborne insects. *Annals of Applied Biology* **43**, 51–61.

Klingenberg, A. & Ulber, B. (1994). Untersuchungen zum Auftreten der Tersilochinae (Hym., Ichneumonidae) als Larvalparasitoide einiger Rapsschädlinge im Raum Göttingen 1990 und 1991 und zu deren Schlupfabundanz nach unterschiedlicher Bodenbearbeitung. *Journal of Applied Entomology* **117**, 287–99.

Ladell, W. R. S. (1936). A new apparatus for separating insects and other arthropods from the soil. *Annals of Applied Biology* **23**, 862–79.

Lancashire, P. D., Bleiholder, H., Boom, T. van den, Langelüddeke, P., Strauss, R., Weber, E. & Witzenberger, A. (1991). A uniform decimal code for growth stages of crops and weeds. *Annals of Applied Biology* **119**, 561–601.

Lehmann, W. (1965). Einfluss chemischer Bekämpfungsmassnahmen auf einige Rapsschädlinge und ihre Parasiten. II. Knospen- und Stengelschädlinge. *Archiv für Pflanzenschutz* **1**, 209–19.

Miczulski, B. (1968). Blonkowski (Hymenoptera) w biocenozie upraw rzepaku. Czesc VI. Bleskotkowate (Chalcidoidea). *Polskie Pismo Entomologiczne* **38** (2), 341–83.

Muirhead-Thomson, R. C. (1991). *Trap Responses of Flying Insects*. Academic Press: London.

Murchie, A. K. (1996). *Parasitoids of Cabbage Seed Weevil and Brassica Pod Midge in Oilseed Rape*. Unpublished PhD thesis, University of Keele, UK.

Murchie A. K. & Williams, I. H. (1998). Effect of host size on the sex of *Trichomalus perfectus* (Walker) (Hymenoptera: Pteromalidae). *IOBC/wprs Bulletin* **21**, 177–81.

Murchie, A. K., Smart, L. E. & Williams, I. H. (1997). Responses of *Dasineura brassicae* Winn. (Diptera: Cecidomyiidae) and its parasitoids *Platygaster subuliformis* Kieffer (Hymenoptera: Platygastridae) and *Omphale clypealis* Thomson (Hymenoptera: Eulophidae) to traps, baited with organic isothiocyanates, in the field. *Journal of Chemical Ecology* **23**, 917–26.

Murchie, A. K., Polaszek, A. & Williams, I. H. (1999a). *Platygaster subuliformis* (Kieffer) (Hym., Platygastridae) new to Britain, an egg-larval parasitoid of the brassica pod midge *Dasineura brassicae* Winnertz (Dipt., Cecidomyiidae). *Entomologist's Monthly Magazine* **135**, 217–22.

Murchie, A. K., Williams, I. H. & Perry, J. (1999b). Edge distributions of *Ceutorhynchus assimilis* Paykull and its parasitoid *Trichomalus perfectus* (Walker) in a crop of winter oilseed rape (*Brassica napus* L.). *BioControl* **44**, 379–90.

Murchie, A. K., Burn, D. J., Kirk, W. D. J. & Williams, I. H. (2001). A novel mechanism for time-sorting insect catches, and its use to derive the diel flight periodicity of brassica pod midge *Dasineura brassicae* (Diptera: Cecidomyiidae). *Bulletin of Entomological Research* **91**, 199–203.

Nilsson, C. (1988a). The pollen beetle (*Meligethes aeneus* F.) in winter and spring rape at Alnarp 1976–1978. I. Migration and sex ratio. *Växtskyddnotiser* **52**, 134–8.

Nilsson, C (1988b). The pollen beetle (*Meligethes aeneus* F.) in winter and spring rape at Alnarp 1976–1978. III. Mortality factors. *Växtskyddnotiser* **52**, 145–50.

Nilsson, C. & Andreasson, B. (1987). Parasitoids and predators attacking pollen beetles (*Meligethes aeneus* F.) in spring and winter rape in southern Sweden. *IOBC/wprs Bulletin* **10**, 64–73.

Nissen, U. (1997). *Ökologische Studien zum Auftreten von Schadinsekten und ihren Parasitoiden an Winterraps norddeutscher Anbaugebiete*. PhD thesis, Christian-Albrechts-University of Kiel, Germany.

Nitzsche, O. (1998). *Auftreten und Effizienz von Parasitoiden als natürliche Gegenspieler von Schadinsekten im Winterraps unter besonderer Berücksichtigung unterschiedlicher Bodenbearbeitungsmaßnahmen nach Winterraps*. PhD thesis, Georg-August-University of Göttingen, Germany.

Nitzsche, O. & Ulber, B. (2002). Studies on parasitism of rape stem weevil, *Ceutorhynchus napi* Gyll. (Col.: Curculionidae), by *Tersilochus fulvipes* Grav. (Hym.: Ichneumonidae) in oilseed rape. (Submitted.)

Powell, W., Walton, M. P. & Jervis, M. A. (1996). *Populations and communities*. In: M. A. Jervis & N. A. C. Kidd (eds). *Insect Natural Enemies.* Chapman & Hall: London.

Šedivý, J. (1983). Tersilochinae as parasitoids of insect pests of winter rape (Hymenoptera: Ichneumonidae). *Contributions of the American Entomological Institute* **20**, 266–76.

Shaw, M. R. (1997). Rearing parasitic Hymenoptera. *The Amateur Entomologist* **25**, 1–46.

Smart, L. E., Blight, M. M. & Ryan, J. (1995). Response of pollen beetles, *Meligethes* spp., to volatiles from *Brassica napus. Proceedings 9th International Rapeseed Congress, Cambridge, UK, 4–7th July 1995*, pp. 1040–2.

Smart, L. E., Blight, M. M. & Lane, A. (1996). Development of a monitoring trap for spring and summer pests of oilseed rape. *Brighton Crop Protection Conference – Pests and Diseases* **1**, 167–72.

Southwood, T. R. E. & Henderson, P. A. (2000). *Ecological Methods.* 3rd edn. Blackwell Science: London.

Sylvester-Bradley, R. & Makepeace, L. J. (1984). A code for stages in development in oilseed rape (*Brassica napus* L.). *Aspects of Applied Biology* **6**, 399–419.

Theron, J. G. (1985). An improved pooter (aspirator). *Antenna* **9** (4), 163.

Ulber, B. (1995). Untersuchungen zur Abundanzdynamik von *Ceutorhynchus napi* Gyll. und *Ceutorhynchus pallidactylus* (Mrsh.) im Raum Göttingen. *Rapssymposium zu Fragen der Phytopathologie und des Pflanzenschutzes, Rostock, 13–14 October 1994,* pp. 95–8.

Ulber, B. (2000). Bibliography of parasitoid species and levels of parasitism of rape stem weevil *Ceutorhynchus napi* Gyll. and cabbage stem weevil *Ceutorhynchus pallidactylus* (Mrsh.). *IOBC/wprs Bulletin* **23** (6), 131–4.

Walczak, B., Kelm, M., Klukowski, Z., Smart, L. E., Ferguson, A. W. & Williams, I. H. (1998). Effect of trap design and 2-phenylethyl isothiocyanate on catches of stem weevils (*Ceutorhynchus pallidactylus* Marsh. and *C. napi* Gyll.) in winter oilseed rape. *IOBC/wprs Bulletin* **21**, 141–6.

Walters, K. F. A. & Lane, A. (1994). Sampling procedures for pests of winter oilseed rape: meeting the needs of the crop consultant. *Aspects of Applied Biology* **37**, 83–94.

Walz, A. J. (1956). Observations on the biologies of some hymenopterous parasites of the cabbage seedpod weevil in northern Idaho. *Annals of the Entomological Society of America* **50**, 219–20.

Warner, D. J., Allen-Williams, L. J., Ferguson, A. W. & Williams, I. H. (2000). Pest–predator spatial relationships in winter rape: implications for ICM. *Pest Management Science* **56**, 977–82.

Williams, I. H. (1989). Pheromone monitoring of brassica pod midge (*Dasineura brassicae* Winn.). *Proceedings of the 7th International Rapeseed Congress, Poznan, Poland, 11–14 May 1987*, pp. 1074–6.

Williams, I. H. (1990). Monitoring *Dasineura brassicae* by means of pheromone traps. *IOBC/wprs Bulletin* **13**, 40–5.

Williams, I. H. & Martin, A. P. (1986). Evidence for a female sex pheromone in the brassica pod midge *Dasineura brassicae. Physiological Entomology* **11**, 353–6.

Williams, I. H., Martin, A. P. & Kelm, M. (1987a). The phenology of the emergence of brassica pod midge (*Dasineura brassicae* Winn.) and its infestation of winter oilseed rape (*Brassica napus* L.). *Journal of Agricultural Science, Cambridge* **108**, 579–89.
Williams, I. H., Martin, A. P. & Kelm, M. (1987b). The phenology of emergence of brassica pod midge (*Dasineura brassicae* Winn.) and its infestation of spring oilseed rape (*Brassica napus* L.). *Journal of Agricultural Science, Cambridge* **109**, 309–14.
Winfield, A. L. (1963). A study of the effects of insecticides on parasites of larvae of blossom beetles (*Meligethes aeneus* F., Coleoptera: Nitidulidae). *Entomologia Experientalis et Applicata* **6**, 309–18.

11

Identification of Hymenopterous Parasitoids Associated with Oilseed Rape Pests

Stefan Vidal
Georg-August Universität, Institut für Pflanzenpathologie und Pflanzenschutz, Grisebachstraße 6, D-37077 Göttingen, Germany

Introduction

As outlined in the earlier chapters in this monograph, insect pests of oilseed rape crops are attacked by a wide range of parasitoids, the vast majority of which belong to the order Hymenoptera. This chapter considers hymenopterous ecto- and endoparasitoids associated with the main pests on European oilseed rape crops, namely: brassica pod midge (*Dasineura brassicae*) (see Chapter 7), cabbage seed weevil (*Ceutorhynchus assimilis*) (see Chapter 6), cabbage stem flea beetle (*Psylliodes chrysocephala*) (see Chapter 8), ceutorhynchid stem weevils (*Ceutorhynchus napi*, *C. pallidactylus* and *C. picitarsis*) (see Chapter 5) and pollen beetles (*Meligethes* spp.) (see Chapter 4).

Understanding of the population dynamics of oilseed rape pests is hampered by the insufficient knowledge of the complexes of natural enemies that may play an important role in the regulation of these herbivorous insects (Thies & Tscharntke, 1999). Although research on parasitoids of herbivorous oilseed rape pests has a long tradition in Europe (Blunck, 1921; Speyer, 1925a, 1925b), and several papers have focused especially on this topic (e.g. Jourdheuil, 1960), our knowledge on the parasitoid species complexes is far from complete. Several factors contribute to these shortcomings. On the one hand, specific life stages of the herbivorous hosts can be sampled in large numbers. Thus, herbivores infesting stems or pods will be easily detected and sampled, along with their accompanying parasitoids. On the other hand, however, overwintering pest pupae in the soil, which could also harbour site-specific or life-stage-specific parasitoid species, are difficult to detect and to sample in sufficient numbers. Therefore, to date, the latter complexes of parasitoid species have only rarely been investigated, if at all. An additional source of bias in our current knowledge of parasitoids of oilseed rape pests results from the differences in the extent and importance of oilseed rape as a crop in various parts of Europe. Numerous papers have been published on the parasitoid species of oilseed rape pests in Germany, Great Britain and Scandinavia (see earlier chapters in this monograph). However, the parasitoid species-pools of oilseed rape pests in eastern Europe have been investigated only recently, and these urgently warrant further studies. Moreover, in the future, if research on the oilseed rape pests intensifies, the well-established relationship between sampling intensity and cumulative species numbers will undoubtedly result in the discovery of additional parasitoid species. Such advances appear likely to result in the wake of the current growing interest in integrated crop management (ICM) on crops such as oilseed rape (see Chapter 3). These interests offer a new chance not only to study the biological antagonists of oilseed rape pests but also to relate parasitism rates (and the efficacy of parasitoids to control pests) to landscape features (Thies & Tscharntke, 1999) or to agricultural cropping systems (Nitzsche & Ulber, 1998).

Although many gaps in knowledge currently exist, it was considered important to complement data provided in Chapters 4–9 by producing a key to adults of hitherto known parasitoid species of common European oilseed rape insect pests. This key (see below) aims to provide simple dichotomus characters for the identification and separation of the various parasitoid species, rather than give the correct phylogenetic relationships between families or subfamilies. The key includes nearly 50 species, but intensified research could result, in the future, in the inclusion of many more.

Finally, it had been hoped to construct a key to parasitoid larvae. However, insufficient material and literature records were available for this to be possible. In the future, as our knowledge base is widened, production of such a key might be a practical proposition.

Key to Adult Parasitoids of Oilseed Rape Insect Pests

Problems and caveats

Although the key is based on a recent literature survey, some caveats have to be addressed. Firstly, papers on the parasitoid species complexes of oilseed rape pests published up to the 1970s are difficult to interpret. Several of the parasitoid genera included in these papers have been revised only recently (e.g. Horstmann, 1971; Buhl, 1999), thus rendering the older species records or species concepts unreliable. Secondly, in cases where parasitoid species groups have not been separated at the time of publication, a species mentioned in a paper may, in a modern sense, comprise two or even more different species. Names of parasitoid species mentioned in such papers have not been included in the key. In other cases, authors simply cited even older references, without critically checking the species names or any of the reared material (e.g. Laborius, 1972). Had such records been included in the key, incorrect names of parasitoid species or erroneous identifications could have been perpetuated. Moreover, it would appear that no published papers actually state that voucher specimens have been deposited in museum or institute collections.

Parasitoid species listed in some published papers have been excluded from the key because their known or potential host relationships would appear to exclude oilseed rape pests. In other instances the determinations seemed to be erroneous or, at the very least, suspect. Finally, species that have been cited in the literature on just one occasion have also been omitted from the key.

By restricting the scope of the key to parasitoids that could be regarded as common and widespread in Europe, the key serves as a starting point for those readers interested in further research on the parasitoid species involved. The key may also serve as a basis for broadening our knowledge on the species complexes. However, being far from complete, in cases where species determinations are questionable or species groups are diverse, the reader should refer to the primary references. For a broader understanding of family concepts in the Hymenoptera or of modern subfamilies concepts in the Ichneumonidae and Braconidae the reader should refer to the book by Goulet & Huber (1993) or to the paper by van Achterberg (1990), where additional characters are given for the understanding of these taxa. The Tersilochinae genera of interest with regard to oilseed rape insect pests were revised by Horstmann (1971) and his paper gives further diagnostic characters to separate difficult species complexes. Chalcidoid families, genera and species are keyed in Tryapitzyn (1988), and Platygastridae could be identified by using the key of Buhl (1999). Pteromalid genera and species are keyed in Graham (1969) and by Bouček & Rasplus (1991), and certain species (with regard to parasitoids of oilseed rape insect pests) are described in more detail in Dessart (1964), Tobias *et al.* (1986), Graham (1987), Zerova (1995) and Zerova & Seryognia (1999).

The following key includes references to 21 species that parasitize brassica pod midge, 10 species that parasitize cabbage seed weevil, seven species that parasitize cabbage stem flea beetle, 11 species that parasitize ceutorhynchid stem weevils and nine species that parasitize pollen beetles. This gives a total of 58 (when the key includes only 49 species), but there are duplications because several species (notably *Diospilus morosus*) are non-specific parasitoids and associated with several host species. Systematic details of the parasitoids are included earlier in the monograph.

Key to the hitherto known parasitoids of common European oilseed rape insect pests (female parasitoids only)

1. Fore wings with many veins, which enclose at least 4 cells (Fig. 11.1: terminology of cells; Fig. 11.2: terminology of veins) . . . . ICHNEUMONOIDEA – 2
– Fore wings with veins reduced, forming fewer than 4 cells (Fig. 11.3), or wings reduced 21
2. Fore wing with vein 1/Rs + M separating median cell 1M and radial cell 1R (Fig. 11.4); if 1/Rs + M not present (genus *Microctonus*) then hind wing with vein 1r – m basal to separation of veins R1 and Rs (Fig. 11.5). . . . . BRACONIDAE – 3
– Fore wing without vein 1/Rs + M, so median cell 1M and radial cell 1R are fused (Fig. 11.6) and hind wing with vein 1r–m apical to separation of veins R1 and Rs (Fig. 11.7) . . . . ICHNEUMONIDAE – 10
3. Labrum exposed and concave; clypeus with apical margin concave (Fig. 11.8) . . . . BRACONINAE – 4
– Labrum either concealed by clypeus or flat and convex and sculptured (Fig. 11.9). 5
4. Fore wing with cell 2R1 large, extending to tip of wing (Fig. 11.10); ovipositor shorter than gaster; surface of tergites 2 and 3 sculptured. . . . . ***Bracon fulvipes*** Nees
– Fore wing with cell 2R1 shorter, not reaching to tip of wing (Fig. 11.11); surface of tergites 2 and 3 smooth . . . . ***Bracon variator*** Nees
5. Fore wing with 1/Rs + M, though sometimes not always complete (Fig. 11.4) 6
– Fore wing without 1/Rs + M (Fig. 11.12) . . . . ***Microctonus melanopus*** Ruthe
6. Hind wing vein 2/Cu distinct (Fig. 11.13) . . . . ***Sigalphus obscurellus*** Nees

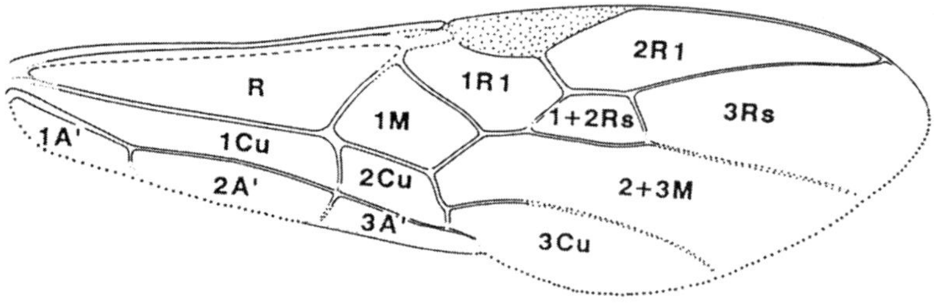

Fig. 11.1 Superfamily ICHNEUMONOIDEA – fore wing to show terminology of cells (after Goulet & Huber, 1993).

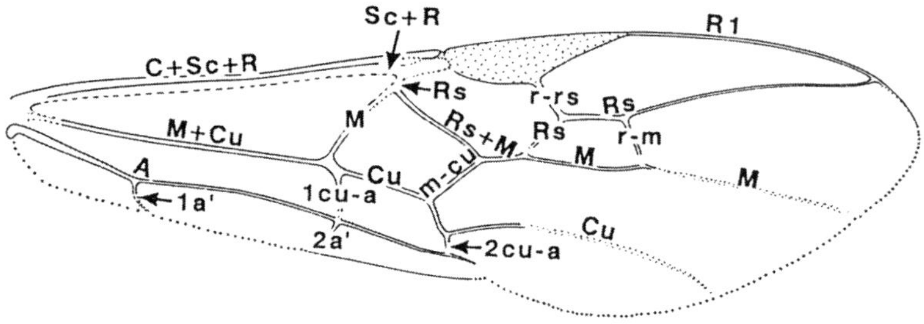

Fig. 11.2 Superfamily ICHNEUMONOIDEA – fore wing to show terminology of veins (after Goulet & Huber, 1993).

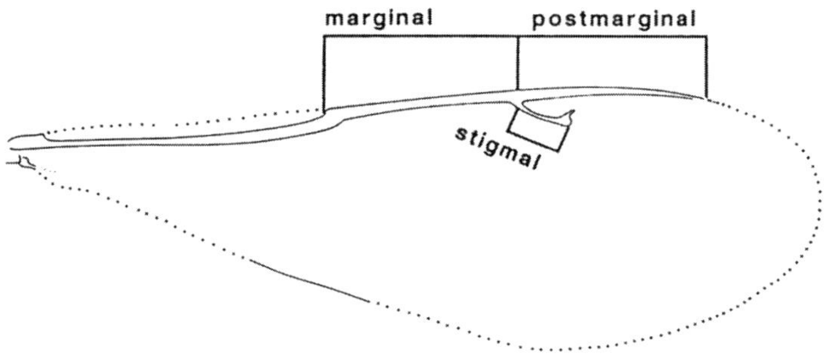

Fig. 11.3 Superfamily CHALCIDOIDEA – fore wing to show terminology of veins (after Goulet & Huber, 1993).

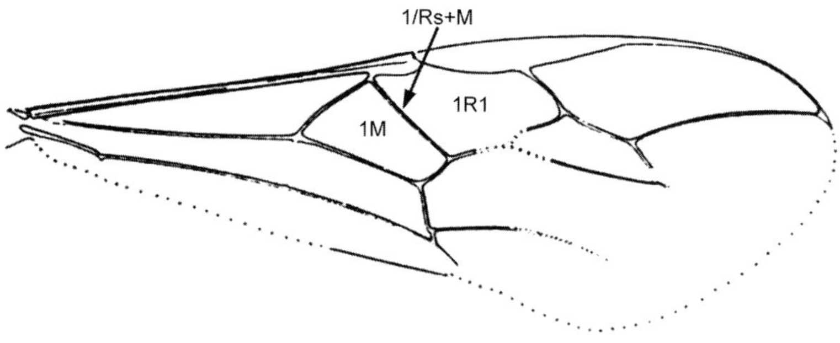

Fig. 11.4 Family BRACONIDAE (other than genus *Microctonus*) – fore wing with 1/Rs + M (arrowed) (after Goulet & Huber, 1993).

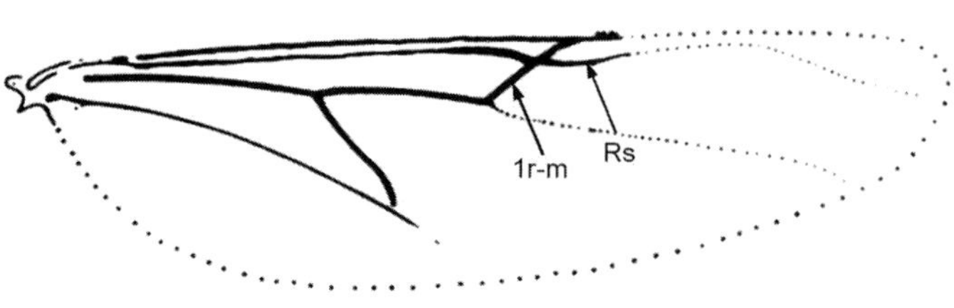

Fig. 11.5 Family BRACONIDAE (genus *Microctonus*) – hind wing with vein 1r–m basal to separation of veins R1 and Rs (after Goulet & Huber, 1993).

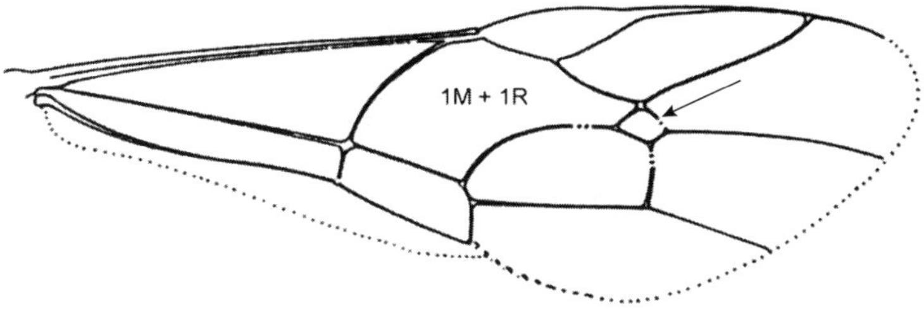

Fig. 11.6 Family ICHNEUMONIDAE – fore wing with median cell 1M and radial cell 1R fused and with areolet (arrowed) (after Goulet & Huber, 1993).

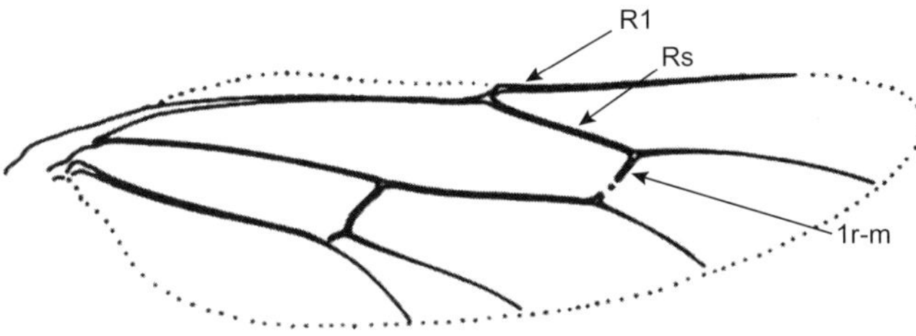

Fig. 11.7 Family ICHNEUMONIDAE – hind wing with vein 1r-m apical to separation of veins R1 and Rs (after Goulet & Huber, 1993).

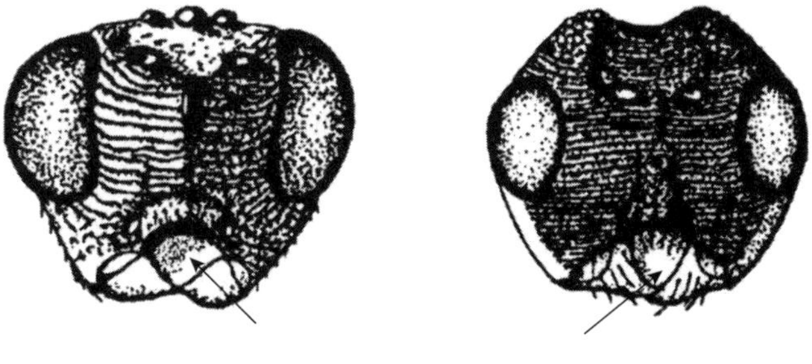

Fig. 11.8 Subfamily BRACONINAE – frontal view of head; labrum (arrowed) exposed and concave (after Achterberg, 1990).

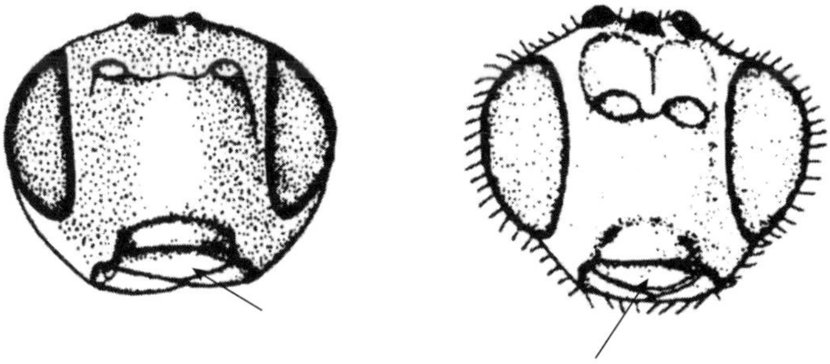

Fig. 11.9 Frontal view of head – labrum (arrowed) either concealed by clypeus or flat or convex and sculptured (after Achterberg, 1990).

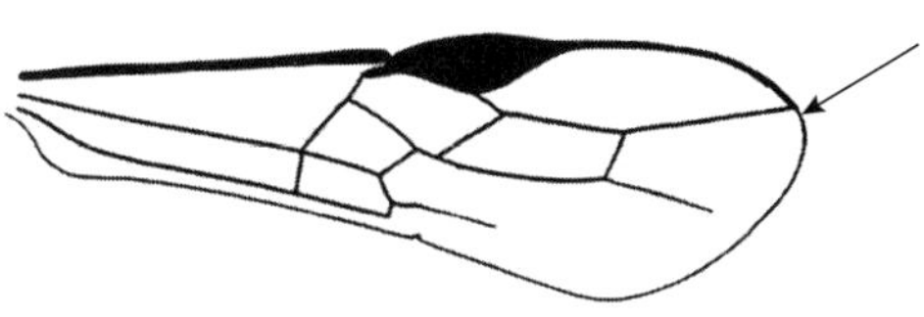

Fig. 11.10 *Bracon fulvipes* – fore wing with cell 2R1 long and reaching to tip of wing (arrowed) (after Tobias *et al.*, 1986).

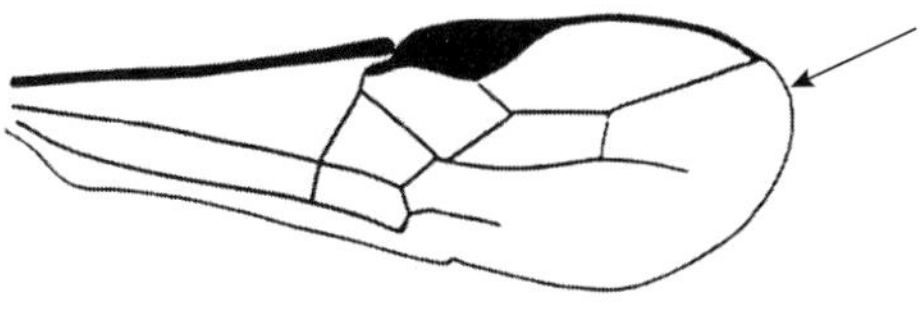

Fig. 11.11 *Bracon variator* – fore wing with cell 2R1 not reaching to tip of wing (arrowed) (after Tobias *et al.*, 1986).

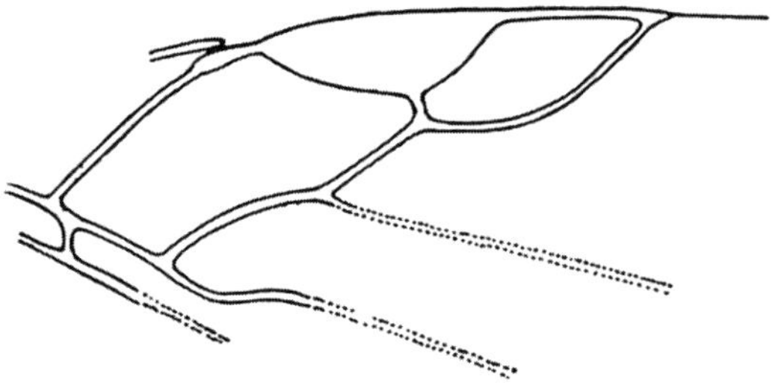

Fig. 11.12 *Microctonus melanopus* – part of fore wing to show absence of c1/Rs + M (after Tobias *et al.*, 1986).

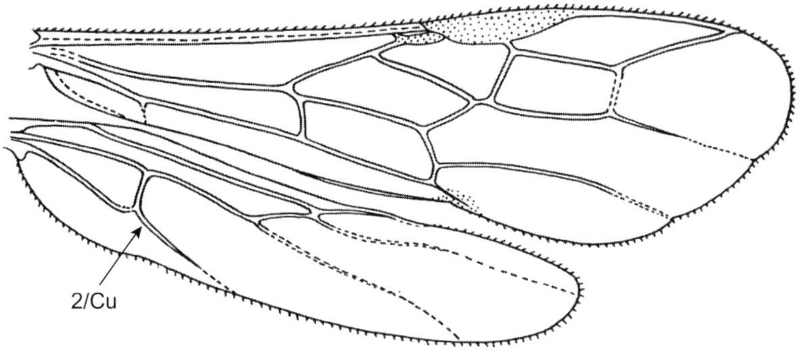

Fig. 11.13 *Sigalphus obscurellus* – fore wing and hind wing, the latter with vein 2/Cu present (after Tobias *et al.*, 1986).

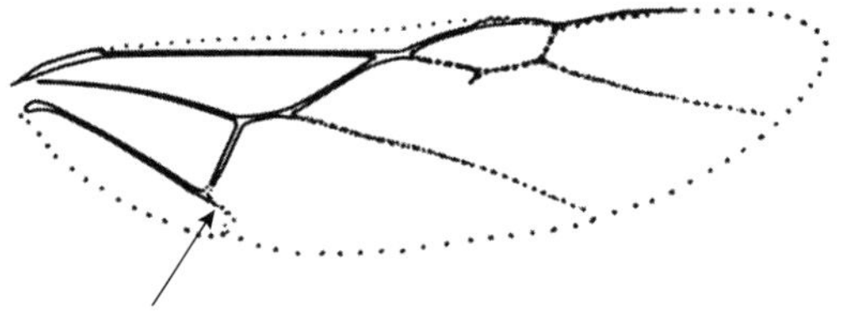

Fig. 11.14 Hind wing with vein 2/Cu absent (arrowed) (after Goulet & Huber, 1993).

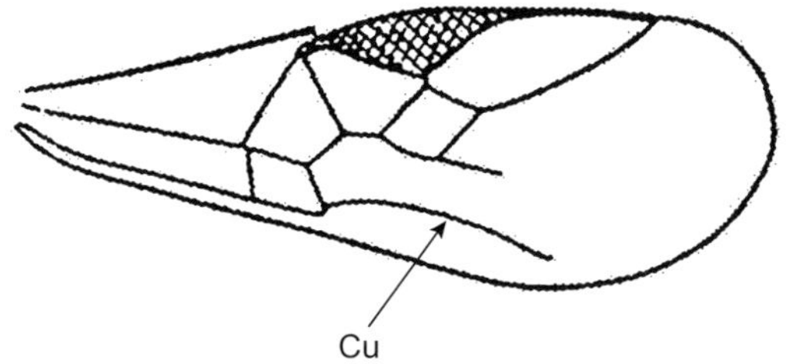

Fig. 11.15 *Diospilus morosus* – fore wing with Cu (arrowed) reaching far beyond cell 1 + 2Rs (after Tobias *et al.*, 1986).

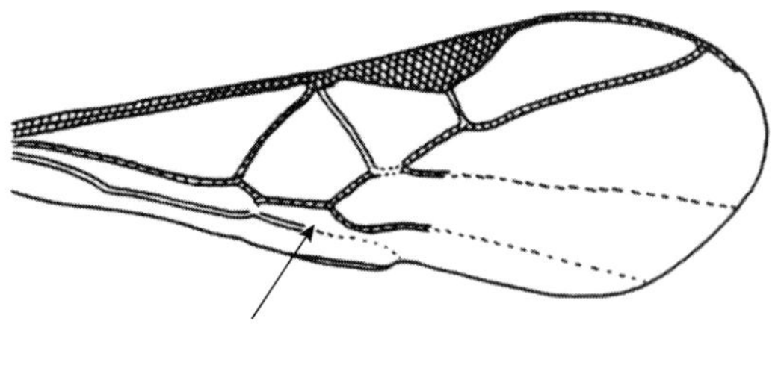

Fig. 11.16 *Blacus nigricornis* – fore wing with cell 2Cu (arrowed) open (after Tobias *et al.*, 1986).

– Hind wing vein 2/Cu absent (Fig. 11.14) 7
7. Fore wing with cell 1Rs present (Fig. 11.15) . . . . *Diospilus* – 8
– Fore wing with cell 1Rs absent (Fig. 11.16) 9
8. Fore wing with Cu reaching far beyond cell 1 + 2Rs (Fig. 11.15) . . . . *Diospilus morosus* Reinhardt
– Fore wing with Cu shorter than cell 1 + 2Rs, distinctly indicated only at the very beginning (Fig. 11.17) . . . . ***Diospilus capito*** (Nees)
9. Fore wing with cell 2Cu open (Fig. 11.16). . . . . . . ***Blacus nigricornis*** Haeselbarth
– Fore wing with cell 2Cu closed. . . . . ***Eubazus sigalphoides*** (Marshall)
10. Fore wing with areolet open (Fig. 11.18); hind wing with basal part of vein M + Cu absent (Fig. 11.19) . . . . TERSILOCHINAE – 11
– Fore wing with areolet closed (Fig. 11.6); hind wing with vein M + Cu complete (Fig. 11.20) . . . . ***Stibeutes curvispina*** (Thomson)
11. Fore wing with second anal cell widely open (Fig. 11.21) . . . . *Aneuclis* – 12
– Fore wing with second anal cell closed (Fig. 11.22). 13
12. Antenna with 16 segments; basal part of antenna: Fig. 11.23. . . . . ***Aneuclis incidens*** (Thomson)
– Antenna with 18 segments; basal part of antenna: Fig. 11.24. . . . . ***Aneuclis melanarius*** (Holmgren)
13. Antenna with fewer than 20 segments; M joining M + Cu at or near 1 cu-a . . . . *Phradis* – 14
– Antenna most often with more than 20 segments; M joining M + Cu beyond 1 cu-a . . . . *Tersilochus* – 15
14. Antenna with 15–16 segments; femur III totally dark; tip of ovipositor: Fig. 11.25 . . . . ***Phradis morionellus*** (Holmgren)
– Antenna with 17–18 segments; femur III darkened; tip of ovipositor: Fig. 11.26 . . . . ***Phradis interstitialis*** (Thomson)
15. Sternaulus distinctly visible, consisting of a curved furrow, extending nearly the whole length of mesopleuron (Fig. 11.27) . . . . ***Tersilochus exilis*** Holmgren
– Sternaulus shorter, less than two-thirds as long as mesopleuron 16
16. Sternaulus weakly indicated on more than half length of mesopleuron, accompanied by wrinkles, and hypostomal carina complete (Fig. 11.28) . . . . ***Tersilochus stenocari*** (Gregor)
– Sternaulus longer or shorter than half length of mesopleuron, and hypostomal carina absent. 17
17. Antennal segment 4 smaller and slighter shorter than the following segments; tip of ovipositor dorsally and ventrally distinctly dentate; basal part of antenna: Fig. 11.29; tip of ovipositor: Fig. 11.30 . . . . ***Tersilochus heterocerus*** Thomson
– Antennal segments gradually decreasing in length, beginning with segment 3 18
18. Ovipositor sheaths × 1.2–1.3 length of gastral segment 1 19
– Ovipositor sheaths more than × 1.4 length of gastral segment 1 20
19. Tip of ovipositor dorsally dentate, ventrally straight; basal part of antenna: Fig. 11.31 . . . . ***Tersilochus tripartitus*** Brischke
– Both sides of tip of ovipositor distinctly dentate; basal part of antenna: Fig. 11.32. . . . . ***Tersilochus microgaster*** (Szépligeti)
20. Ovipositor sheaths × 1.4 length of gastral segment 1; legs darkened; basal part of antenna: Fig. 11.33 . . . . ***Tersilochus obscurator*** Aubert

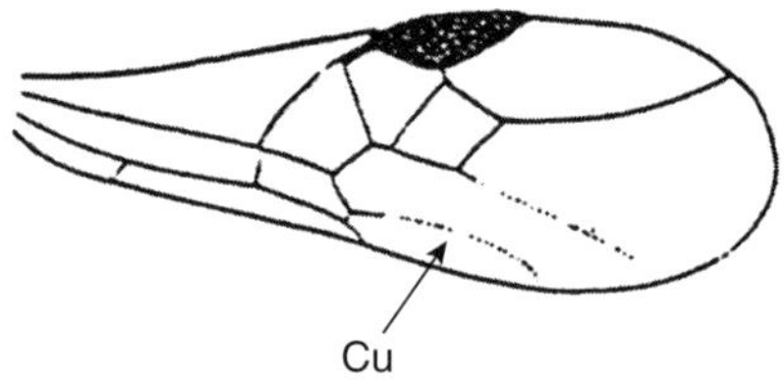

Fig. 11.17 *Diospilus capito* – fore wing with Cu (arrowed) shorter than cell 1 + 2Rs, distinctly indicated only at the very beginning, i.e. basally (after Tobias *et al.*, 1986).

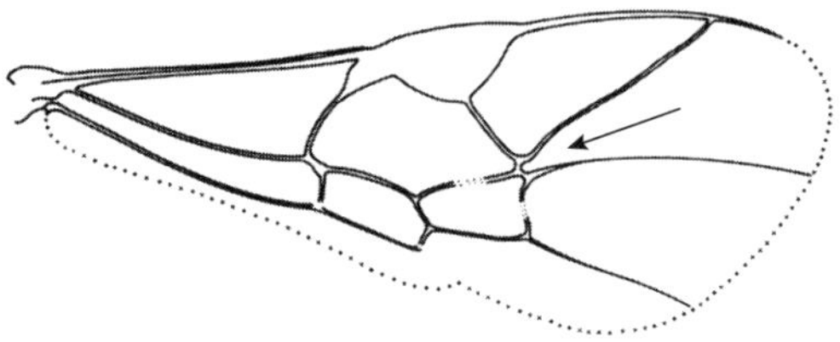

Fig. 11.18 Subfamily TERSILOCHINAE – fore wing with areolet open (arrowed) (after Goulet & Huber, 1993).

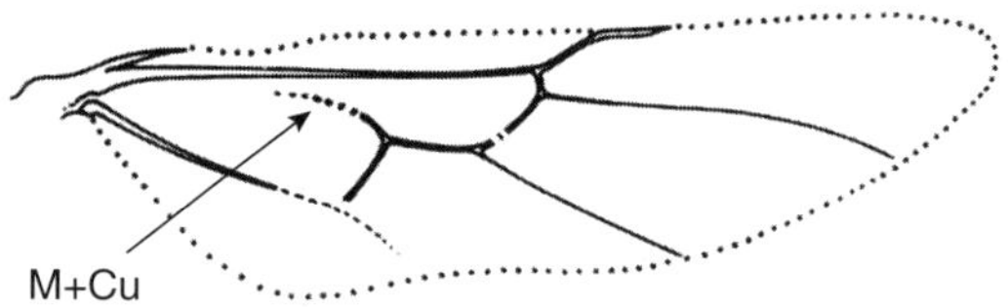

Fig. 11.19 Subfamily TERSILOCHINAE – hind wing with basal part of vein M + Cu absent (after Goulet & Huber, 1993).

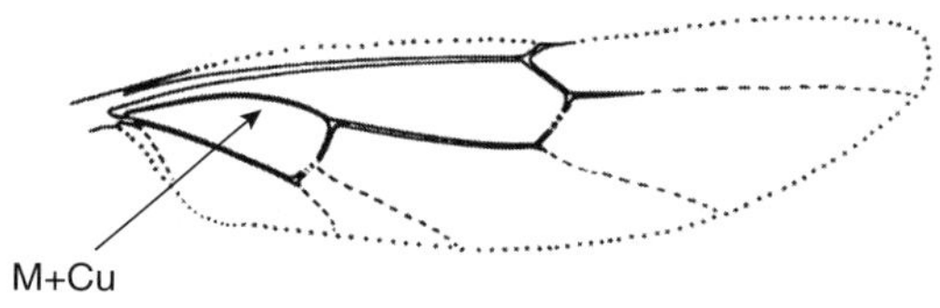

Fig. 11.20 *Stibeutes curvispina* – hind wing with vein M + Cu complete (after Goulet & Huber, 1993).

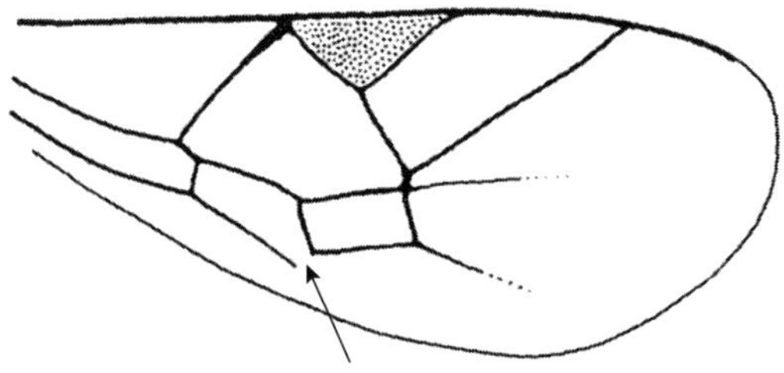

Fig. 11.21 *Aneuclis* – fore wing with second anal cell widely open (arrowed) (after Horstmann, 1971).

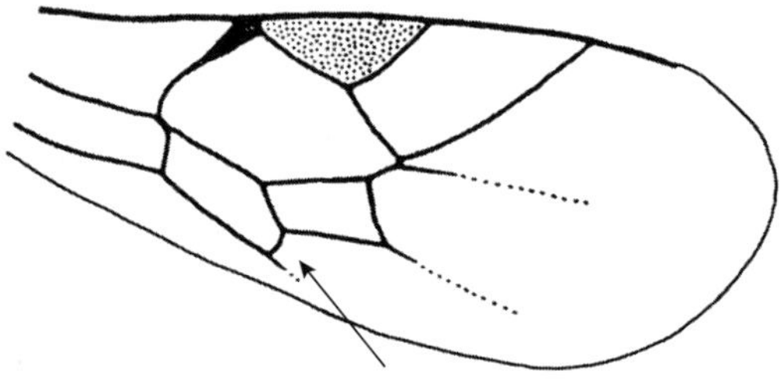

Fig. 11.22 Fore wing with second anal cell closed (arrowed) (after Horstmann, 1971).

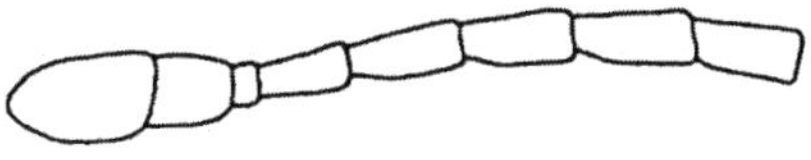

Fig. 11.23 *Aneuclis incidens* – basal part of antenna (after Horstmann, 1971).

Fig. 11.24 *Aneuclis melanarius* – basal part of antenna (after Horstmann, 1971).

Fig. 11.25 *Phradis morionellus* – tip of ovipositor (after Horstmann, 1971).

Fig. 11.26 *Phradis interstitialis* – tip of ovipositor (after Horstmann, 1971).

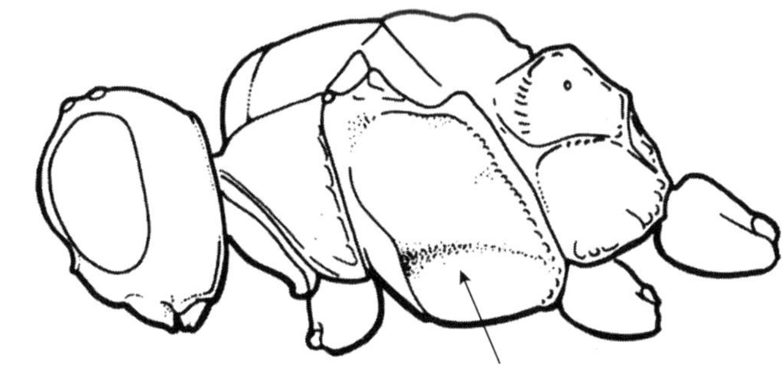

Fig. 11.27 *Tersilochus exilis* – sternaulus (arrowed) distinctly visible, forming a curved furrow along nearly the whole length of the mesopleuron (after Goulet & Huber, 1993).

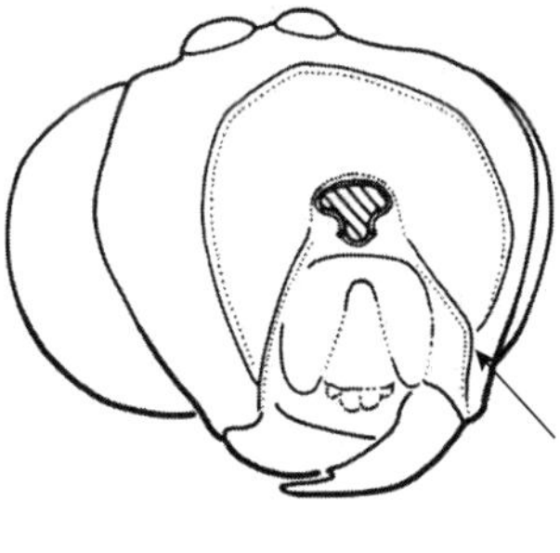

Fig. 11.28 *Tersilochus stenocari* – latero-frontal view of head to show complete hypostomal carina (arrowed) (after Goulet & Huber, 1993).

Fig. 11.29 *Tersilochus heterocerus* – basal part of antenna (after Horstmann, 1971).

Fig. 11.30 *Tersilochus heterocerus* – tip of ovipositor (after Horstmann, 1971).

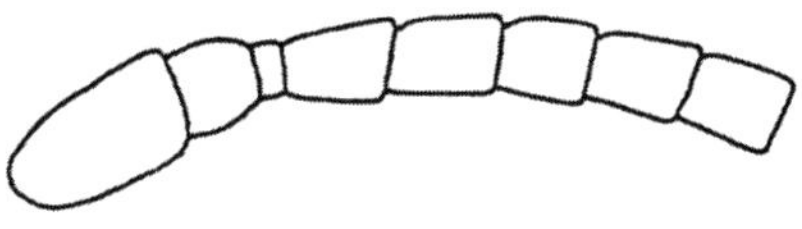

Fig. 11.31 *Tersilochus tripartitus* – basal part of antenna (after Horstmann, 1971).

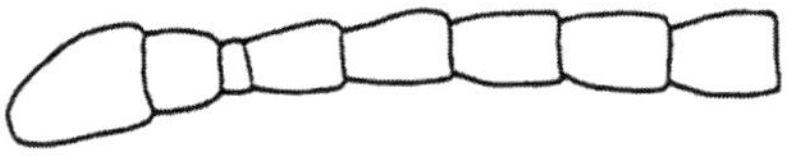

Fig. 11.32 *Tersilochus microgaster* – basal part of antenna (after Horstmann, 1971).

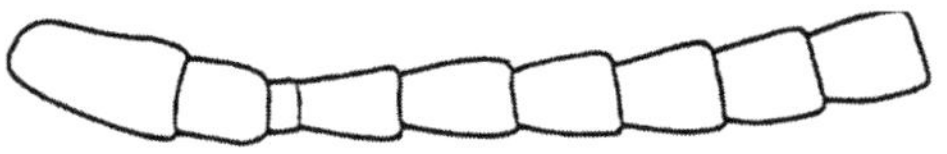

Fig. 11.33 *Tersilochus obscurator* – basal part of antenna (after Horstmann, 1971).

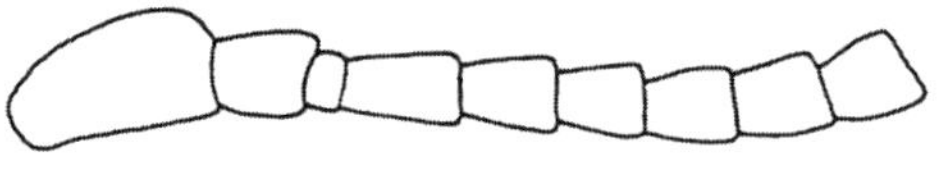

Fig. 11.34 *Tersilochus fulvipes* – basal part of antenna (after Horstmann, 1971).

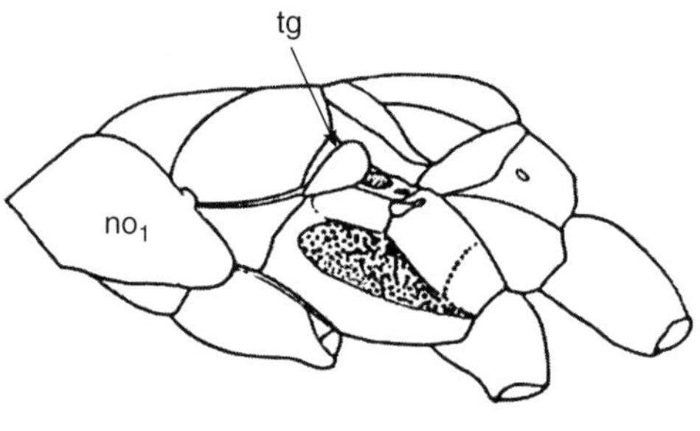

Fig. 11.35 Superfamily CHALCIDOIDEA – lateral view of thorax with pronotum (no_1) separated from tegula (tg) (after Goulet & Huber, 1993).

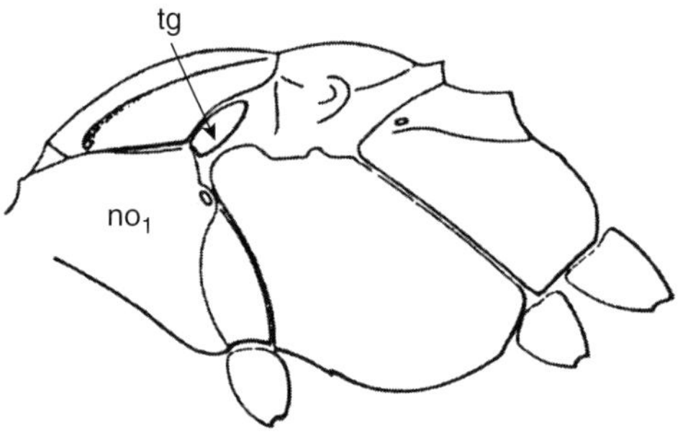

Fig. 11.36 Family EULOPHIDAE – lateral view of thorax with pronotum (no_1) extending back to tegula (tg) (after Goulet & Huber, 1993).

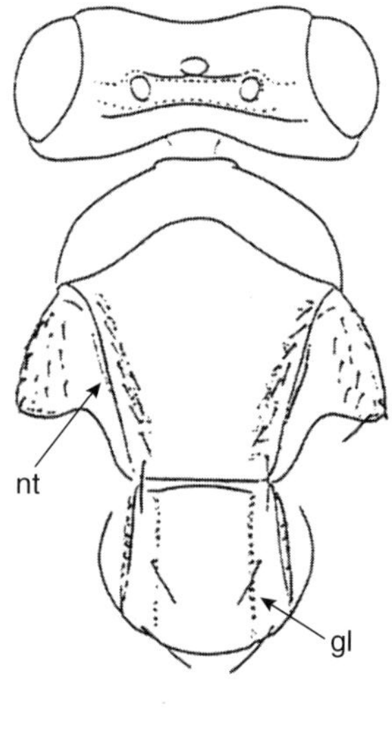

Fig. 11.37 Head and thorax to show complete notauli (nt) on mesoscutum and grooved line (gl) on scutellum (after Graham, 1987).

– Ovipositor sheaths × 2.0 length of gastral segment 1; legs yellow-red; basal part of antenna: Fig. 11.34 . . . . ***Tersilochus fulvipes*** (Gravenhorst)
21. Apex of fore tibiae with one spur 22
– Apex of fore tibiae with two spurs, fore wing with up-curved Rs 46
22. Pronotum (Fig. 11.35: no_1) in lateral view separated from tegula (Fig. 11.35: tg) . . . . CHALCIDOIDEA – 23
– Pronotum (Fig. 11.36: no_1) in lateral view extending posteriorly to tegula (Fig. 11.36: tg) 36
23. Tarsi with 4 tarsomeres . . . . EULOPHIDAE – 24
– Tarsi with 5 tarsomeres. 27
24. Mesoscutum with notauli complete (Fig. 11.37: nt) 25
– Mesoscutum without notauli . . . . ***Necremnus leucarthros*** (Nees)
25. Scutellum with two sublateral, longitudinal, grooved lines (Fig. 11.37: gl), postmarginal vein absent. . . . . TETRASTICHINAE – 26
– Scutellum without grooved lines, postmarginal vein present (Fig. 11.3); clypeus yellow, antenna with 3 flagellomeres . . . . ***Omphale clypealis*** (Thomson)
26. Vertex with a sharp transverse carina behind posterior ocelli (Fig. 11.37), body extensively yellow, not metallic . . . . ***Sigmophora brevicornis*** (Panzer)
– Vertex without a carina behind posterior ocelli, body with weak olive-bluish tints, sometimes richly yellow-marked . . . . ***Aprostocetus epicharmus*** (Walker)
27. Mesopleuron enlarged, undivided and convex (Fig. 11.38) 28
– Mesopleuron not enlarged; differentiated (Fig. 11.35). 29
28. Wings rudimentary . . . . ***Eupelmus vesicularis*** (Retzius)
– Wings developed; marginal vein at most as long as wide. . . . . ***Cerchysiella planiscutellum*** (Mercet)
29. Ovipositor sheaths far exerted, longer than hind tibiae (Fig. 11.39), femur III with a tubercle near. . . . . ***Pseudotorymus napi*** (Amerling & Kirchner)
– Ovipositor sheaths only slightly exerted, shorter than hind tibiae 30
30. Body black, pronotum in dorsal view large and rectangular (Fig. 11.40) . . . . EURYTOMIDAE – 31
– Body with metallic colour, pronotum small (Fig. 11.41: general appearance) . . . . PTEROMALIDAE – 32
31. Lower margin of mesepisternum straight, without a process (Fig. 11.42) . . . . ***Eurytoma curculionum*** Mayr
– Lower margin of mesepisternum (occasionally masked by fore coxae) with a triangular process (Fig. 11.43) . . . . ***Eurytoma dentata*** Mayr
32. Antennae with 2 anelli and 6 flagellomeres. 33
– Antennae with 3 anelli and 5 flagellomeres (Fig. 11.44); head in frontal view: Fig. 11.45 (males with fore wing venation enlarged) . . . . ***Mesopolobus morys*** (Walker)
33. Anterior margin of clypeus *with* a median tooth (Fig. 11.46). . . . . ***Stenomalina gracilis*** (Walker)
– Anterior margin of clypeus *without* a tooth, truncate or emarginate 34
34. Dorsal surface of hind coxa hairy proximally; sides of basal tergite of gaster conspicuously hairy laterally 35
– Dorsal surface of hind coxa bare in basal half; basal tergite with only few hairs . . . . ***Pteromalus* spp.**

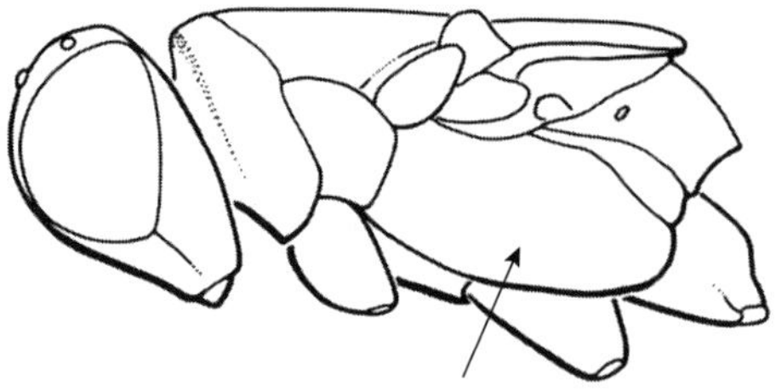

Fig. 11.38 Mesopleuron (arrowed) enlarged, undivided and convex (after Goulet & Huber, 1993).

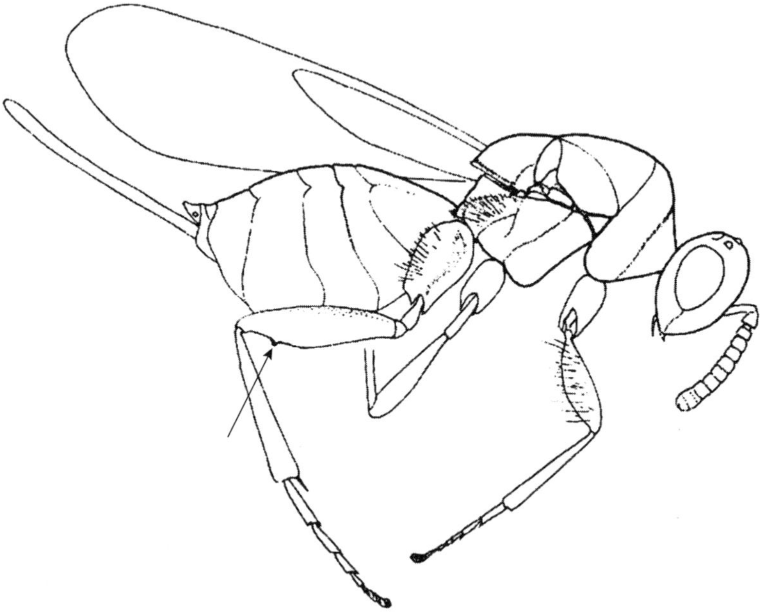

Fig. 11.39 *Pseudotorymus napi* – ovipositor sheaths far exerted, longer than hind tibiae, femur III with a tubercle near apex (arrowed) (after Zerova & Seryognia, 1999).

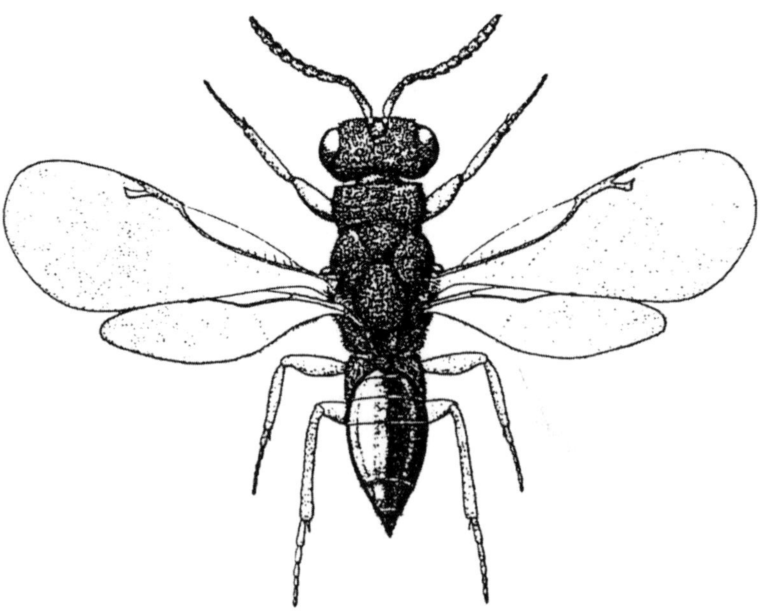

Fig. 11.40 Family EURYTOMIDAE – general appearance; pronotum in dorsal view large and rectangular (after Gauld & Bolton, 1988).

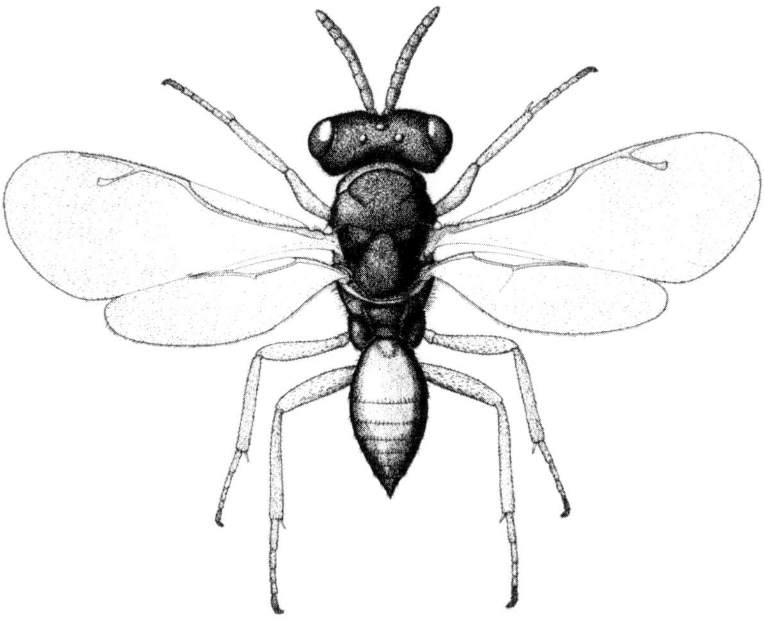

Fig. 11.41 Family PTEROMALIDAE – general appearance (after Gauld & Bolton, 1988).

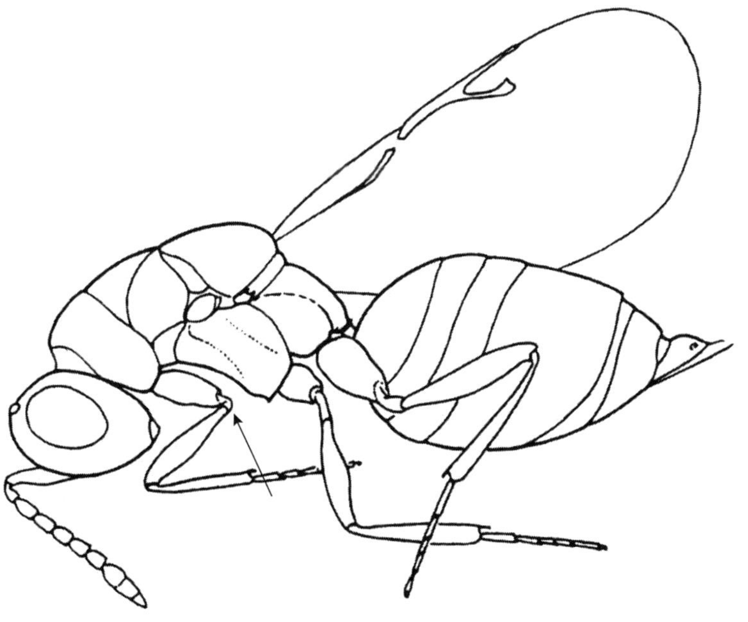

Fig. 11.42 *Eurytoma curculionum* – lower margin of mesepisternum (arrowed) without a process (after Zerova, 1995).

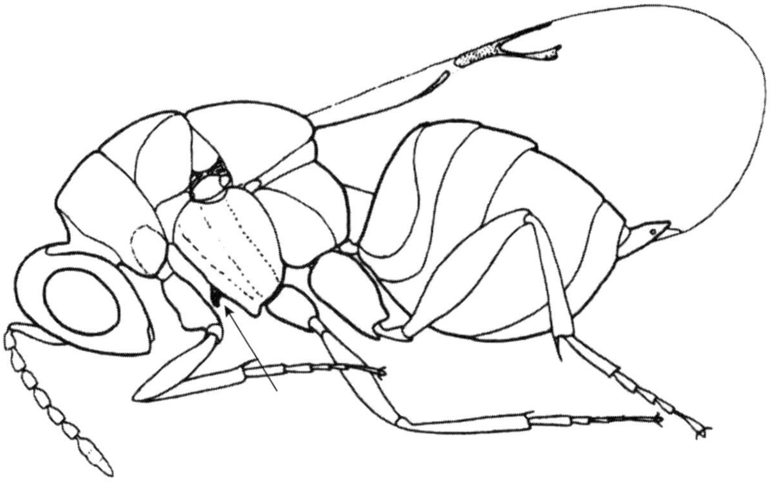

Fig. 11.43 *Eurytoma dentata* – lower margin of mesepisternum (arrowed) (occasionally masked by fore coxae) with triangular process (after Zerova, 1995).

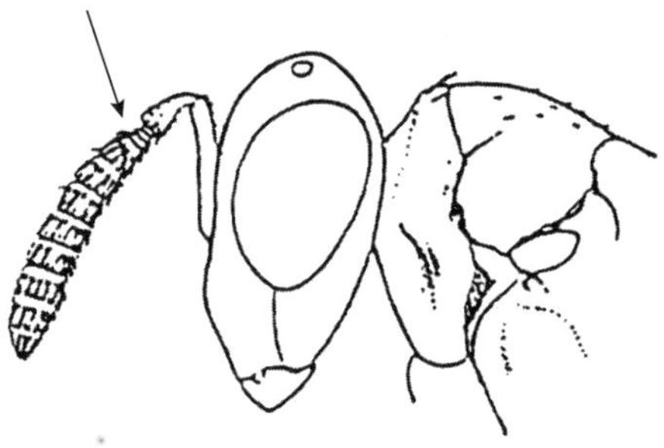

Fig. 11.44 *Mesopolobus morys* – antenna with 3 anelli (arrrowed) and 5 flagellomeres (after Bouček & Rasplus, 1991).

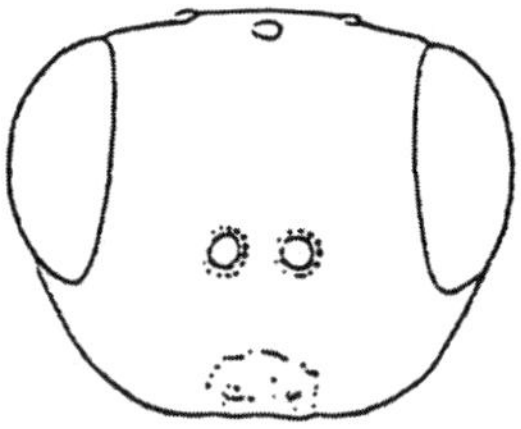

Fig. 11.45 *Mesopolobus morys* – frontal view of head (after Graham, 1969).

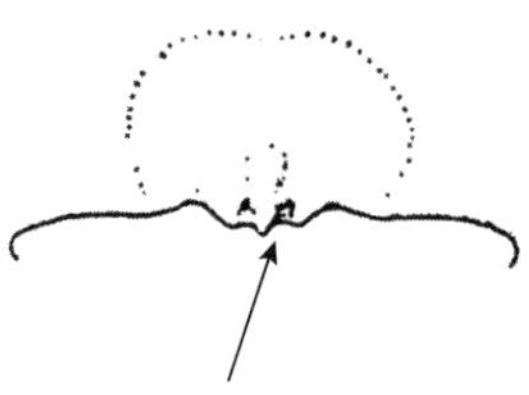

Fig. 11.46 *Stenomalina gracilis* – anterior margin of clypeus to show median tooth (arrowed) (after Graham, 1969).

35. Median area of propodeum with a distinct transverse costula; antennae slender, proximally not or hardly stouter than pedicellus, and widening somewhat towards tip; colour of body dark green. . . . . ***Trichomalus lucidus*** (Walker)
– Median area of propodeum without a distinct costula; antennae distinctly stouter than pedicellus, hardly widening towards tip; colour of body often coppery tinged . . . . ***Trichomalus perfectus*** (Walker)
36. Antennal sockets separated from dorsal margin of clypeus by more than diameter of socket (general appearance: Fig. 11.47). . . . . ***Brachyserphus parvulus*** (Nees)
– Antennal sockets continuous with dorsal margin of clypeus 37
37. Fore wing with stigmal vein . . . . CERAPHRONOIDEA – 38
– Fore wing without stigmal or postmarginal vein; antennae with 10 or fewer segments (general appearance: Fig. 11.48). . . . . PLATYGASTRIDAE – 42
38. Fore wing with large, semicircular pterostigma . . . . ***Conostigmus rufescens*** Kieffer
– Fore wing without large pterostigma. 39
39. Mesonotum and scutellum convex; thorax laterally compressed and notably higher than wide. . . . . *Aphanogmus* – 40
– Mesonotum and scutellum flat; thorax broader than high . . . . *Ceraphron* – 41
40. Mesonotum with distinct median groove throughout (Fig. 11.49) . . . . ***Aphanogmus abdominalis*** (Thomson)
– Mesonotum without median groove (Fig. 11.50) . . . . ***Aphanogmus tenuicornis*** Thomson
41. Wings fully developed . . . . ***Ceraphron longipennis*** (Kieffer)
– Wings reduced (Fig. 11.51) . . . . ***Ceraphron xanthosoma*** (Kieffer)
42. Gastral petiolus with distinct cornutus (Fig. 11.52); fore wing with subcostal vein . . . . *Inostemma* – 43
– Gastral petiolus without cornutus; fore wing without subcostal vein. 45
43. Vertex with distinct notch; cornutus of petiolus longitudinally striated 44
– Vertex with hardly visible notch; cornutus of petiolus reticulate . . . . ***Inostemma reticulatum*** (Szelényi)
44. Tergite 6 at least twice as long as basal width; cornutus overreaching head . . . . ***Inostemma boscii*** (Jurine)
– Tergite 6 at most slightly longer than basal width; cornutus reaching only to ocelli . . . . ***Inostemma walkeri*** Kieffer
45. Scutellum posteriorly with a semitransparent lamella; gastral petiolus ensely haired . . . . ***Synopeas lugubris*** Thomson
– Scutellum posteriorly without a lamella; gastral petiolus with only sparse hairs 46
46. Scutellum somewhat angled, with tuft of hairs at apex; body length about 3 mm. . . . . ***Isocybus thomsoni*** Kieffer
– Scutellum semicircular, without tuft of hairs at apex; body length at most about 2 mm . . . . *Platygaster* – 47
47. Tergite 4 strongly transverse 48
– Tergite 4 hardly transverse 49
48. Antennae with flagellomeres 8 and 9 transverse; notauli complete . . . . ***Platygaster oebalus*** Walker
– Antennae with flagellomeres 8 and 9 each at least as long as wide; notauli incomplete. . . . . ***Platygaster iolas*** Walker

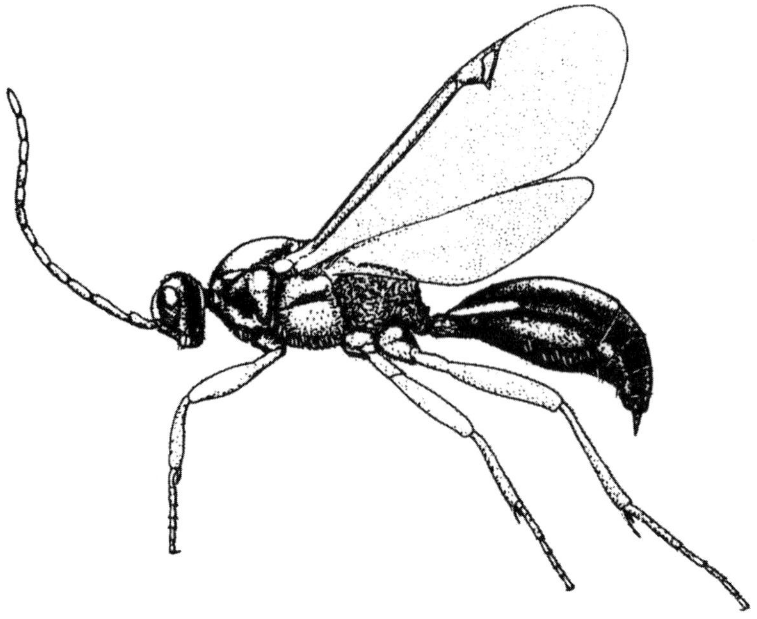

Fig. 11.47 *Brachyserphus parvulus* – general appearance (after Gauld & Bolton, 1988).

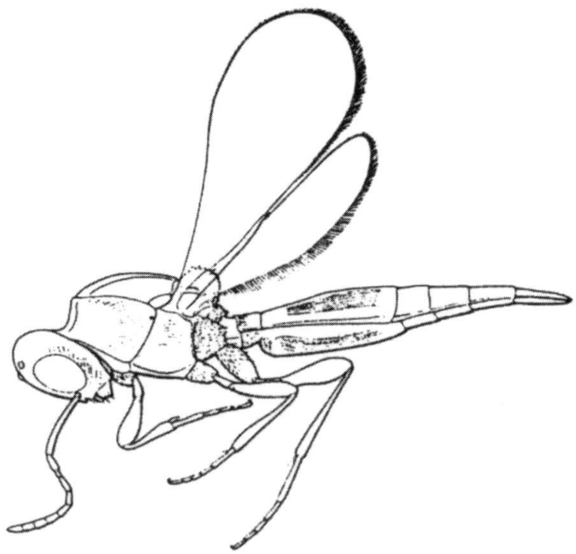

Fig. 11.48 Family PLATYGASTRIDAE – general appearance (after Goulet & Huber, 1993).

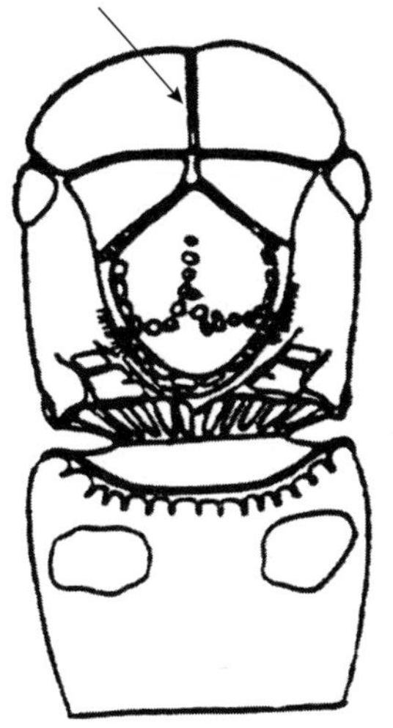

Fig. 11.49 *Aphanogmus abdominalis* – mesonotum with a median groove (arrowed) (after Dessart, 1964).

Fig. 11.50 *Aphanogmus tenuicornis* – mesonotum without a median groove (after Tryapitzyn, 1988).

Fig. 11.51 *Ceraphron xanthosoma* – fore wing (after Dessart, 1964).

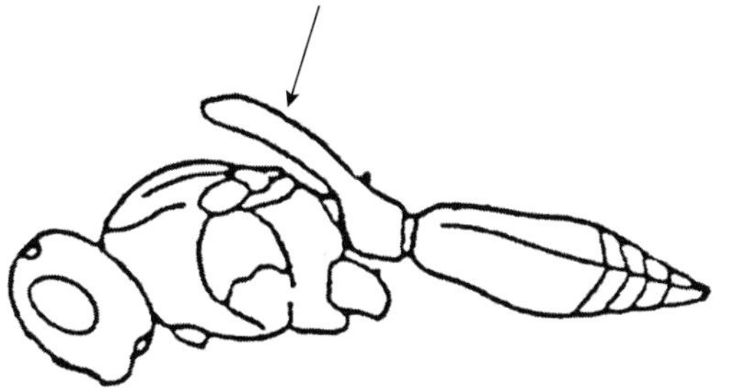

Fig. 11.52 *Inostemma* – gastral petiolus with distinct cornutus (arrowed) (after Tryapitzyn, 1988).

49. Occiput striated 50
- Occiput not striated. . . . . ***Platygaster subuliformis*** (Kieffer)
50. Antennae with flagellomeres 8 and 9 transverse; tergites 3 to 6 smooth; tergites thickened at interfaces. . . . . ***Platygaster tisias*** Walker
- Antennae with flagellomeres 8 and 9 at least slightly longer than wide; tergites 3 to 6 with faint longitudinal microsculpturing, and tergites not thickened at interfaces. . . . . ***Platygaster gladiator*** Zetterstedt

Acknowledgements

I am indebted to Peter Neerup Buhl for his valuable comments on the platygastrid part of the key and to David V. Alford for helpful suggestions and reviews which substantially improved the whole chapter.

References

Achterberg, C. van (1990). Illustrated key to the subfamilies of the Holarctic Braconidae (Hymenoptera: Ichneumonoidae). *Zoologische Mededelingen Leiden* **64**, 1–20.

Blunck, H. (1921). Die Schmarotzerwespen des Rapslglanzkäfers im Jahre 1920. *Arbeiten aus der Biologischen Reichsanstalt für Land- und Forstwirtschaft* **10**, 429–33.

Bouček, Z. & Rasplus, J. Y. (1991). *Illustrated Key to West-Palearctic Genera of Pteromalidae (Hymenoptera: Chalcidoidea).* INRA: Paris.

Buhl, P. N. (1999). A synopsis of the Platygastridae of Fennoscandia and Denmark. *Entomofauna* **20**, 17–52.

Dessart, P. (1964). Contribution à l'étude des Hyménoptères Proctotrupoidea (IV). Trois Ceraphronidae parasites de la Cecidomyie du colza: *Dasyneura brassicae* (Winnertz), en France. *Bulletin et Annales de la Société Royale d'Entomologie de Belgique* **100**, 109–30.

Gauld, I. & Bolton, B. (eds) (1988). *The Hymenoptera*. British Museum of Natural History and Oxford University Press: London.

Goulet, H. & Huber, J. T. (1993). Hymenoptera of the world: an identification guide to families. *Agriculture Canada, Research Branch, Publication 1894/E*, 668 pp.

Graham, M. W. R. de V. (1969). The Pteromalidae of north-western Europe (Hymenoptera: Chalcidoidea). *Bulletin of the British Museum (Natural History) Entomology Series, Supplement* **16**, 908 pp.

Graham, M. W. R. de V. (1987). A reclassification of the European Tetrastichinae (Hymeoptera: Eulophidae), with a revision of certain genera. *Bulletin of the British Museum (Natural History) Entomology Series* **55**, 1–392.

Horstmann, K. (1971). Revision der europäischen Tersilochinen (Hymenoptera, Ichneumonidae) Teil I. *Veröffentlichungen der Zoologischen Staatsammlung München* **15**, 45–138.

Jourdheuil, P. (1960). Influence de quelques facteurs écologiques sur les fluctuations de population d'une biocénose parasitaire: étude relative à quelques Hyménoptères (Ophioninae, Diospilinae, Euphorniae) parasites de divers Coléoptères inféodés aux Crucifères. *Annales des Épiphyties* **11**, 445–658.

Laborius, A. (1972). Untersuchungen über die Parasitierung des Kohlschotenrüsslers (*Ceutorrhynchus assimilis* Payk.) und der Kohlschotenmücke (*Dasyneura brassicae* Winn.) in Schleswig Holstein. *Zeitschrift für Angewandte Entomologie* **72**, 14–31.

Nitzsche, O. & Ulber, B. (1998). Influence of different tillage treatments following the harvest of oilseed-rape on the mortality of pollen beetle (*Meligethes* spp.) parasitoids. *Zeitschrift für Pflanzenkrankheiten und Pflanzenschutz* **105**, 417–21.

Speyer, W. (1925a). *Perilitus melanopus* Ruthe (Hym., Braconidae) als Imaginalparasitoid von *Ceutorrhyncus quadridens* Panz. Zugleich eine kurze Zusammenfassung unserer bisherigen Kenntnisse von Schlupfwespen als Parasiten der Käfer-Imagines. *Zeitschrift für Angewandte Entomologie* **11**, 132–46.

Speyer, W. (1925b). Kohlschotenrüssler (*Ceutorrhynchus assimilis* Payk.) und Kohlschotenmücke (*Dasyneura brassicae* Winn.) und ihre Parasiten. *Arbeiten aus der Biologischen Reichsanstalt für Land- und Forstwirtschaft* **12**, 79–108.

Thies, C. & Tscharntke, T. (1999). Landscape structure and biological control in agroecosystems. *Science* **285**, 893–5.

Tobias, W. J., Belokobylskii, S. A. & Kotenko, A. G. (eds) (1986). [*Keys to the Fauna of the European Part of the USSR, Volume III, Hymenoptera, Part IV.*] Nauka: Leningrad. [In Russian.]

Tryapitzyn, V. A. (1988). *Keys to the Fauna of the European Part of the USSR, Volume III, Hymenoptera, Part II.* E. J. Brill: Leiden.

Zerova, M. D. (1995). [*Parasitic Hymenoptera – Eurytominae and Eudecatominae of Palaearctics.*] National Academy of Sciences of Ukraine, Kiev Naukova Dumka, Kiev. [In Russian.]

Zerova, M. D. & Seryognia, L. Y. (1999). [Torymid chalcidoid wasps (Hymenoptera, Cahclidoidea, Torymidae) of the tribes Podagrionini and Monodontomerini of the Ukrainian fauna.] *Vestnik Zoologii: Supplement N* **13**, 130 pp. [In Russian.]

12

Predators of Oilseed Rape Pests

Wolfgang Büchs
Federal Research Centre for Agriculture and Forestry, Messeweg 11/12, D-38104 Braunschweig, Germany
David V. Alford
Orchard House, Oakington Road, Dry Drayton, Cambridge CB3 8DD, UK

Introduction

A wide range of arthropods prey upon oilseed rape pests. However, unlike parasitoids, such predators are usually non-specific and their attacks on pests within oilseed rape fields tend to be fortuitous rather than targeted. Nevertheless, there are occasions when specific prey items – for example, fully grown larvae of pests such as brassica pod midge (*Dasineura brassicae*), cabbage seed weevil (*Ceutorhynchus assimils*) and pollen beetle (*Meligethes aeneus*) which drop to the ground prior to pupating in the soil – are likely to feature as important components of the diet of predators.

In this monograph (Chapters 4–9), parasitoids are considered in direct relation to their main hosts. However, this approach is inappropriate for predators. Instead, predators are discussed here on a group by group basis, with particular emphasis on the major taxa that occur in arable crop situations, namely ground beetles (Coleoptera: Carabidae), rove beetles (Coleoptera: Staphylinidae) and spiders (Araneae). Where appropriate, taxa of lesser importance are also discussed, but, inevitably, available information on these (particularly with regards their impact or potential impact on oilseed rape pests) is relatively scarce.

In general, predators are either stenophagous or polyphagous organisms, the former being specialist feeders and the latter more or less catholic in their choice of prey. Stenophagous predators in oilseed rape are represented mostly by lacewings (Neuroptera) (= Planipennia), ladybirds (Coleoptera: Coccinellidae) and hover flies (Diptera: Syrphidae). Polyphagous predators in oilseed rape are represented mainly by ground beetles, rove beetles and spiders, but also vertebrates such as birds, bats and shrews. Vertebrates, however, are not considered further here. Other invertebrate predators with a wide range of prey, and which may be encountered in oilseed rape fields, include larvae of soldier beetles (Coloptera: Cantharidae), various dipterans (suborder Brachycera), predacious bugs (Heteroptera) and centipedes (Chilopoda).

Especially in oilseed rape, stenophagous predators are of minor importance, because most of them are adapted to feed on aphids, of which (usually) in Europe only cabbage aphid (*Brevicoryne brassicae*) plays a role as a pest in oilseed rape during the vegetative period. Peach/potato aphid (*Myzus persicae*) is an important virus vector on oilseed rape (see Chapter 2) but is present mainly in the autumn, when these aphidophagous predators are not active or, at best, are present in only small numbers; their population peaks of abundance occur during the summer when peach/potato aphid is not a problem on oilseed rape.

Compared with stenophagous predators, polyphagous predators are considered to have the following advantages:

- they are able to switch over to an alternative prey in periods (or years) when the pests are scarce, without detriment to their reproductive success;
- if environmental conditions are appropriate, they are present in fields in considerable numbers in all seasons;
- owing to their all-year-round presence, they are able to suppress pest populations at an early stage, either before or at the onset of pest population build-up;
- they are able to develop temporal preferences for certain prey items, such as pests in oilseed rape crops, as and when these become available;
- several species can coexist, so that the absence of one species of predator (e.g. owing to stochastic catastrophes or unsuitable environmental conditions) can lead to its replacement by another.

Of all the groups of epigaeic (i.e. ground-dwelling) predators, ground beetles have the greatest biomass (Goltermann, 1994). Pawlizki (1984), for example, measured the biomass of rove beetles, spiders and ground beetles on the basis of number of individuals present, and this gave a relationship of 1 : 1.3 : 3, respectively.

The Main Groups of Predators

According to data from pitfall traps (Goltermann, 1994) the epigaeic predator community in oilseed rape fields is dominated by ground beetles and rove beetles in spring and early summer (the main period when larvae of many oilseed rape pests drop to the ground from the flower stands); spiders dominate in the autumn and early winter, when aphid virus vectors (primarily peach/potato aphid) and adults and larvae of, for example, cabbage stem flea beetle (*Psylliodes chrysocephala*) and turnip sawfly (*Athalia rosae*) are present.

Beetles (Insecta: Coleoptera)

Ground beetles (Carabidae)

Without doubt, ground beetles (Fig. 12.1) are amongst the most important invertebrate predators in arable cropping situations. A wide range of species occur, several of which can be abundant (see, for example, Thiele, 1977; Luff, 1987; Kromp, 1989). Species composition and dominance vary from site to site, but in arable situations within Europe there tend to be 20–30 key species. Adults and larvae of some ground beetles are phytophagous, often feeding on seeds, and those of *Zabrus tenebrioides* are pests of cereals and grasses; others, however, are voracious carnivores.

The literature on ground beetles as predators of crop pests is vast, with many on-farm studies of their beneficial role within arable cropping systems. Studies often include

Cabbage stem flea beetle (*Psylliodes chrysocephala*) – adult.

Cabbage stem flea beetle (*Psylliodes chrysocephala*) – larva.

Cabbage aphid (*Brevicoryne brassicae*) – colony.

Pollen beetle (*Meligethes aeneus*) – larva.

Rape stem weevil (*Ceutorhynchus napi*) – larva in stem.

Plate 1. Examples of oilseed rape pests.

Brassica pod midge (*Dasineura brassicae*) – larvae in pod.

Cabbage seed weevil (*Ceutorhynchus assimilis*) – larval exit hole in pod wall.

Cabbage stem weevil (*Ceutorhynchus pallidactylus*) – adult.

Cabbage stem weevil (*Ceutorhynchus pallidactylus*) – larva in stem.

Rape winter stem weevil (*Ceutorhynchus picitarsis*) – larvae in stem.

Baris coerulescens – adult.

Plate 2. Examples of oilseed rape pests.

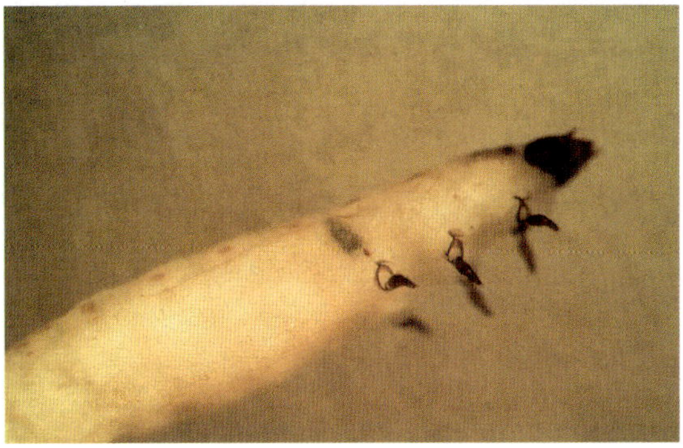

Tersilochus heterocerus (family ICHNEUMONIDAE) – egg with black chorion beneath cuticle of a pollen beetle larva.

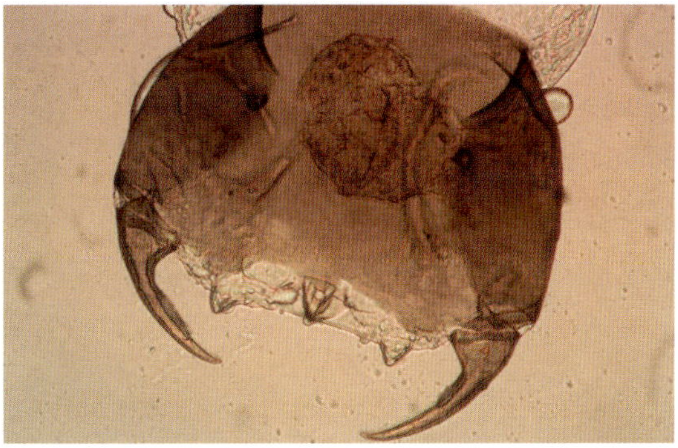

Tersilochus sp. (family ICHNEUMONIDAE) – larval head capsule.

Tersilochus fulvipes (family ICHNEUMONIDAE).

Trichomalus perfectus (family PTEROMALIDAE) – female ovipositing through pod wall into a cabbage seed weevil larva.

Trichomalus perfectus (family PTEROMALIDAE) – larva feeding on a cabbage seed weevil larva.

Omphale clypalis (family EULOPHIDAE) – female searching for brassica pod midge larvae.

Plate 3. Examples of parasitoids of oilseed rape pests.

Tersilochus heterocerus (family ICHNEUMONIDAE).

Phradis morionellus (family ICHNEUMONIDAE).

Stenomalina gracilis (family PTEROMALIDAE).

Mesopolobus morys (family PTEROMALIDAE).

Trichomalus perfectus (family PTEROMALIDAE).

Platygaster subuliformis (family PLATYGASTRIDAE).

Plate 4. Examples of parasitoids of oilseed rape pests.

Collecting samples from a beating tray, using a pooter.

Sweep netting.

Malaise trap fitted with a time-sorting device.

Omni-directional flight trap.

Yellow water-trap at top of crop canopy.

Yellow water-trap, baited with isothiocyanates.

Plate 5. Examples of collecting or monitoring equipment used in oilseed rape.

Funnel emergence trap, with collecting bottle on top.

Johnson and Taylor suction trap.

Funnel trap to collect larvae dropping from the crop canopy.

Window trap.

Plate 6. Examples of collecting or monitoring equipment used in oilseed rape.

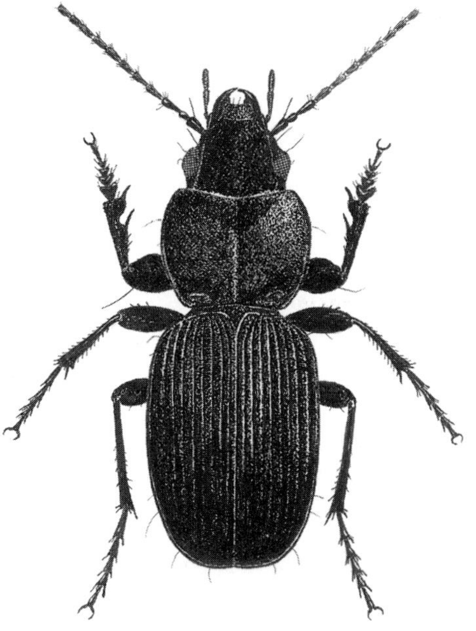

Fig. 12.1 *Pterostichus melanarius*, a typical ground beetle – family CARABIDAE (×4).

assessments of the enhancement in predation to be obtained following the establishment of 'beetle banks' and the maintenance of other on-farm structural element that will aid their season-to-season survival (see, for example, Sotherton, 1984, 1985). However, observations on predatory ground beetles in oilseed rape fields are relatively few, most studies being directed towards species or populations in cereal fields, or towards ground beetles and arable land in general. For an in-depth, up-to-date review, see Kromp (1999); this publication also examines the association of ground beetles with on-farm landscape structural elements, which are considered in this monograph in Chapter 15.

Persistence of species within a habitat is determined particularly by larval survival, as this is the most vulnerable stage in the lifecycle (Lövei & Sunderland, 1996). This is also true for various other groups of predators.

Over 20 species of ground beetle occur regularly in European oilseed rape fields. In detail, however, species composition within communities will inevitably vary from country to country and also from site to site, according to regional or local conditions; changes have also occurred over time, with larger species (e.g. *Carabus* spp.) nowadays being less abundant than formerly (see Chapter 15). Currently, at least in central Europe, the generally most frequently encountered species in oilseed rape fields (cited in descending order of frequency) are as follows: *Pterostichus melanarius*, *Anchomenus dorsalis*, *Amara similata*, *Nebria brevicollis*, *Poecilus cupreus*, *Trechus quadristriatus*, *Harpalus affinis*, *Bembidion lampros*, *Pseudoophonus rufipes*, *Bembidion tetracolum*, *Amara aenea*, *Loricera pilicornis*, *Amara ovata*, *Amara familiaris*, *Calathus cinctus*, *Harpalus tardus*, *Calathus fuscipes*, *Clivina fossor*, *Agonum muelleri*, *Notiophilus biguttatus*, *Calathus erratus*, *Carabus auratus*, *Pterostichus madidus* and *Bembidion properans*. Systematic details of these species, and also of various other predatory species associated with oilseed rape fields, are given in Table 12.1. To avoid confusion, the recent (Assmann *et al.*, 1998) revisions of the genus *Pterostichus* are ignored in the text of this monograph.

Table 12.1 Examples of ground beetles (Coleoptera: Carabidae) inhabiting European oilseed rape fields.

Tribe CARABINI
Carabus Linnaeus
– *auratus* Linnaeus†#
– *cancellatus* Illiger#
– *granulatus* Linnaeus#
– *nemoralis* Müller†#

Tribe NEBRIINI
Nebria Latreille
– *brevicollis* (Fabricius)†

Tribe NOTIOPHILINI
Notiophilus Duméril
– *biguttatus* (Fabricius)†#

Tribe LORICERINI
Loricera Latreille
– *pilicornis* (Fabricius)†

Tribe SCARITINI
Clivina Latreille
– *fossor* (Linnaeus)†

Tribe BROSCINI
Broscus Panzer
– *cephalotes* (Linnaeus)

Tribe TRECHINI
Trechus Clairville
– *quadristriatus* (Schrank)†

Tribe BEMBIDIINI
Asaphidion des Gozis
– *flavipes* (Linnaeus)†#
Bembidion Latreille
– *femoratum* Sturm
– *lampros* (Herbst)†
– *obtusum* Serville†
– *properans* Stephens†
– *quadrimaculatum* (Linnaeus)
– *tetracolum* Say†#

Tribe PTEROSTICHINI
Abax Bonelli
– *ater* (Villers) (= *parallelepipedus* (Piller & Mitterpacher))#
Agonum Bonnelli
– *bipustulatus* (Fabricius)

Tribe LEBIINI
Dromius Bonelli
– *muelleri* (Herbst)†#
Anchomenus Bonelli (= ***Platynus*** auct. partim)
– *assimilis* (Paykull) [Note 1]
– *dorsalis* (Pontoppidan)† [Note 2]
Argutor Dejean [Note 3]
– *vernalis* (Panzer)†#
Calathus Bonelli
– *cinctus* Motschulsky†
– *erratus* Sahlberg†#
– *fuscipes* (Goeze)†
– *melanocephalus* (Linnaeus)
Melanius Bonelli [Note 3]
– *anthracinus* (Illiger)†#
Morphnosoma Lutshnik [Note 3]
– *melanarius* (Illiger)†
Phonias Des Gozis [Note 3]
– *strenuus* (Panzer)†#
Poecilus Bonelli
– *cupreus* (Linnaeus)†#
– *lepidus* (Leske)†#
– *versicolor* (Sturm)†#
Pterostichus Bonelli
– *macer* (Marsham)
– *niger* (Schaller)†
– *nigrita* (Paykull)
– *oblongopunctatus* (Fabricius)†#
Steropidius Jeanne [Note 3]
– *madidus* (Fabricius)†
Stomis Clairville
– *pumicatus* (Panzer)† [Note 4]

Tribe AMARINI
Amara Bonelli†
– *aenea* (Degeer)†
– *communis* (Panzer)†
– *eurynota* (Panzer)†
– *familiaris* (Dufschmid)†
– *ovata* (Fabricius)†
– *plebeja* (Gyllenhal)†
– *similata* (Gyllenhal)†

Tribe HARPALINI
Harpalus Latreille
– *affinis* (Schrank) (= *aeneus* (Fabricius))†
– *tardus* (Panzer)†#
Ophonus Dejean
– *signaticornis* (Duftschmidt)†#
Pseudoophonus Motschulsky
– *rufipes* (Degeer)† [Note 5]

Tribe LICININI
Badister Clairville
– *melanocephalus* Dejean
Demetrias Bonelli
– *atricapillus* (Linnaeus)

† Dominant and subdominant species (according to publications considered, not only on species set listed).
# Uncommon in oilseed rape in northern Europe (at least in Sweden) (C. Nilsson, *pers. comm.*).

Note 1. Often cited as *Agonum assimile.*
Note 2. Often cited as *Agonum dorsale.*
Note 3. Formerly *Pterostichus* (see Assmann *et al.*, 1998), and cited as such elsewhere in this monograph.
Note 4. Spring rape only.
Note 5. Often cited as *Harpalus rufipes.*

Rove beetles (Staphylinidae)

Rove beetles (Fig. 12.2) form a very large group of predatory insects. There are, for example, almost 1,000 species in the British Isles, and many more within the whole of Europe. Of these, relatively few species occur regularly in oilseed rape fields, with no more than about 30 species being considered generally abundant in such situations (Table 12.2). However, except for large species (e.g. *Ocypus* spp., *Philonthus* spp. and *Quedius* spp.), the nutritional status of most rove beetle species is unclear. Most species (e.g. *Tachyporus* spp., Aleocharinae spp.) are said to be partly or temporarily fungivorous/saprophagous or at least mycetophilous, preferring habitats dominated by fungi or decaying plant parts. Further, some species of the genus *Aleochara* (e.g. *Aleochara bilineata*) are parasitoids (see Chapter 9), but they occur only in very small numbers in arable fields. Information on the nutritional status of a range of rove beetles is available as follows:

- *Amischa analis* — Hartmann (1979)
- *Anotylus rugosus* — Andersen *et al.* (1983); Tomlin *et al.* (1985); Eghtedar (1970)
- *Lesteva pubescens* — Good & Giller (1991)
- *Liogluta* spp. — Hartmann (1979)
- *Omalium* spp. — Steel (1970); Topp *et al.* (1982)
- *Oxypoda exoleta* — Zimmermann & Büchs (1999)
- *Philonthus cognatus* — Eghtedar (1970)
- *Tachyporus* spp. — Coaker & Williams (1963); Basedow (1973); Potts & Vickerman (1974); Sunderland (1975); Vickerman & Sunderland (1975); Andersen *et al.* (1983); Poehling *et al.* (1985); Janssens & de Clercq (1990); Dennis *et al.* (1991).

Very little information is available on the species composition of rove beetles in oilseed rape fields. The only concise work is that of Krooß (1996), done in the region of Göttingen, Germany. Except for the most abundant group (the Aleocharinae), which was neglected owing to taxomonic difficulties, the most frequently encountered rove beetles in winter oilseed rape fields included the following: *Anotylus inustus*, *A. rugosus*, *A. sculpturatus*, *A. tetracarinatus*, *Lathrobium fulvipenne*, *Lesteva longoelytrata*, *Mycetoporus splendidulus*,

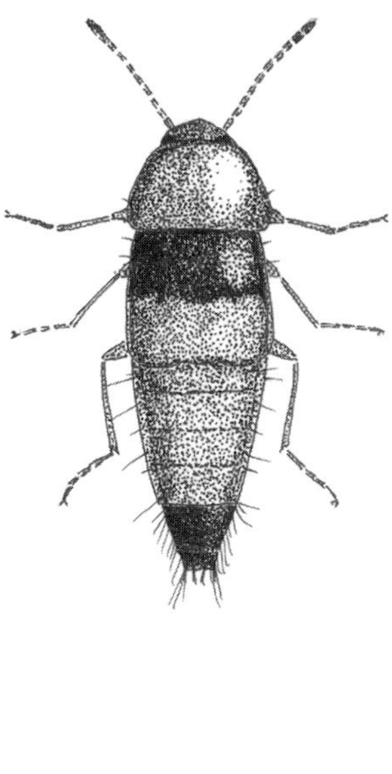

Fig. 12.2 *Tachyporus hypnorum*, a typical rove beetle – family STAPHYLINIDAE (×10).

Table 12.2 Examples of rove beetles (Coleoptera: Staphylinidae) inhabiting European oilseed rape fields.

Subfamily PROTEININAE
Proteinus Latreille†*

Subfamily OMALIINAE
Acrolocha Thomson
– *minuta* (Olivier)†*
Lesteva Latreille
– *longoelytrata* (Goeze) (= *longelytrata*, variant name)†#*
Olophrum Erichson
– *assimile* (Paykull)†*
Omalium Gravenhorst
– *caesum* Gravenhorst†#*
– *rivulare* (Paykull)†#*

Subfamily OXYTELINAE
Anotylus Thomson (= **Oxytelus** in part)
– *insecatus* (Gravenhorst)†
– *inustus* (Gravenhorst)†#
– *rugosus* (Fabricius)†#
– *sculpturatus* (Gravenhorst)†#*
– *tetracarinatus* (Block)†#*
Coprophilus Latreille
– *striatulus* (Fabricius)#
Platystethus Mannerheim
– *nitens* (Sahlberg)†*

Subfamily STENINAE
Stenus Latreille#

Subfamily PAEDERINAE
Lathrobium Gravenhorst
– *fulvipenne* (Gravenhorst)†#
– *longulum* (Gravenhorst)#

Subfamily STAPHYLININAE
Gabrius *Stephens*
– *nigritulus* (Gravenhorst)
Philonthus *Stephens*
– *carbonarius* (Gyllenhal)†
– *cognatus* Stephens (= *fuscipennis* (Mannerheim))†#
– *ebeninus* (Gravenhorst)†*
– *rotundicollis* (Ménétriés)†#*
Staphylinus Linnaeus (= **Ocypus** Samouelle)
– **nero** Faldermann (= *similis* (Fabricius))#
Xantholinus Dejean
– *linearis* (Olivier)†#
– *jarrigei* Coiffait (= *semirufus* Steel)†#*

Subfamily TACHYPORINAE
Mycetoporus Mannerheim
– *splendidulus* (Gravenhorst)†
Tachinus Gravenhorst
– *signatus* Gravenhorst (= *rufipes* Degeer)†#
Tachyporus Gravenhorst
– *chrysomelinus* (Linnaeus)†# [Note 1]
– *hypnorum* (Fabricius)†#
– *nitidulus* (Fabricius)†
– *obtusus* (Linnaeus)#
– *solutus* Erichson#

Subfamily ALEOCHARINAE
Aleochara Gravenhorst#
– *bilineata* Gyllenhall†
Atheta Thomson# [Note 2]
Gyrophaena Mannerheim#
Lomechusa Gravenhorst (= **Atemeles** Dillwyn)#*
Oxypoda Mannerheim#*

† Dominant or subdominant species (according to publications considered not only on species set listed).
# Included in key in Box 13.2.
* Uncommon in oilseed rape in northern Europe (at least in Sweden) (C. Nilsson, *pers. comm.*).

Note 1. Recently identified as a complex of three species.
Note 2. *Atheta* spp. are very common in oilseed rape fields in Sweden; other common species include: *Xantholinus tricolor* (Fabricius) and *Gyrohypnus augustulus* Stephens (Staphylininae), *Amischa analis* (Gravenhorst) and *Drusila* (= *Astilbus*) *canaliculatus* (Fabricius) (Aleocharinae) (C. Nilsson, *pers. comm.*).

Omalium caesum, *O. rivulare*, *Philonthus cognatus*, *P. ebeninius*, *P. rotundicollis*, *Proteinus* spp., *Tachyporus hypnorum*, *T. nitidulus*, *T. rufipes*, *T. solutus* and *Xantholinus linearis*. Of these, *Tachyporus* spp. and, to a lesser extent, *Philonthus* spp. were most abundant. Other species sometimes reported as numerous in oilseed rape crops include *Philonthus carbonarius* and *Platystethus nitens*.

In spring rape, Volkmar *et al.* (1999) found *Anotylus insecatus*, *A. inustus*, *A. rugosus*, *Philonthus carbonarius*, *P. cognatus*, *Platystethus nitens* and *Tachyporus hypnorum* as dominant species. Therefore, the rove beetle fauna of spring rape fields appears to be similar to that of winter rape fields but without species such as *Lesteva longoelytrata*, *Omalium caesum* and *O. rivulare*, which are active mainly in winter.

As indicated above, members of the subfamily Aleocharinae (although often numerous in arable crops such as oilseed rape) tend to be neglected owing to difficulties in determining species. The same is true of the larval stages of rove beetles in general.

Ladybirds (Coccinellidae)

Ladybirds (Fig. 12.3) (Table 12.3) are well-known predators of aphids but are of limited importance in oilseed rape (see Chapter 16). Tischler (1958) recorded adults and larvae of *Coccinella septempunctata* and *Propylea quattuordecimpunctata* as frequently occurring in an oilseed rape crop rotation.

Soldier beetles (Cantharidae)

Soldier beetles (Fig. 12.4) (Table 12.3) are predatory in both their adult and larval stages. Various species occur in agro-ecosystems, and larvae of *Cantharis* spp. are often abundant in arable fields (Langenstück *et al.*, 1998). Tischler (1958) found adults and larvae of *Cantharis fusca*, *Cantharis livida* and *Rhagonycha fulva* in fields of an oilseed rape crop rotation, as well as in adjacent forest edges, at Schleswig-Holstein, Germany. More recently, W. Büchs (*unpublished*) reared adult soldier beetles from several larvae collected from oilseed rape fields in the autumn; *Cantharis lateralis* was the most numerous species.

Other predatory beetles

Tischler (1958) mentioned the presence of carrion beetles (Silphidae) (Table 12.3) in cultivated areas that include oilseed rape in the crop rotation. Although the family includes well-known scavengers, adults of many species are predatory. On loamy soils, he recorded *Silpha tristis*; on more-sandy soils, *Silpha obscura* was one of the most dominant of all beetles. Both species are to some extent carnivorous, but they feed also on decaying animals. Tischler also found *Nicrophorus vespillo* and *Thanatophilus sinuatus* in the fields; these species are known to prey on insects up to the size of larger members of the family Scarabaeidae (chafer beetles). In addition to carrion beetles, two histerids (*Grammostethus marginatus* and *Paralister purpurascens*) (Histeridae) (Table 12.3) occur regularly in arable fields treated with stable manure, where they feed on dipterous larvae and pupae. However, owing to today's crop production techniques (e.g. no manure applications and, often, no swathing before harvesting) these beetles are no longer common in oilseed rape fields.

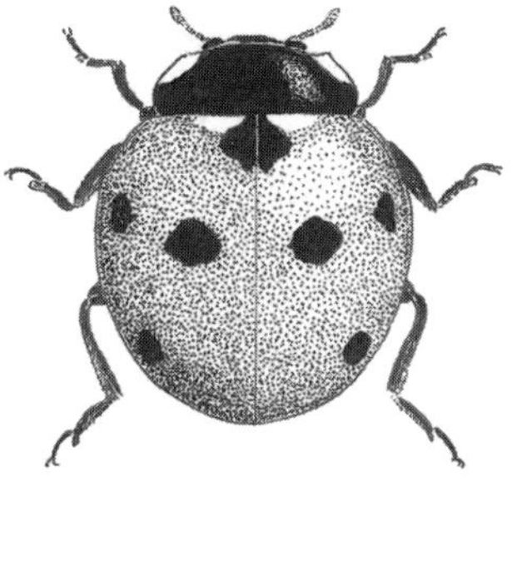

Fig. 12.3 *Coccinella septempunctata*, a typical ladybird – family COCCINELLIDAE (×4).

Table 12.3 Examples of miscellaneous predatory beetles (Coleoptera) inhabiting European oilseed rape fields.

Family CANTHARIDAE
- ***Cantharis*** Linnaeus
  - – *fusca* Linnaeus
  - – *lateralis* Linnaeus
  - – *livida* Linnaeus
- ***Rhagonycha*** Eschscholz
  - – *fulva* (Scopoli)

Family COCCINELLIDAE
- Subfamily COCCINELLINAE
  - ***Coccinella*** Linnaeus
    - – *septempunctata* Linnaeus
  - ***Propylea*** Mulsant
    - – *quattuordecimpunctata* Linnaeus

Family HISTERIDAE
- ***Grammostethus*** Lewis
  - – *marginatus* (Erichson) [Note 1]
- ***Paralister*** Bickhardt
  - – *purpurascens* (Herbst) [Note 2]

Family MELYRIDAE
- ***Malachius*** Fabricius

Family SILPHIDAE
- ***Nicrophorus*** Fabricius (= ***Necrophorus***: misspelling)
  - – *vespillo* (Linnaeus)
- ***Phosphuga*** Leach
  - – *atrata* Linnaeus [Note 3]
- ***Silpha*** Linnaeus
  - – *obscura* Linnaeus
  - – *tristis* Illiger
- ***Thanatophilus*** Leach
  - – *sinuatus* (Fabricius)

Note 1. Also cited as *Margarinotus marginatus.*
Note 2. Also cited as *Margarinotus purpurascens.*
Note 3. Also cited as *Silpha atrata* Linnaeus.

True flies (Insecta: Diptera)

Large numbers of flies occur in oilseed rape crops, but most are of no economic importance, either as pests or as natural enemies (i.e. as biocontrol agents). Nevertheless, a few parasitoids (Tachinidae – see Chapter 9) and a wider range of predators are represented. Brief details of the main predatory groups to be found in oilseed rape crops are given below, and systematic details are given in Table 12.4.

Dance flies (Empidoidea, e.g. Empididae and Hybotidae)

Larvae of dance flies are almost exclusively predacious, preying on various insects but mainly on the larvae of other Diptera. The adults (Fig. 12.5) either are predacious or feed on nectar or pollen obtained from flowers (Chvala, 1983); their prey comprises mainly

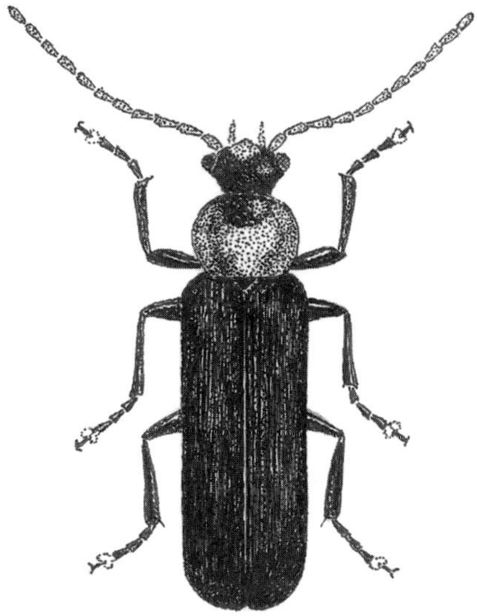

Fig. 12.4 *Cantharis fusca*, a typical soldier beetle – family CANTHARIDAE (×4).

Table 12.4 Examples of dominant and subdominant predatory flies (Diptera) in European oilseed rape fields.

Suborder BRACHYCERA
- Family RHAGIONIDAE
  - ***Rhagio*** Fabricius
    - – *scolopacea* (Linnaeus)
    - – *tringarius* (Linnaeus)
- Family THEREVIDAE
  - ***Dialineura*** Rondani
    - – *anilis* (Linnaeus)
- Family EMPIDIDAE
  - ***Hilara*** Meigen
    - – *anglodanica* Lundbeck
    - – *monedula* Collin
  - ***Microphorus*** Macquart (= ***Microphor*** Macquart)
    - – *holosericeus* (Meigen)
  - ***Phyllodromia*** Zetterstedt
- Family HYBOTIDAE
  - ***Crossopalpus*** Bigot
  - ***Drapetis*** Meigen
  - ***Elaphropeza*** Macquart
  - ***Platypalpus*** Macquart
    - – *annulipes* (Meigen)
    - – *interstinctus* (Collin)
    - – *minutus* (Meigen)
    - – *pallidicornis* (Collin)
    - – *pallidiventris* (Meigen)
  - ***Tachydromia*** Meigen
  - ***Tachypeza*** Meigen
- Family DOLICHOPODIDAE
  - ***Campsicnemus*** Haliday
    - – *curvipes* (Fallén)
  - ***Chrysotus*** Meigen
    - – *collini* Parent
  - ***Dolichopus*** Latreille
    - – *acuticornis* Wiedemann
    - – *longicornis* Stannius
  - ***Medetera*** Fischer von Waldheim
    - – *micacea* Loew
    - – *truncorum* Meigen

Suborder CYCLORRHAPHA
- Family SYRPHIDAE
  - ***Episyrphus*** Matsumura & Adachi
    - – *balteatus* (Degeer)
- Family MUSCIDAE
  - Subfamily MUSCINAE
    - ***Muscina*** Robineau-Desvoidy
      - – *stabulans* (Fallén)
  - Subfamily PHAONIINAE
    - ***Helina*** Robineau-Desvoidy
      - – *setiventris* Ringdahl
      - – *quadrum* (Fabricius)
    - ***Phaonia*** Robineau-Desvoidy
      - – *trimaculata* (Bouché)
  - Subfamily COENOSIINAE
    - ***Coenosia*** Meigen
      - – *intermedia* (Fallén)
      - – *tigrina* (Fabricius)

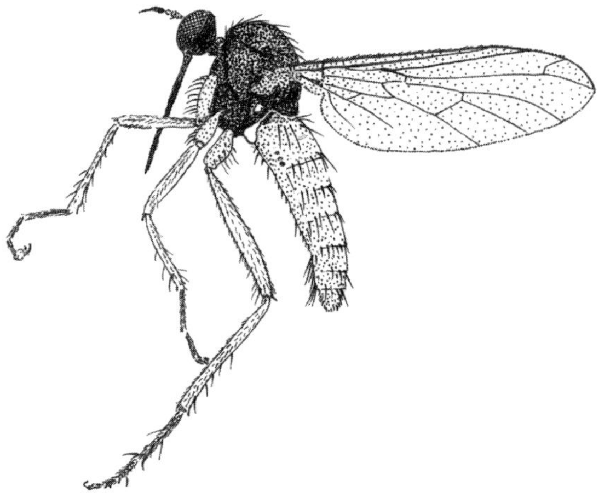

Fig. 12.5 A dance fly – family HYBOTIDAE (×5).

small dipterans (e.g. Agromyzidae and Cecidomyiidae), including adults of brassica pod midge (Prescher & Büchs, 1998).

Hover flies (Syrphidae)

Adult hover flies feed on nectar and pollen that they collect from flowers. The larvae of several species, however, are predatory and important enemies of aphids, including cabbage aphid (van Emden, 1961; Hughes, 1963). The main species to attack cabbage aphid is *Episyrphus balteatus* (Pollard, 1969, 1971). However, because numbers of larvae tend to be greatest in summer, long after cabbage aphid colonies have become established (see Chapter 2), their impact on populations in oilseed rape crops (whether winter rape or spring rape) is limited. Nevertheless, *E. balteatus* is regarded as an important aphidophagous predator in oilseed rape crops, where it often occurs in abundance (Büchner, 1995).

Long-legged flies (Dolichopodidae)

Long-legged flies (Fig. 12.6) (e.g. *Dolichopus acuticornis*, *D. longicornis*, *Medetera* and *Sciapus*) are predacious as adults and larvae, their victims including mainly other Diptera, springtails (Collembola), thrips (Thysanoptera) and psocids (Psocoptera) (Assis-Fonseca, 1978). According to Laurence (1951), they also include Hemiptera in their diet. With regards oilseed rape pests, their main prey is assumed to be brassica pod midge (S. Prescher, *pers. comm.*).

Muscid flies (Muscidae)

Most muscid flies inhabiting oilseed rape fields are saprophagous, with only adults of the genus *Coenosia* being known to be predacious on other insects. However, larvae of many members of the Muscidae are carnivorous in their second, third and fourth instars. With regards oilseed rape crops, larvae of the false stable fly (*Muscina stabulans*) occur within

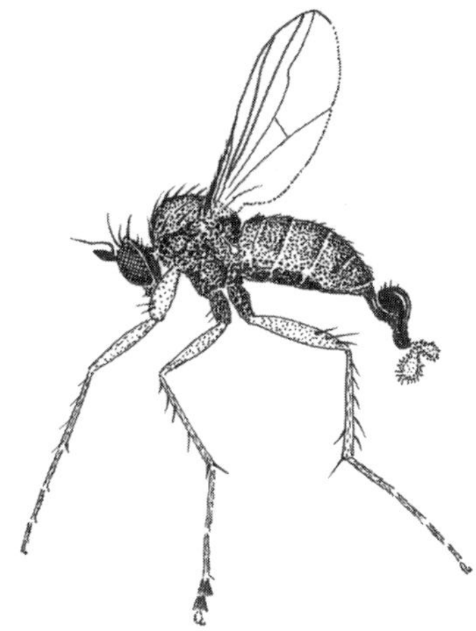

Fig. 12.6 A male long-legged fly – family DOLICHOPODIDAE (×5).

the stems of plants, where they are predators of cabbage stem weevil (*Ceutorhynchus pallidactylus*) larvae. Larvae of the muscid *Phaonia trimaculata* have also been reported (Fritzsche, 1955; Jourdheuil, 1960) to attack larvae of cabbage stem weevil within the stems of oilseed rape plants (and also those of rape stem weevil, *Ceutorhynchus napi*), but the stems must first have split open to allow access to the egg-laying flies. The larvae of *Phaonia trimaculata* can also occur at the basal parts of the stems and around the roots of brassicaceous plants (including oilseed rape), where they feed on cabbage root fly (*Delia radicum*) larvae (Shaw, 1970; Stubbs & Chandler, 1978). Adults of *Coenosia tigrina* are also known to attack the larvae of cabbage root fly (Perron *et al.*, 1956).

Snipe flies (family Rhagionidae)

Larvae of snipe flies (e.g. *Rhagio* spp.) (Fig. 12.7) have been recorded in oilseed rape fields, where they feed on small insects and earthworms; adults of *Rhagio* are also predacious. Effects on oilseed rape pests, however, are unknown.

Stiletto flies (Therevidae)

Stiletto flies (e.g. *Dialineura* spp. and *Thereva* spp.) (Fig. 12.8) are well-known predators of other insects. The soil-inhabiting larvae, which are often numerous in oilseed rape fields, are also partly carnivorous, feeding (in their third and fourth instars) on beetle larvae (including wireworms – Elateridae) and pupae of Lepidoptera (Kröber, 1930).

Miscellaneous predatory insects

Lacewings (Insecta: Neuroptera = Planipennia)

Lacewings are carnivorous as both adults and larvae. Members of two families (Chrysopidae and Hemerobiïdae) occur in oilseed rape fields. However, they play only

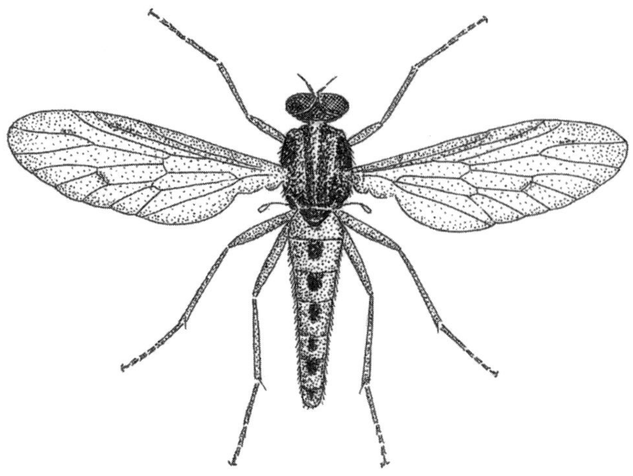

Fig. 12.7 *Rhagio scolopacea*, a typical snipe fly – family RHAGIONIDAE (×3).

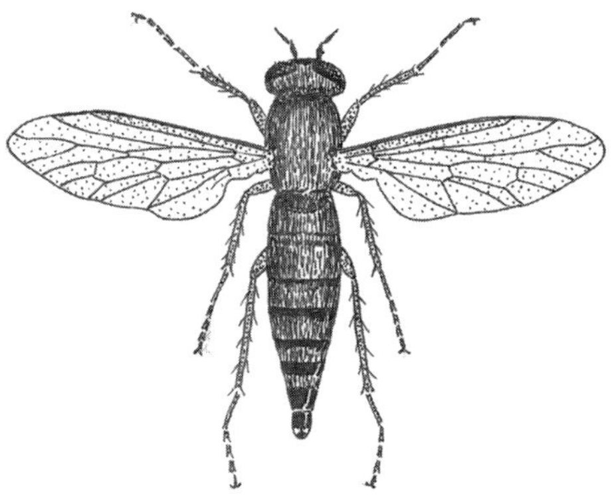

Fig. 12.8 *Thereva lunulata*, a typical stiletto fly – family THEREVIDAE (×3).

a minor part as predators in oilseed rape crops, with *Chrysoperla carnea* (Chrysopidae), as a predator of aphids, being most frequently reported. For further information on the feeding habits of lacewings and their role as predators in crops see McEwan *et al.* (2001).

Predatory bugs (Insecta: Heteroptera)

Plant bugs in various families (e.g. Anthocoridae, Geocoridae, Miridae, Nabidae, Pentatomidae and Saldidae) are well-known predators; see Wheeler (2001) for an up-to-date survey of their feeding habits. However, although often important natural enemies of pests in perennial crops (such as orchards), predatory bugs are of lesser importance in arable situations.

Other predatory insects

Among thirteen species of thrips (Thysanoptera) found on winter and spring rape in Bulgaria, members of two predatory genera (Aeolothripidae) were recorded (Dochev & Kircheva, 1990). Predatory thrips, however, are not known to play an important role as predators of oilseed rape pests; nor are predatory ants (Hymenoptera: Formicidae) or earwigs (Dermaptera: Forficulidae).

Spiders (Arachnida: Araneae)

Spiders are well-known predators yet, compared with insects such as ground beetles, they have received relatively little attention as natural enemies of crop pests. A wide range of species can occur in arable fields, of which money spiders (Linyphiidae) and wolf spiders (Lycosidae) are most abundant. Harenberg (1997) recorded the following as the most numerous species in oilseed rape fields in Germany:

- Linyphiidae — *Araeoncus humilis*, *Bathyphantes gracilis*, *Centromerita bicolor*, *Erigone atra*, *E. dentipalpis*, *Lepthyphantes tenuis*, *Meioneta rurestris*, *Oedothorax apicatus*, *O. retusus*, *Pelecopsis parallela*, *Porrhomma microphthalmum* and *Stemonyphantes lineatus*
- Tetragnathidae — *Pachygnatha degeeri*
- Theridiidae — *Theridion impressum* [the most abundant spider reported in oilseed rape by Nyfeller & Benz (1979)]
- Lycosidae — *Pardosa palustris*
- Salticidae — *Salticus scenicus*

Brief information on a range of families represented in oilseed rape fields is given below (see Table 12.5 for systematic details).

Crab spiders (Thomisidae)

During the flowering period, crab spiders (Fig. 12.9) sometimes ambush insects that visit the open blossoms of oilseed rape plants; the pale-coloured and generally common species *Misumena vatia* is probably the most frequently encountered species. Crab spiders do not produce webbing but instead await the arrival of their prey and then jump upon them.

Money spiders (Linyphiidae)

Money spiders are the most frequent spider taxon in arable crops in general, and are the most abundant group to inhabit oilseed rape. Many are typical 'pioneer' species, which primarily recolonize disturbed or newly generated habitats. This is valid also for fields after tillage. Money spiders are very small and, typically, they construct tiny horizontal webs on the soil surface, spun between soil clumps, crevices or plant debris. The most frequent species in arable fields (including oilseed rape) belong to the genera *Bathyphantes*, *Erigone*, *Lepthyphantes*, *Oedothorax* and *Porrhomma*.

Wolf spiders (Lycosidae)

Wolf spiders (e.g. members of the genera *Lycosa*, *Pardosa* and *Trochosa*) (Fig. 12.10) are typically epigaeic predators. Within oilseed rape crops, their prey includes pollen beetle larvae and brassica pod midge larvae that have dropped to the ground prior to pupation.

Table 12.5 Examples of spiders (Araneae) inhabiting European oilseed rape fields.

Family DICTYNIDAE
- ***Argenna*** Thorell
  - – *subnigra* (Pickard-Cambridge)

Family GNAPHOSIDAE
- ***Zelotes*** Gistel
  - – *lutetianus* (Koch)
  - – *pusillus* (Koch)

Family CLUBIONIDAE
- ***Clubiona*** Latreille
  - – *reclusa* Pickard-Cambridge

Family THOMISIDAE
- ***Misumena*** Latreille
  - – *vatia* (Clerck)
- ***Xysticus*** Koch

Family SALTICIDAE
- ***Salticus*** Latreille
  - – *scenicus* (Clerck)†#

Family LYCOSIDAE
- ***Alopecosa*** Simon (= ***Tarentula*** Sundevall)
  - – *pulverulenta* (Clerck)
- ***Lycosa*** Latreille
- ***Pardosa*** Koch
  - – *agrestis* Westring
  - – *amentata* (Clerck)†
  - – *palustris* (Linnaeus)†
  - – *pullata* (Clerck)
- ***Trochosa*** Koch
  - – *terricola* Thorell
- ***Xerolycosa*** Dahl
  - – *miniata* (Koch)

Family TETRAGNATHIDAE
- ***Pachygnatha*** Sundevall
  - – *clercki* Sundevall
  - – *degeeri* Sundevall†

Family THERIDIIDAE
- ***Theridion*** Walckenaer
  - – *impressum* Koch†#

Family ARGIOPIDAE (= ARANEIDAE)
- ***Araneus*** Clerck
  - – *cornutus* Clerck
  - – *cucurbitinus* Clerck

Family LINYPHIIDAE

Subfamily ERIGONINAE
- ***Araeoncus*** Simon
  - – *humilis* (Blackwall)†
- ***Dicymbium*** Menge
  - – *nigrum* (Blackwall)
- ***Erigone*** Audouin
  - – *atra* (Blackwall)†
  - – *dentipalpis* (Wider)†
  - – *longipalpis* (Sundevall)
- ***Micrargus*** Dahl
  - – *herbigradus* (Blackwall)†#
- ***Mioxena*** Simon
  - – *blanda* (Simon)†#
- ***Oedothorax*** Bertkau
  - – *agrestis* (Blackwall)
  - – *apicatus* (Blackwall)†
  - – *fuscus* (Blackwall)
  - – *retusus* (Westring)†
- ***Pelecopsis*** Simon
  - – *parallela* (Wider)†
- ***Savignia*** Blackwall
  - – *frontata* Blackwall
- ***Walckenaera*** Blackwall
  - – *acuminata* Blackwall
  - – *capito* (Westring)† [Note 1]
  - – *nudipalpis* (Westring)† [Note 2]
  - – *vigilax* (Blackwall)†# [Note 3]

Subfamily LINYPHIINAE
- ***Bathyphantes*** Menge
  - – *gracilis* (Blackwall)†
- ***Centromerita*** Dahl
  - – *bicolor* (Blackwall)†#
- ***Centromerus*** Dahl
  - – *sylvaticus* (Blackwall)†#
- ***Diplostyla*** Emerton
  - – *concolor* (Wider)†# [Note 4]
- ***Lepthyphantes*** Menge
  - – *tenuis* (Blackwall)†
- ***Meioneta*** Hull
  - – *rurestris* (Koch)†
- ***Porrhomma*** Simon
  - – *microphthalmum* (Pickard-Cambridge)†
  - – *pygmaeum*†#
- ***Stemonyphantes*** Menge
  - – *lineatus* (Linnaeus)

† Dominant or subdominant species (according to publications considered, not only on species set listed).
# Uncommon in oilseed rape in northern Europe (at least in Sweden) (C. Nilsson, *pers. comm.*).

Note 1. Also cited as *Widera capito*.
Note 2. Also cited as *Trachynella nudipalpis*.
Note 3. Also cited as *Cornicularia vigilax*.
Note 4. Also cited as *Bathyphantes concolor*.

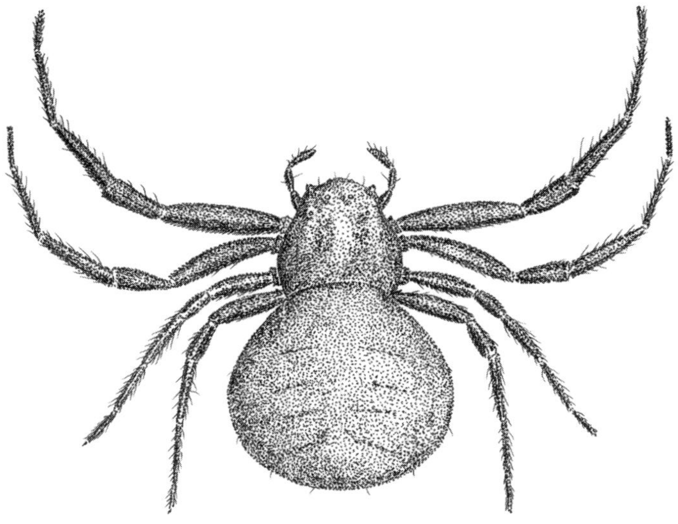

Fig. 12.9 A crab spider – family THOMISIDAE (×10).

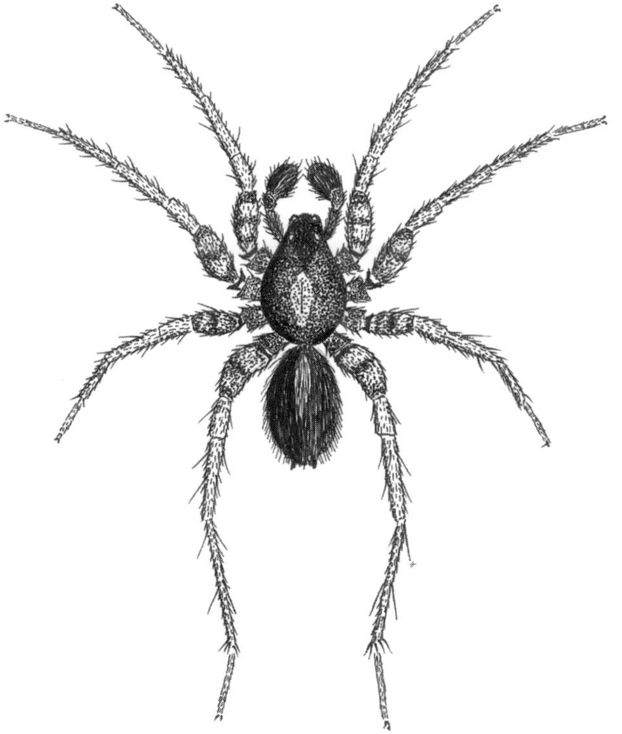

Fig. 12.10 A male wolf spider – family LYCOSIDAE (×5).

Other spiders

Orb-web producers (e.g. *Araneus* spp. – Argiopidae; *Pachygnatha* spp. – Tetragnathidae) sometimes inhabit arable crop plants, but they are of only minor significance as predators of oilseed rape pests.

The theridiid *Theridion impressum* (Theridiidae) forms webs amongst the inflorescences or terminal clusters of developing pods of oilseed rape plants. Here, they are able to capture a range of insects, including aphids, midges, thrips and, to a lesser extent, adult pollen beetles. The web density of *T. impressum* is determined by structure of the available vegetation. This species prefers 'spacey' plants with vertical branches as sites for web construction; oilseed rape plants (unlike grasses), therefore, are ideal.

Jumping spiders (Salticidae) (e.g. *Salticus scenicus*) often occur in oilseed rape fields, especially in sunny situations; they do not construct webs but stalk their prey and then pounce upon them. Other spiders that catch prey without using webs include *Clubiona* (Clubionidae) and *Zelotes* (Gnathosidae); these spiders often hide during the daytime in silken retreats spun beneath stones or other shelter but are generally uncommon in oilseed rape fields.

Miscellaneous invertebrates

Centipedes (Chilopoda)

Although centipedes occur in oilseed rape crops (e.g. Büchs *et al.*, 1999), no species have been highlighted as of particular significance with regards any impact on oilseed rape pests. The only published records of actual species in oilseed rape fields are those of Tischler (1958). In cultivated areas (including oilseed rape) he found three species in small numbers (*Geophilus electricus*, *G. longicornis* and, occasionally, *Lithobius forficatus*); in addition, but without actual reference to oilseed rape, he found *Lithobius calcaratus* and *L. crassipes*. The only centipede that occurred regularly (nearly 75% of all centipedes found) in all sampled crops – between May and October as adults and in June/July as sub-adults – was *Lamyctes* (= *Hemicops*) *fulvicornis*.

Harvestmen (Arachnida: Opiliones)

Several species of harvestmen (e.g. *Laccinius ephipiatus*, *Mitopus morio* and *Platybunus triangularis*) have been found in oilseed rape crops (Tischler, 1958; Büchs *et al.*, 1999). Harvestmen are partly saprophagous, but they also prey upon small insects (Immel, 1955). Their role as carnivores in oilseed rape fields, however, is not known.

Acknowledgements

This work was part-funded by the European Commission Concerted Action: *Minimizing pesticide use and environmental impact by the development and promotion of bio-control strategies for oilseed rape pests, FAIR 96-1314.*

References

Andersen, A., Hansen, A. G., Rydland, N. & Oeyre, G. (1983). Carabidae and Staphylinidae (Col.) as predators of eggs of the turnip root fly *Delia floralis* Fallén. *Zeitschrift für Angewandte Entomologie* **95**, 499–506.

Assis-Fonseca, E. C. M. de (1978). Diptera: Dolichopodidae. *Handbooks for the Identification of British Insects* **9** (5), 1–90.

Assmann, T., Balkenohl, H., Daffner, H., Gebert, J., Hieke, F., Lohse, G. A., Lompe, A., Müller-Motzfeld, G., Schmidt, J., Trautner, J. & Wrase, D. W. (1998). *Ergänzungen und Berichtigungen zu 'Die Käfer Mitteleuropas' 1. Familie Carabidae.* In: W. Lucht & C. Klausnitzer (eds). *Die Käfer Mitteleuropas. 4. Supplementband Vol. 15.* Goecke & Evers: Krefeld.

Basedow, T. (1973). Der Einfluß epigäischer Raubarthropoden auf die Abundanz phytophager Insekten in der Agrarlandschaft. *Pedobiologia* **13**, 410–22.

Büchner, S. (1995). *Die Dipterenfauna unterschiedlich bewirtschafteter Ackerflächen.* PhD thesis, University of Göttingen, Germany.

Büchs, W., Harenberg, A., Prescher, S., Weber, G. Hattwig, F. (1999). Entwicklung von Evertebratenzönosen bei verschiedenen Formen der Flächenstilllegung und Extensivierung. *Mitteilungen aus der Biologischen Bundesanstalt für Land- und Forstwirtschaft Berlin-Dahlem* **368**, 9–38.

Chvala, M. (1983). The Empoidea (Diptera) of Fennoscandia and Denmark II. General Part. The families Hybotidae, Atelestidae and Microphoridae. *Fauna Entomologica Scandinavica* **12**, 1–279.

Coaker, T. H. & Williams, D. A. (1963). The importance of some Carabidae and Staphylinidae as predators of the cabbage root fly *Erioischia brassicae* (Bouche). *Entomologia Experimentalis et Applicata* **6**, 156–64.

Dennis, P., Wratten, S. D. & Sotherton, N. W. (1991). Mycophagy as a factor limiting predation of aphids (Hemiptera: Aphididae) on staphylinid beetles (Coleoptera: Staphylinidae) in cereals. *Bulletin of Entomological Research* **81**, 25–31.

Dochev, K. & Kircheva, M. (1990). Thrips (Thysanoptera) on the rape in Bulgaria. *Proceedings of the 3rd International Symposium on Thysanoptera, Kazimierz Dolny, Poland, 11–16 June 1990*, pp. 97–100.

Eghtedar, E. (1970). Zur Biologie und Ökologie der Staphyliniden *Philonthus fuscipennis* Mannh. und *Oxytelus rugosus* Grav. *Pedobiologia* **10**, 169–79.

Emden, H. F. van (1961). The effect of uncultivated land on the distribution of cabbage aphid (*Brevicoryne brassicae*) and on an adjacent crop. *Journal of Applied Ecology* **1**, 171–96.

Fritzsche, R. (1955). *Phaonia trimaculata* Bouché als Parasit des Großen Kohltriebrüßlers *Ceuthorrynchus napi* Gyll. und des Gefleckten Kohltriebrüßlers *Ceuthorrynchus quadridens* Panz. *Nachrichtenblatt für den Deutschen Pflanzenschutzdienst (Berlin)* **9**, 35–6.

Goltermann, S. (1994). *Das Auftreten von Laufkäfern (Col., Carabidae) auf Winterrapsfeldern und deren Einfluß auf den Massenwechsel von* Meligethes aeneus *F. (Col., Nitidulidae).* PhD thesis, University of Rostock, Germany.

Good, J. A. & Giller, P. S. (1991). The diet of predatory staphylinid beetles. *Entomologist's Monthly Magazine* **127**, 77–89.

Harenberg, A. (1997). *Auswirkungen extensiv geführter Anbausysteme in verschiedenen Fruchtfolgen (Raps-, Zuckerrübenfruchtfolge) und einer selbstbegrünenden Dauerbrache auf Spinnen (Arachnida: Araneae).* PhD thesis, Technical University of Braunschweig, Germany.

Hartmann, P. (1979). *Biologisch-ökologische Untersuchungen an Staphylinidenpopulationen verschiedener Ökosysteme des Solling.* PhD thesis, University of Göttingen, Germany.

Hughes, R. D. (1963). Population dynamics of the cabbage aphid *Brevicoryne brassicae* (L.). *Journal of Animal Ecology* **32**, 393–424.

Immel, V. (1955). Einige Bemerkungen zur Biologie von *Platybunus bucephalus* (Opiliones, Eupnoi). *Zoologische Jahrbücher für Systematik* **83**, 475–84.

Janssens, P. & Clercq, R. de (1990). Observations on Carabidae, Staphylinidae and Araneae as predators of cereal aphids in winter wheat. *Mededelingen van de Faculteit Landbouwwetenchappen Rijksuniversiteit Gent* **55**, 471–5.

Jourdheuil, P. (1960). Influence de quelques facteurs écologiques sur les fluctuations de population d'une biocénose parasitaire: étude relative à quelques Hyménoptères (Ophioninae,

Diospilinae, Euphorinae) parasites de divers Coléoptères inféodés aux Crucifères. *Annales des Épiphyties* **11**, 445–658.

Kröber, O. (1930). *Familie Therevidae (Stilettfliegen)*. In: *Die Tierwelt Deutschlands und der angrenzenden Meeresteile. 26. Teil – Zweiflügler oder Diptera*. Gustav Fischer: Jena.

Kromp, B. (1989). Carabid beetle communities (Carabidae, Coleoptera) in biologically and conventionally farmed agroecosystems. *Agriculture, Ecosystems and Environment* **27**, 241–51.

Kromp, B. (1999). Carabid beetles in sustainable agriculture: a review on pest control efficacy, cultivation impacts and enhancement. *Agriculture, Ecosystems and Environment* **74**, 187–228.

Krooß, S. (1996). *Staphyliniden in abgestuft extensiv bewirtschafteten Agrarökosystemen: Populationsdynamik, Gemeinschaftsstruktur und Bedeutung als Prädatoren*. PhD thesis, University of Göttingen, Germany.

Langenstück, C., Heimbach, U. & Larink, O. (1998). Larven der Cantharidae (Insecta: Coleoptera) auf Ackerflächen in SO-Niedersachsen und Aspekte ihrer Biologie. *Braunschweiger Naturkundliche Schriften* **5**, 551–68.

Laurence, B. R. (1951). The prey of some tree trunk frequenting Empididae and Dolichopodidae (Dipt.). *Entomologist's Monthly Magazine* **87**, 166–9.

Lövei, G. L. & Sunderland, K. D. (1996). Ecology and behaviour of ground beetles (Coleoptera: Carabidae). *Annual Review of Entomology* **41**, 231–6.

Luff, M. L. (1987). Biology of polyphagous ground beetles in agriculture. *Agricultural Zoology Reviews* **2**, 237–78.

McEwan, P., New, T. R. & Whittington, A. E. (eds) (2001). *Lacewings in the Crop Environment*. Cambridge University Press: Cambridge.

Nyfeller, M. & Benz, G. (1979). Zur ökologischen Bedeutung der Spinnen der Vegetationsschicht von Getreide- und Rapsfeldern. *Zeitschrift für Angewandte Entomologie* **87**, 348–76.

Pawlizki, K.-H. (1984). Auswirkungen abgestufter Produktionsintensitäten auf die Aktivitätsabundanz von Feldcarabiden (Coleopterea, Carabidae) sowie auf die Selbstregulation von Agrarökosystemen. *Bayerische Landwirtschaftliche Jahrbücher* **61**, Sonderheft **2**, 11–40.

Perron, J. P., Leroux, E. J. & Lafrance, J. (1956). Notes on *Coenosia tigrina* (F.) (Diptera: Anthomyiidae), mainly on habits and rearing. *Canadian Entomologist* **138**, 608–11.

Poehling, H. M., Dehne, H. W. & Sprick, P. (1985). Untersuchungen zur Bedeutung von Carabiden und Staphyliniden als Blattlausantagonisten in Winterweizenfeldern und deren Beeinträchtigung durch insektizide Wirkstoffe. *Mededelingen van de Faculteit Landbouwwetenchappen Rijksuniversiteit Gent* **50**, 519–30.

Pollard, E. (1969). The effect of removal of arthropod predators on an infestation of *Brevicoryne brassicae* (Hemiptera, Aphididae) on Brussels sprouts. *Entomologia Experimentalis et Applicata* **12**, 118–24.

Pollard, E. (1971). Hedges VI. Habitat diversity and crop pests: a study of *Brevicoryne brassicae* and its syrphid predators. *Journal of Applied Ecology* **8**, 751–80.

Potts, G. R. & Vickerman, G. P. (1974). Studies on cereal ecosystems. *Advanced Ecological Research* **8**, 107–97.

Prescher, S. & Büchs, W. (1998). Der Einfluß abgestufter Extensivierungsmaßnahmen im Raps- und Erbsenanbau auf nützliche Fliegen (Diptera, Brachycera). *Gesunde Pflanzen* **50**, 213–18.

Shaw, M. W. (1970). Egg-laying by some dipterous pests of cultivated cruciferae in north-east Scotland. *Journal of Applied Ecology* **7**, 353–91.

Sotherton, N. W. (1984). The distribution and abundance of predatory arthropods overwintering on farmland. *Annals of Applied Biology* **105**, 423–9.

Sotherton, N. W. (1985). The distribution and abundance of predatory Coleoptera overwintering in field boundaries. *Annals of Applied Biology* **106**, 17–21.

Steel, W. O. (1970). The larvae of the genera of the Omaliinae (Coleoptera: Staphylinidae) with particular reference to the British fauna. *Transactions of the Royal Entomological Society London* **122**, 211–23.

Stubbs, A. & Chandler, P. (1978). *A Dipterist's Handbook. The Amateur Entomologist 15.* The Amateur Entomologist's Society: Hanworth.

Sunderland, K. D. (1975). The diet of some predatory arthropods in cereal crops. *Journal of Applied Ecology* **12**, 507–15.

Thiele, H. U. (1977). Carabid beetles in their environments. A study on habitat selection by adaptation in physiology and behaviour. Springer: Berlin.

Tischler, W. (1958). Synökologische Untersuchungen an der Fauna der Felder und Feldgehölze. *Zeitschrift für Morphologie und Ökologie Tiere* **47**, 54–114.

Tomlin, A. D., Miller, J. J., Harris, C. R. & Tolman, J. R. (1985). Arthropod parasitoids and predators of the onion maggot (Diptera: Anthomyiidae) in southwestern Ontario. *Journal of Economic Entomology* **78**, 975–81.

Topp, W., Hansen, K. & Brandl, R. (1982). Artengemeinschaften von Staphyliniden an Aas. *Entomologia Generalis* **7**, 347–64.

Vickerman, G. P. & Sunderland, K. D. (1975). Arthropods in cereal crops: nocturnal activity, vertical distribution and aphid predation. *Journal of Applied Ecology* **12**, 755–66.

Volkmar, C., Wetzel, T., Lübke-Al Hussein, M., Jany, D. & Richter, L. (1999). Ökologische Begleituntersuchungen beim Anbau transgener Rapspflanzen am Standort Friemar (Freistaat Thüringen) in den Jahren 1994 bis 1996. *Archiv für Phytopathologie und Pflanzenschutz* **32**, 89–130.

Wheeler, A. G. Jr. (2001). *Biology of the Plant Bugs.* Cornell University Press: Cornell.

Zimmermann, J. & Büchs, W. (1999). Kurzflügelkäfer (Coleoptera: Staphylinidae) in unterschiedlich intensiv bewirtschafteten Ackerflächen. *Agrarökologie* **32**, 1–154.

13

Taxonomy and Identification of Predators

David V. Alford
Orchard House, Oakington Road, Dry Drayton, Cambridge CB3 8DD, UK

Wolfgang Büchs & Sabine Prescher
Federal Research Centre for Agriculture and Forestry, Messeweg 11/12, D-38104 Braunschweig, Germany

Bernhard Kromp
Ludwig Boltzmann-Institute for Biological Agriculture and Applied Ecology, Rinnböckstraße 15, A-1110 Wien, Austria

Bernd Ulber
Institut für Pflanzenpathologie und Pflanzenschutz, Universität Göttingen, Grisebachstraße 6, D-37077 Göttingen, Germany

Introduction

Most predators associated with arable crops, including oilseed rape, are insects (class Insecta) and spiders (class Arachnida: order Araneae). Identification of insects and spiders is often a specialist task, requiring the detailed examination (and sometimes dissection) of specimens. Within this chapter, the main groups of invertebrate predators associated with oilseed rape fields (see Chapter 12) are described and, in some cases, details given of how to identify members of important families, genera or species.

Although many keys are available for the identification of specific groups of insects and spiders, these inevitably include a vast number of species from habitats other than agro-ecosystems and species that are not natural enemies of oilseed rape pests. Also, such keys are usually designed for the specialist and often require detailed microscopical examination (and perhaps dissection) of specimens, thereby limiting their use as an aid for practitioners in crop protection. The current monograph focuses on the biocontrol of oilseed rape pests. Therefore, an attempt has been made in this chapter to present a simplified taxonomic overview of predators likely to be encountered within oilseed rape fields (see Chapter 12). Simplified keys to several groups or species are included, as are group-by-group descriptions. Specialist scientific terminology cannot be avoided entirely (especially in the keys); nor is it always possible to avoid mentioning microscopical characters, although these have been kept to a minimum. Terms likely to be unfamiliar to the general reader are explained in the Glossary.

Insects

Features of the main groups of predatory insects of significance in agricultural cropping systems (including features of the various orders of insects) are outlined by Alford (1999).

Beetles (order Coleoptera)

Several families are important predators on oilseed rape fields (see Chapter 12), of which ground beetles (Carabidae) and rove beetles (Staphylinidae) are of greatest significance.

A general key to all species of beetles of central Europe is provided by Freude *et al.* (1965 *et seq.*); see also the catalogue by Lucht (1987). This most professional key for beetles in Europe is expanded and revised continuously by supplementary volumes (e.g. Lohse & Lucht, 1989–1993; Lucht & Klausnitzer, 1998). The work is also augmented by keys to coleopteran larvae, for which information is generally scarce (even in the case of well-known predatory groups such as ground beetles). Indeed, even basic taxonomic knowledge on most coleopteran larvae is lacking. For example, with regards the central European beetle fauna, larvae are probably known for only 30–40% of species. However, the percentage of known larvae differs considerably from family to family, even in those with mainly carnivorous larvae (e.g. Carabidae – 41.8%; Cantharidae – 27.4%; Coccinellidae – 72.0%; Histeridae – 27.3%). For Staphylinidae, no estimation is possible (see Klausnitzer, 1991, 1994, 1996, 1997a, 1999, 2001). Other important (but mostly older) works on predacious coleoptera larvae include Perris (1877), Böving (1929), Böving & Craighead (1931), van Emden (1942), Peterson (1951), Hayashi *et al.* (1959), Ghilarov (1964), Mamaev (1972), Klausnitzer (1978), Morimoto & Hayashi (1986 *et seq.*), Costa *et al.* (1988), Paulian (1990) and Stehr (1991).

Recently, beetle taxonomy has undergone major changes (Lawrence & Newton, 1995). However, for practical reasons, these revisions are not considered here. For systematic details of species cited in this chapter see Tables 12.1–12.3.

Ground beetles (family Carabidae)

Gound beetles are very active, fast-moving beetles with a characteristic appearance (see Fig. 12.1). Although most are black, many have a metallic blue, green or coppery lustre. The pronotum is shield-like and often of characteristic shape or form. The elytra each usually have nine longitudinal ridges (intervals), separated by distinct furrows (striae) or series of punctures. The antennae are 11-segmented and filiform, and clearly inserted on the side of the head, between the eyes and the mandibles.

Various publications deal in depth with the identification of ground beetles, but these do not distinguish between species of significance or otherwise within agro-ecosystems. The European fauna extends to over 2,600 species (Wachmann *et al.*, 1995); in the UK alone there are over 350 named species of ground beetle, subdivided into more than 60 genera (Pope, 1977). Recently, Assmann *et al.* (1998) reported major changes in the systematics of the genus *Pterostichus*, which is one of the major genera of abundant predator species in oilseed rape fields (see Chapter 12). However, to avoid confusion, such changes have not been employed within this monograph. Specialist keys to Carabidae are provided by Trautner & Geigenmüller (1987) for western and central Europe and by Freude *et al.* (1976); see also Lindroth (1974). For a quick and easy determination of genera, and of the most common species, the German key of Trautner *et al.* (1984) is recommended. To determine ground beetle larvae to species, only the advanced keys of Klausnitzer (1978) and Arndt (1991, 1997) are available.

Within the EC-funded project BORIS (see Preface), where the emphasis was placed on predatory species of relevance to oilseed rape crops, an attempt was made to produce a simplified non-dichotymous 'key' to ground beetles that could be used to separate adults of the more frequently encountered species in European oilseed rape fields on the basis of the overall body size of adults; typical eastern and southern European species, however, were not included. This key (B. Kromp, *unpublished*) was based on Freude *et al.* (1976), Trautner & Geigenmüller (1987), Lohse & Lucht (1989), de Snoo & van der Poll (1990 – *unpublished manuscript*) and Hurka (1996). A modified version of this 'key' is presented in Boxes 13.1a–d.

Box 13.1a
General descriptions of 'very large' ground beetles (Carabidae) frequently present in European oilseed rape fields – body length 16–32 mm

1. ***Carabus cancellatus*** Illiger – antennal segment 1 usually red; elytra longitudinally ridged and granular, and strongly incurved near apex; upper surface metallic, mostly coppery or green; body length 17–32 mm.
2. ***Carabus auratus*** Linnaeus – at least antennal segments 1–3 yellowish-red; elytra golden-green and metallic, with flat, broad, longitudinal ridges (Fig. 13.1a); 17–30 mm long.
3. ***Carabus granulatus*** Linnaeus – elytra longitudinally ridged and granular, and at most weakly incurved near apex (Fig. 13.1b); upper surface mostly dark coppery; body length 16–23 mm.

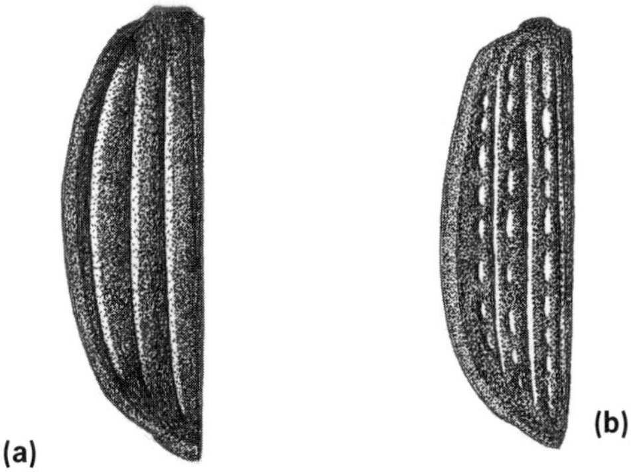

Fig. 13.1 Elytron – (a) *Carabus auratus*; (b) *Carabus granulatus*.

Box 13.1b
General descriptions of 'large' ground beetles (Carabidae) frequently present in European oilseed rape fields – body length 11–17 mm

1. ***Pterostichus*** spp. – body smooth and black; legs black; head with two supra-orbital punctures (Fig. 13.2); antennae pubescent from segment 4 onwards.
   1.1. ***Pterostichus melanarius*** (Illiger) – 3rd elytral interval with two punctures; raised lateral beading of pronotum very broad, with a distinct roll between it and the basal depressions (Fig. 13.3); elytra elongate; body length 13–17 mm.
2. ***Pseudoophonus*** spp. – body pubescent and black; legs red; head with one supra-orbital puncture (Fig. 13.4); antennae pubescent from segment 3 onwards.
   2.1. ***Pseudoophonus rufipes*** (Degeer) – elytra entirely punctate and with a yellowish pubescence (*best seen in dry specimens*); sides of pronotum weakly concave towards base, and hind angles sharp (Fig. 13.5); body length 11–16 mm.

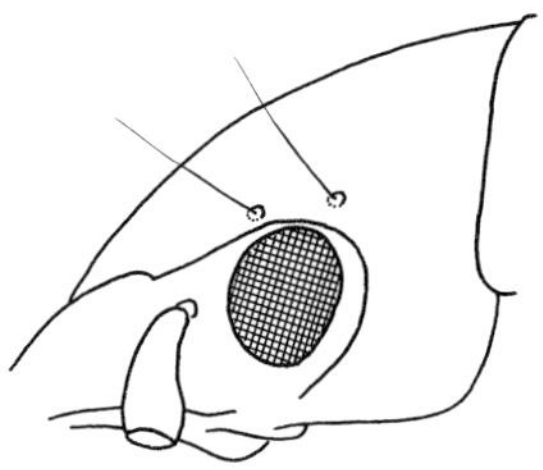

Fig. 13.2 Side of head of a ground beetle to show two supra-orbital punctures – e.g. *Pterostichus*.

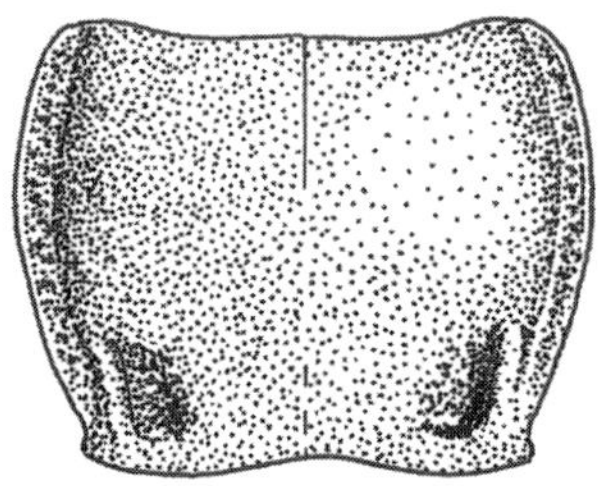

Fig. 13.3 Pronotum – *Pterostichus melanarius*.

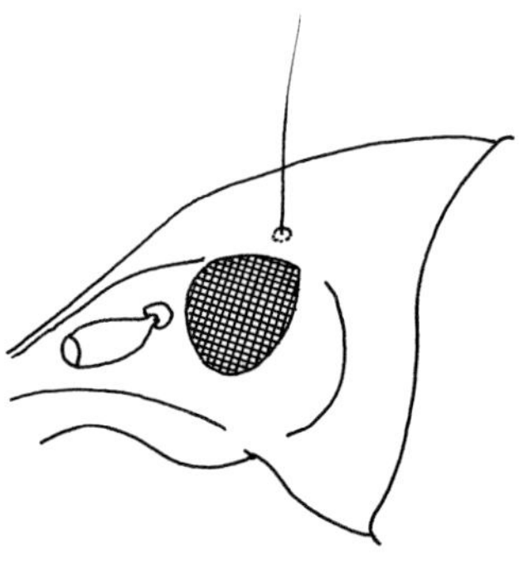

Fig. 13.4 Side of head to show single supra-orbital puncture – e.g. *Harpalus* and *Pseudoophonus*.

Fig. 13.5 Outline of pronotum – *Pseudoophonus rufipes*.

Box 13.1c
General descriptions of 'medium-sized' ground beetles (Carabidae) frequently present in European oilseed rape fields – body length 6–13 mm

1. ***Nebria brevicollis*** (Fabricius) – head with one supra-orbital puncture (see Fig. 13.4); pronotum heart-shaped, with a thick marginal beading (Fig. 13.6); brown to black; body length 10–13 mm.
2. ***Poecilus*** – head with two supra-orbital punctures (see Fig. 13.2); antennae pubescent from segment 4 onwards; antennal segments 1–3 with a distinct longitudinal keel (Fig. 13.7).
   2.1. ***Poecilus cupreus*** (Linnaeus) – elytra smooth, metallic green, coppery, bronzy or blue; antennal segments 1 and 2 red; femora, and sometimes rest of legs, also often red; head and, especially, base of pronotum with extensive punctation; inner side of hind tibiae with a row of 8–10 pale, fine setae (*best viewed posteriorly from above*) (Fig. 13.8a); body length 9–13 mm.
   2.2. ***Poecilus versicolor*** (Sturm) – similarly coloured to *P. cupreus* but legs always black, and inner side of hind tibiae with a row of 5–8 dark, thicker and shorter setae (Fig. 13.8b); punctation on head and, especially, centrally on pronotal base less extensive; body length 8–12 mm (*this species more abundant in grassland and organically cultivated fields*).
3. ***Harpalus*** spp. – head with one supra-orbital puncture (see Fig. 13.4); antennae pubescent from segment 3 onwards.
   3.1. ***Harpalus affinis*** (Schrank) – upper surface usually metallic green or coppery; antennae unicolourous yellowish-red; legs sometimes blackened; outer two or three elytral intervals punctate and pubescent (*view dried specimens from behind*); body length 9–12 mm.
4. ***Calathus*** spp. – elongate-oval; base of pronotum distinctly broader than front margin (Fig. 13.9), broadly adjoining elytra; tarsal claws denticulate (Fig. 13.10).
   4.1. ***Calathus fuscipes*** (Goeze) – black, with antennae and legs mainly reddish-brown; 3rd and 5th elytral intervals with several punctures dorsally; body length 9–14 mm.
   4.2. ***Calathus melanocephalus*** (Linnaeus) – head and elytra black; pronotum reddish-yellow; body length 6–9 mm.
5. ***Agonum*** and ***Anochomenus*** spp. – pronotum rounded or heart-shaped, with greatest width in middle; mentum with a median tooth; tarsal claws smooth (i.e. not dentate).
   5.1. ***Agonum muelleri*** (Herbst) – metallic green, elytra bronzy or dark copper-coloured; base of pronotum rounded (Fig. 13.11a); antennae and legs blackish-brown; tibiae and antennal segment 1 mostly conspicuously brighter-coloured; antennae pubescent from segment 4 onwards; 5th elytral striae each with a lon-

gitudinal impression situated close to the end (*view posteriorly, from above*); body length 6.9–9.0 mm.

5.2. *Anchomenus dorsalis* (Pontoppidan) – head and pronotum metallic green; elytra yellowish-brown, each with a large, dark-green apical spot; antennae pubescent from segment 3 onwards; hind angles of pronotum angular (Fig. 13.11b); body length 5.8–7.5 mm.

6. ***Amara*** spp. – body more or less egg-shaped (hence the vernacular name 'VW beetles') (Fig. 13.12); pronotum with two pairs of lateral setae, one of which is situated at the hind angles; head with two supra-orbital punctures; body length 6–9 (*species are difficult to identify*).
7. ***Stomis pumicatus*** (Panzer) – body brown and slender; mandibles extremely long (Fig. 13.13); antennal segment 1 long and shaft-like, and at least as long as segments 2 + 3 (Fig. 13.14a); body length 6.5–8.3 mm.
8. ***Loricera pilicornis*** (Fabricius) – black, usually with a greenish metallic lustre; antennal segments 1–6 with long setae (Fig. 13.14b); body length 6–8 mm.

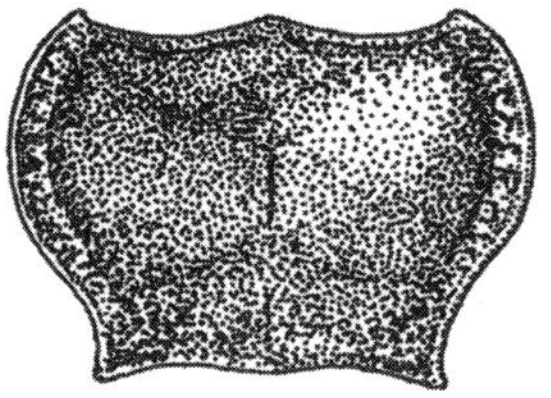

Fig. 13.6 Pronotum – *Nebria brevicollis*.

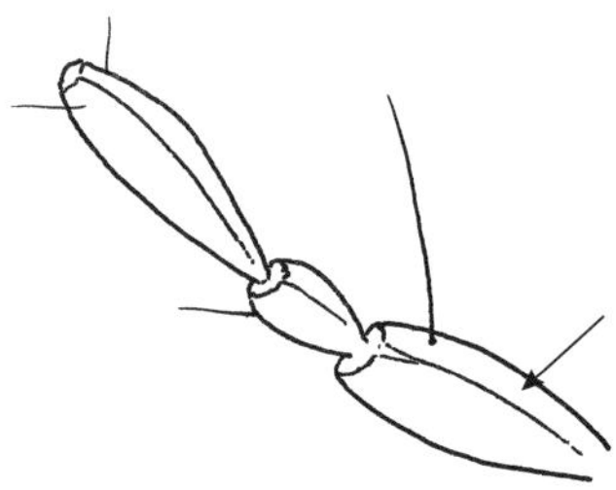

Fig. 13.7 Basal segments of antenna to show keel (arrowed) – *Poecilus cupreus*.

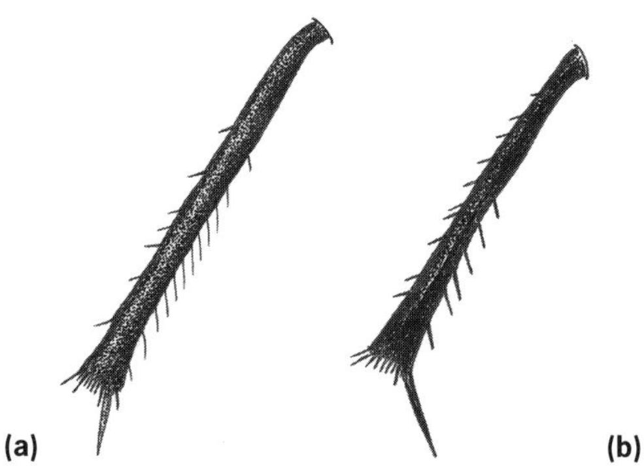

Fig. 13.8 Hind tibia – (a) *Poecilus cupreus*; (b) *Poecilus versicolor*.

Fig. 13.9 Outline of pronotum – *Calathus fuscipes*.

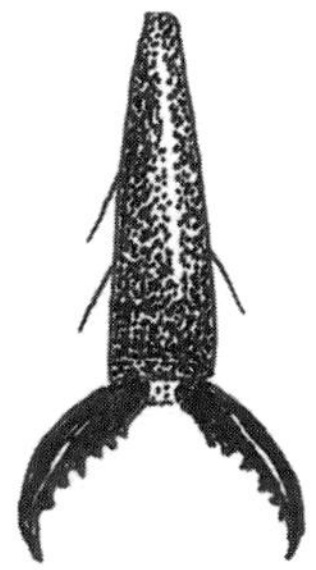

Fig. 13.10 Tarsus – *Calathus fuscipes*.

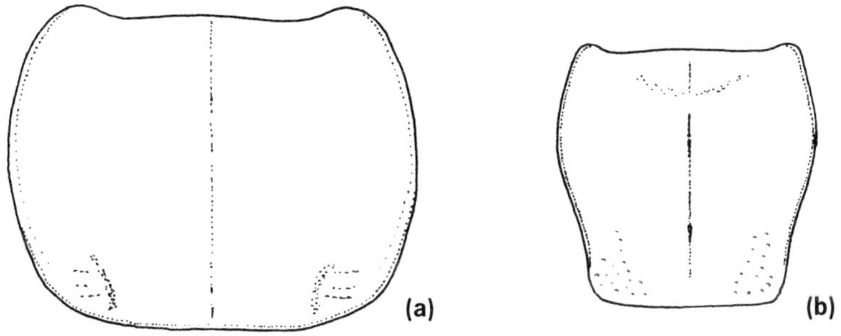

Fig. 13.11 Pronotum – (a) *Agonum muelleri*; (b) *Anchomenus dorsalis*.

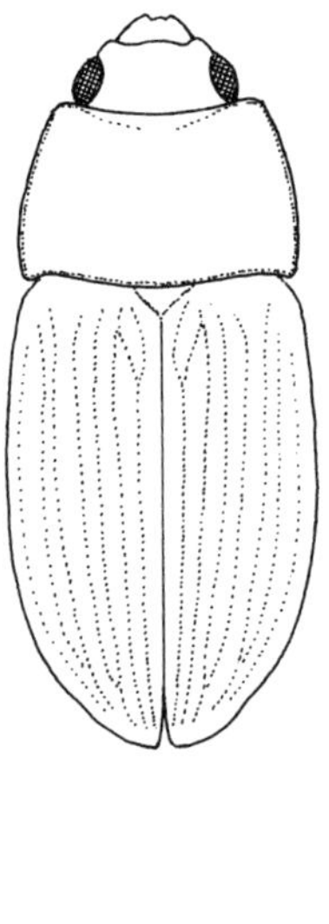

Fig. 13.12 Outline of body – *Amara plebeja*.

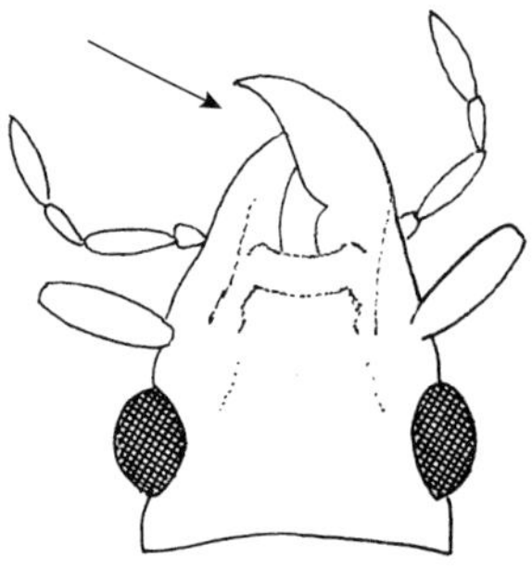

Fig. 13.13 Head of *Stomis pumicatus* to show long mandibles (arrowed).

(a) 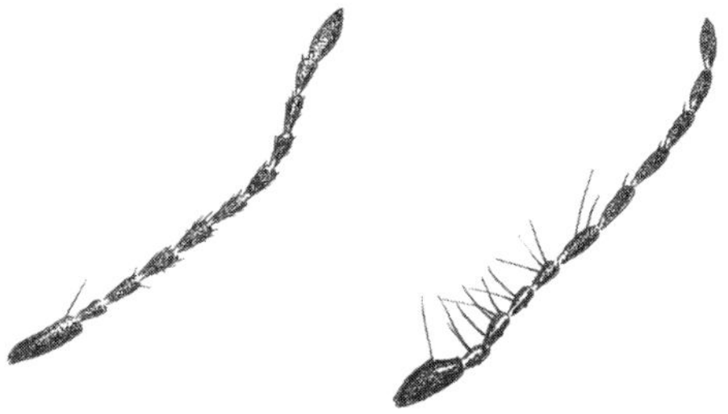 (b)

Fig. 13.14 Antenna – (a) *Stomis pumicatus*; (b) *Loricera pilicornis*.

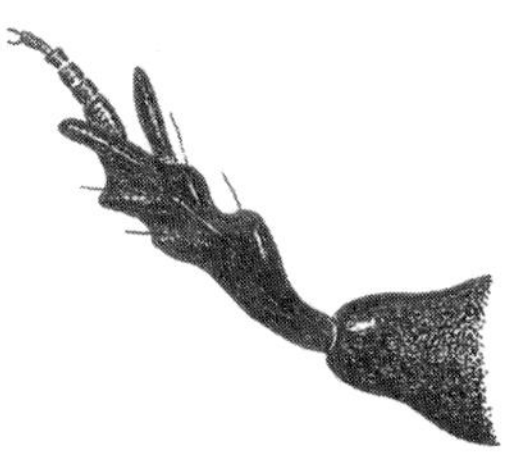

Fig. 13.15 Fore leg of *Clivina fossor*.

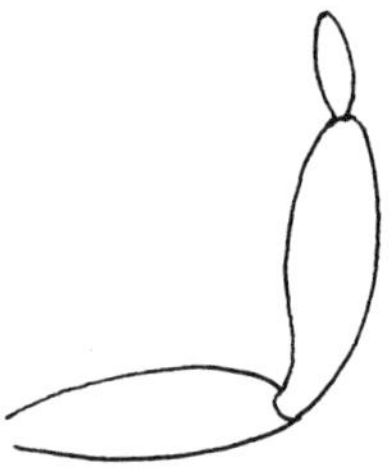

Fig. 13.16 Maxillary palp – *Bembidion*.

Box 13.1d
General descriptions of 'small' ground beetles (Carabidae) frequently present in European oilseed rape fields – body length usually <6 mm

1. ***Clivina fossor*** (Linnaeus) – unicolourous, red to brownish-black; fore legs fossorial (Fig. 13.15); pronotum with a complete lateral beading, reaching to the base; elytra with a row of setae laterally; body length 5.0–6.5 mm.
2. ***Bembidion*** spp. – last segment of maxillary palp much shorter and narrower than segment 2 (Fig. 13.16); elytra glabrous, with distinct (but often extensively obsolete) striae or rows of punctures.
   - 2.1. ***Bembidion tetracolum*** Say – elytra each with two pale yellow patches (Fig. 13.17); antennae dark from segment 4 onwards; femora mostly yellowish-red; body length 4.9–6.1 mm.
   - 2.2. ***Bembidion femoratum*** Sturm – as for *B. tetracolum* but elytral patches larger and body length shorter.
   - 2.3. ***Bembidion lampros*** (Herbst) – pronotum heart-shaped (Fig. 13.18a), metallic dark bronze; body length 3–4 mm.
   - 2.4. ***Bembidion properans*** Stephens – as for *B. lampros*.
   - 2.5. ***Bembidion quadrimaculatum*** (Linnaeus) – head and pronotum black, with a metallic lustre; elytra dark brown and with four yellow spots; base of pronotum distinctly roll-shaped along its whole width (Fig. 13.18b); legs and antennal segments 1–4 yellowish; body length 2.8–3.5 mm.
3. ***Asaphidion flavipes*** (Linnaeus) – dark metallic, the elytra speckled with a silvery to coppery pubescence and irregular punctation; legs and antennae reddish-yellow; eyes protruding (Fig. 13.19); last segment of maxillary palp much shorter and narrower than segment 2; body length 3.9–5.4 mm.
4. ***Trechus quadristriatus*** (Schrank) – head blackish-brown; elytra brown to reddish-brown and glabrous; legs and antennae reddish-yellow; 3rd elytral interval with three setiferous punctures, with 2nd interval broadened apically; 1st elytral stria bending forward apically (the so-called *Trechus* bow) (Fig. 13.20); body length 2.8–4.5 mm.

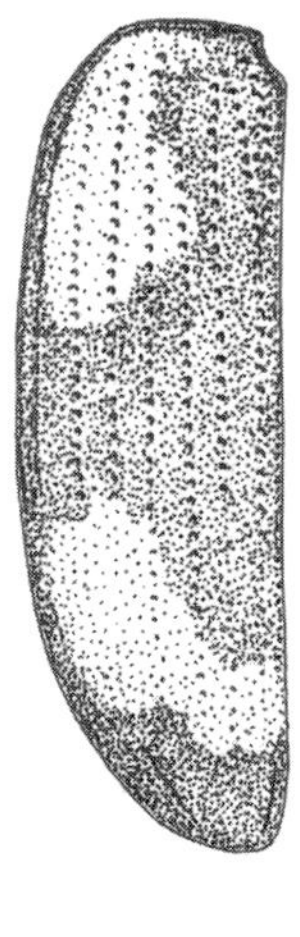

Fig. 13.17 Elytron – *Bembidion tetracolum*.

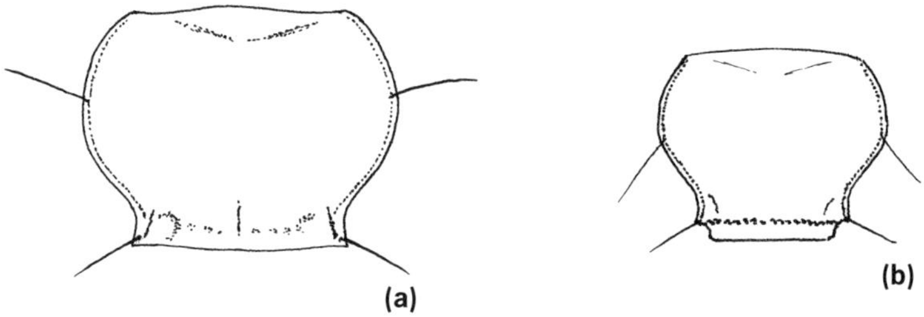

Fig. 13.18 Pronotum – (a) *Bembidion lampros*; (b) *Bembidion quadrimaculatum*.

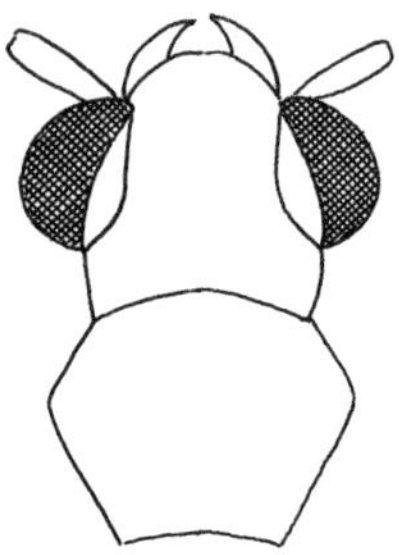

Fig. 13.19 Head – *Asaphidion flavipes*.

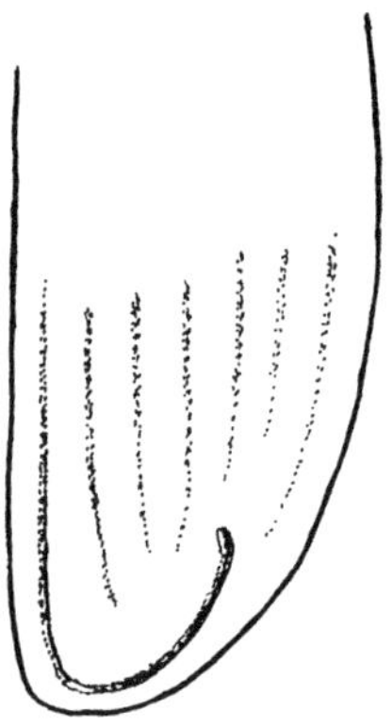

Fig. 13.20 Apical part of elytron – *Trechus quadristriatus*.

Rove beetles (family Staphylinidae)

Rove beetles belong to a very large family of mainly small, usually black beetles; some species, however, are brightly coloured. The elytra are characteristically truncated, exposing at least three abdominal segments (see Fig. 12.2).

Various specialist keys to the identification of rove beetles are available, including Reitter (1909), Joy (1932), Tottenham (1954), Lohse (1964, 1974 – the latter completed in a supplement by Assing *et al.*, 1998) and Schaefer (2000). A simplified key to adult rove beetles commonly occurring in oilseed rape fields in Europe, based on Schaefer (2000), is presented in Box 13.2.

Box 13.2
Simplified key to rove beetles (Staphylinidae) frequently present in European oilseed rape fields

1. Head large; eyes prominent, occupying whole side of head and extending laterally beyond width of pronotum (Fig. 13.21) . . . . STENINAE, e.g. ***Stenus*** spp.
– Head and eyes otherwise 2
2. Antennae inserted on anterior margin of the head, in front of the eyes (Fig. 13.22) 3
– Antennae inserted on anterior part of frons between the eyes (Fig. 13.23); head small and inclined . . . . ALEOCHARINAE – 6
3. Distance between the outer margins of the mandibles greater than the distance between the antennal bases (see Fig. 13.22) 4
– Distance between the antennal bases greater than (or as great as) the distance between the outer margins of the mandibles at their bases (Fig. 13.24). 9
4. Bases of antennae equidistant or only slightly further from each other than from the eyes; 1st segment of metatarsi hardly longer than 2nd . . . . XANTHOLININAE – 5
– Bases of antennae further from each other than from the eyes; 1st segment of metatarsi often extended . . . . STAPHYLININAE – 11
5. Body black or brownish-black, 6–8 mm long . . . . . . ***Xantholinus linearis*** (Olivier)
– Pronotum entirely yellowish-red . . . . ***Xantholinus jarrigei*** Coiffait
6. Tarsal formula 4, 4, 5 . . . . ***Gyrophaena*** spp.
– Tarsal formula 4, 5, 5 7
– Tarsal formula 5, 5, 5; terminal segment of maxillary palps stylet-like. 8
7. Hind part of body with golden dorsal patches of pubescence . . . . . ***Lomechusa*** spp.
– Hind part of body without golden dorsal patches of pubescence . . . . . . ***Atheta*** spp.
8. Maxillary palps 5-segmented; labial palps 4-segmented . . . . ***Aleochara*** spp.
– Maxillary palps 4-segmented; labial palps 3-segmented . . . . ***Oxypoda*** spp.
9. Antennae inserted beneath the extending anterior angles of the frons, above the mandibular bases (Fig. 13.24b); head large and mostly wider than pronotum, from which it is separated by a distinct neck; body thin and parallel-sided. . . . . PAEDERINAE, *Lathrobium* spp. – 10
– Other combination of characters 13
10. Elytra red, body 7.5–9.0 mm long . . . . ***Lathrobium fulvipenne*** (Gravenhorst)
– Front of body shiny, 4.5–5.5 mm long . . . . . . ***Lathrobium longulum*** (Gravenhorst)
11. Head and pronotum bare and mostly shiny, with only single long setae; pronotum with two lines of punctures . . . . *Philonthus* spp. – 12
– Head and pronotum with long setae and covered with a short pubescence. 16
12. Black, 8.0–11.5 mm long. . . . . ***Philonthus cognatus*** Stephens
– Mainly black but with front of body shiny greenish, 8–11 mm long. . . . . ***Philonthus rotundicollis*** (Ménétriés)
13. Body boat-shaped; hind part of body elongate-conical, strongly tapering towards the apex and with long setae; head retracted under pronotum, up to the eyes; pronotum shiny and mostly glabrous; antennae filiform and inserted uncovered on the anterior, descending part of the head . . . . TACHYPORINAE – 14
– Body not boat-shaped; antennae inserted beneath lateral margin of frons, which is dilated by round or linear projections. . . . . OXYTELINAE and OMALIINAE – 17

14. Terminal segment of maxillary palp minute and very short (Fig. 13.25a); head and pronotum glabrous; body 2–5 mm long. . . . . *Tachyporus* spp. – 15
– Terminal segment of maxillary palp as long as, or longer than, penultimate segment (Fig. 13.25b); legs and hind margin of elytra reddish; body 5.0–6.5 mm long . . . . **Tachinus signatus** Gravenhorst
15. Abdomen black, apart from the reddish-brown hind margins of segments; elytra reddish-brown, apart from the black front margins; head and pronotum black; body 3–4 mm long . . . . **Tachyporus hypnorum** (Linnaeus)
– Head black; pronotum and elytra reddish-yellow; body 3.5–4.0 mm long. . . . . **Tachyporus chrysomelinus** (Linnaeus)
– Head, pronotum and hind parts of elytra reddish-brown; body 3.5–4.0 mm long . . . . **Tachyporus obtusus** (Linnaeus)
– Pronotum reddish-yellow; body 3.5–5.0 mm long. . . . **Tachyporus solutus** Erichson
16. Hind part of body with lateral patches of golden-yellow hairs; body black but elytra reddish-brown, 15–25 mm long. . . . . **Staphylinus** spp.
– Hind part of body without lateral patches of golden-yellow hairs; body black but elytra reddish-brown or black, 12–32 mm long . . . . **Staphylinus** spp.
– Hind part of body without lateral patches of golden-yellow hairs; body entirely black, pronotum with distinct punctures, 12–20 mm long . . . . **Staphylinus nero** Faldermann
17. Hind part of body laterally with a broad and flat rim (Fig. 13.26), the outer margin of which is razor-sharp. . . . . OMALIINAE – 18
– Margin of body laterally turned upwards and with very narrow, separate dorsal plates (Fig. 13.27); if outer margin razor sharp, then body >6 mm long or of a metallic colour . . . . OXYTELINAE – 19
18. Body black or brownish-black, <4 mm long. . . . . . . **Omalium caesum** Gravenhorst
– Elytra usually yellowish-brown; body <3 mm long. . . . **Omalium rivulare** (Paykall)
– Elytra often reddish-brown; body 3.5–4.5 mm long . . . . **Lesteva longoelytrata** (Goeze)
19. Tarsi 5-segmented; body 5–7 mm long, shiny brownish-black . . . . **Coprophilus striatulus** (Fabricius)
– Tarsi 3-segmented; pronotum and elytra with long notches. . . . . *Anotylus* spp. – 20
20. Body 3.0–3.5 mm long, black . . . . **Anotylus sculpturatus** (Gravenhorst)
– Body 3.0–3.5 mm long, black; legs yellowish-red . . **Anotylus inustus** (Gravenhorst)
– Body 4.5–5.5 mm long; black; legs yellow. . . . . **Anotylus rugosus** (Fabricius)
– Body 1.7–2.2 mm long, dull black; legs reddish-yellow . . . . **Anotylus tetracarinatus** (Block)

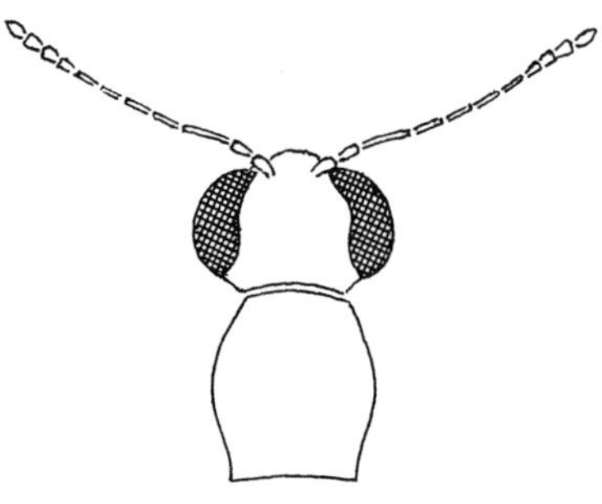

Fig. 13.21 Head of *Stenus* – subfamily STENINAE.

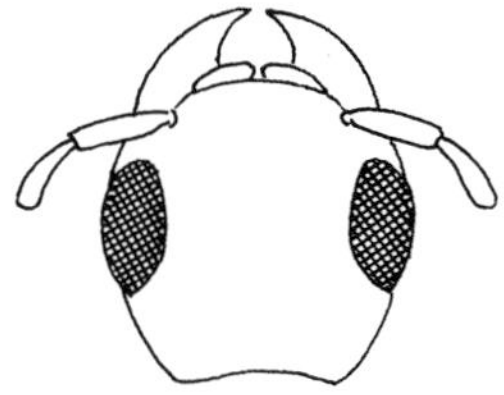

Fig. 13.22 Head of *Philonthus cognatus* to show insertion of antennae.

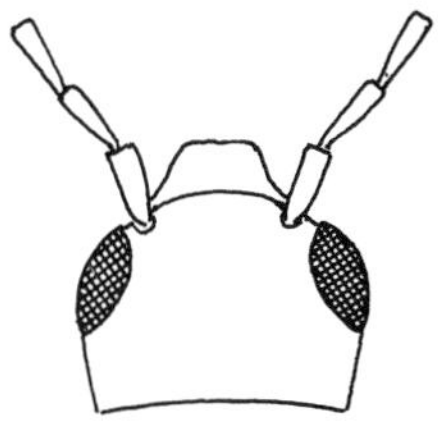

Fig. 13.23 Head of *Oxypoda* – subfamily ALEOCHARINAE.

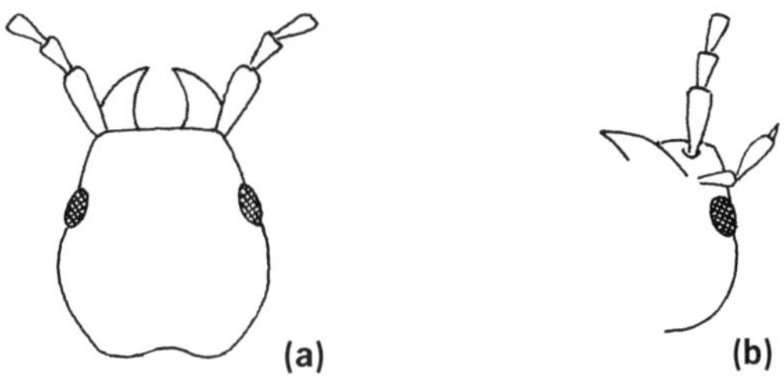

Fig. 13.24 Head of *Lathrobium* to show insertion of antennae – (a) dorsal view; (b) ventro-lateral view.

Fig. 13.25 Maxillary palp – (a) *Tachyporus*; (b) *Tachinus*.

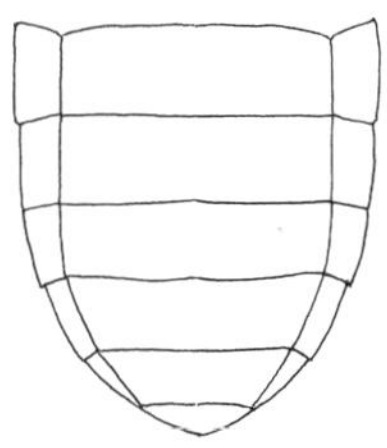

Fig. 13.26 Abdomen of *Omalium* to show arrangement of tergites – subfamily OMALIINAE.

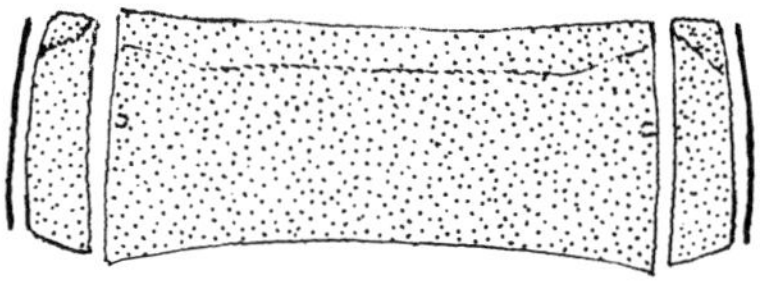

Fig. 13.27 Abdominal tergite of *Coprophilus* to show arrangement of dorsal plates – subfamily OXYTELINAE.

Box 13.3
Key to ladybirds (Coccinellidae) frequently present in European oilseed rape crops

1. Elytra bright red to orange-red, each with four large black spots (see Fig. 12.3); pronotum black, with the anterior angles each marked with a creamish-white patch; eyes fully exposed; tarsal claws with a basal projection; overall body length 5.5–8.0 mm . . . . **Coccinella septempunctata** Linnaeus
– Elytra yellow to yellowish-orange, each with seven large black spots which often coalesce (Fig. 13.28); pronotum yellow or orange, more or less spotted or suffused with black; eyes partly covered by pronotum; tarsal claws without a basal projection; overall body length 3.5–4.5 mm . . . . **Propylea quattuordecimpunctata** (Linnaeus)

Ladybirds (family Coccinellidae)

Ladybirds are well-known, mostly predatory insects, with a more or less hemispherical appearance, and the head retracted beneath the pronotum (see Fig. 12.3); the antennae end in a 3-segmented club. For keys and taxonomic details see, for example, Reitter (1911), Pope (1953) and Fürsch (1967, 1992).

All predatory ladybirds are members of the subfamily Coccinellinae, in which the mandibles have bifid tips. The two most commonly found species in oilseed rape crops are readily distinguished by their distinctive elytral patterns; see also the additional (in part generic) features outlined in Box 13.3.

Larvae of *Coccinella septempunctata* are mainly dull slate-blue, with creamish-yellow spots on abdominal segments 1 and 3 (Fig. 13.29); those of *Propylea quattuordecimpunctata* are mainly blackish, with prominent yellowish markings (Fig. 13.30).

Soldier beetles (family Cantharidae)

Soldier beetles are narrow, elongate and soft-bodied, with smooth, soft and downy elytra and filiform antennae (see Fig. 12.4). Adults are both phytophagous and carnivorous, but, in oilseed rape, the predatory larvae are of greater importance (see Chapter 16). Keys to adult soldier beetles are available in, for example, Dahlgren & Wittmer (1979) and Lohse & Lucht (1992). Keys to larvae are available in Verhoeff (1917, 1923), Larsson (1938), Striganova (1964), Fitton (1975), Klausnitzer (1978, 1997b), and LeSage (1991). The main diagnostic features of adults and larvae commonly found in oilseed rape crops are summarized in Boxes 13.4 and 13.5, respectively.

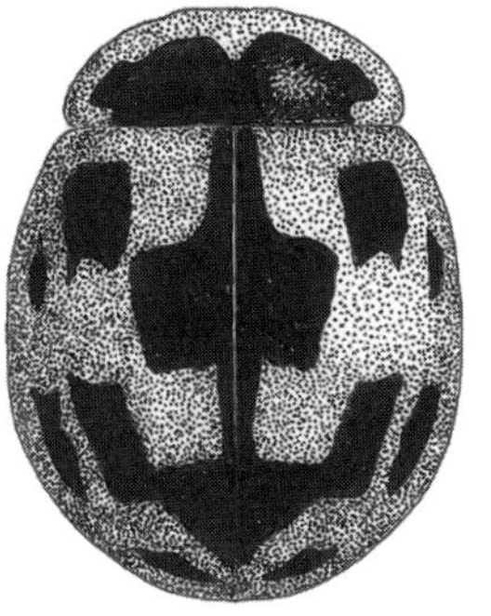

Fig. 13.28 Elytra and pronotum of fourteen-spot ladybird (*Propylea quatturodecimpunctata*).

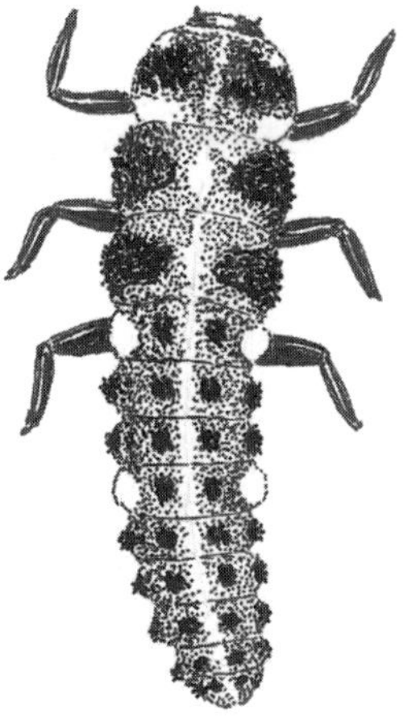

Fig. 13.29 Larva of seven-spot ladybird (*Coccinella septempunctata*) (×7).

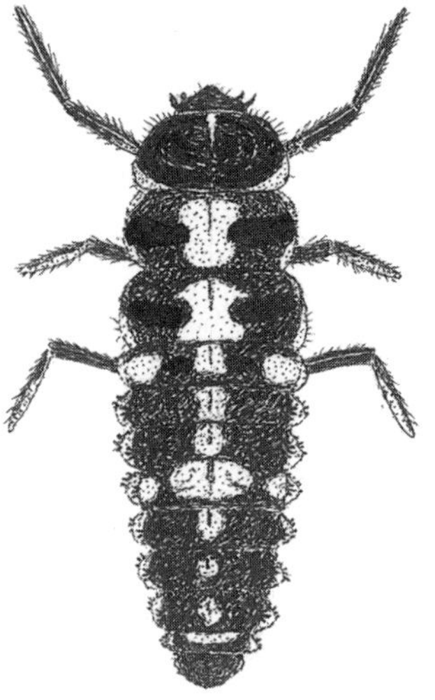

Fig. 13.30 Larva of fourteen-spot ladybird (*Propylea quattuordec-impunctata*).

Box 13.4
Key to soldier beetles (Cantharidae) frequently present in European oilseed rape crops

1. Head with an indentation between the antennae; pronotum bulbous, with rounded sides. 2
– Head without an indentation between the antennae; pronotum relatively straight-sided and narrowing anteriorly; head and thorax yellowish-red; elytra yellowish-red with black tips; base of antennae black; overall body length 7–10 mm . . . . ***Rhagonycha flava*** (Scopoli)
2. Head, pronotum and elytra reddish-yellow; overall body length 10.0–13.5 mm . . . . ***Cantharis livida*** Linnaeus
– Head mainly black; elytra dark. 3
3. Head mainly black; pronotum orange, marked anteriorly and posteriorly with black; elytra black; overall body length 11–15 mm . . . . ***Cantharis fusca*** Linnaeus
– Head mainly black; pronotum yellowish-orange; elytra greyish-black; overall body length 5.5–7.0 mm . . . . ***Cantharis lateralis*** Linnaeus

Box 13.5
Key to fourth-instar larvae of the most common species of soldier beetle (Cantharidae) in European oilseed rape fields

1. Body dark (Fig. 13.31), up to 24 mm long; clypeus (viewed from above) with a blunt central projection. . . . . ***Cantharis fusca*** Linnaeus
– Body mainly pale, up to 20.5 mm long; clypeus (viewed from above) with a more or less pointed central projection 2
2. Dorsal sclerites on thoracic segments 2 and 3 broadly lamellate (Fig. 13.32a); outline of clypeus (viewed from above) slightly concave, with central projection minute; apex of final abdominal segment rounded, with a pair of parallel sclerites; body up to 11 mm long. . . . . ***Cantharis lateralis*** Linnaeus
– Dorsal sclerites on thoracic segments 2 and 3 narrowly lamellate (Fig. 13.32b); central projection on clypeus small; final abdominal segment narrowing posteriorly to a blunt apex, with a pair of converging sclerites; body mainly pale, up to 20.5 mm long . . . . ***Cantharis livida*** Linnaeus

Other predatory beetles

Carrion beetles (Silphidae) (also known as 'burying beetles' and 'sexton beetles', owing to their habit of burying cadavers) are mainly black, but some species are marked prominently with orange; the antennae are strongly clubbed and the elytra short, exposing the three hindmost abdominal segments – see, for example, Reitter (1909). Histerid beetles (Histeridae) are mainly shiny-black, sometimes red-marked, with geniculate, strongly clubbed antennae and strongly denticulate tibiae; the elytra are also relatively short, exposing the two hindmost abdominal segments – see, for example, Reitter (1909) and Halstead (1963). Keys to members of the Melyridae, including colour illustrations of *Malachius* spp., are available in Reitter (1911).

True flies (order Diptera)

Oilseed rape fields are inhabited by a wide range of flies, including members of various predatory taxa (see Table 12.4 for systematic details). Features of the main groups of predatory flies are summarized as follows:

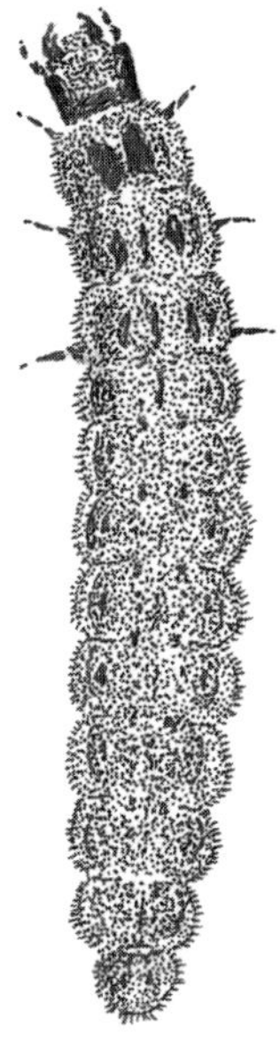

Fig. 13.31 Larva of a soldier beetle (*Cantharis fusca*) – family CANTHARIDAE (×6).

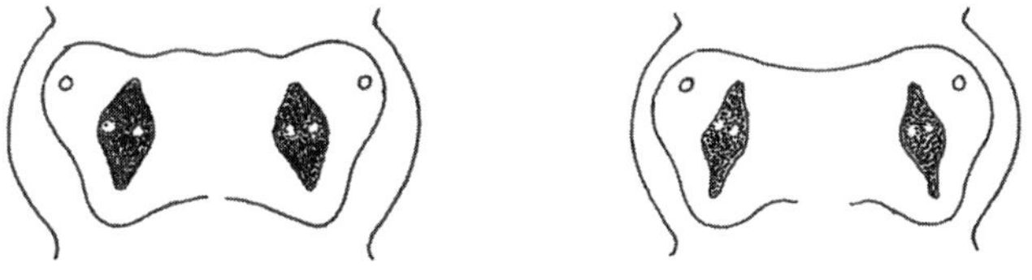

Fig. 13.32 Thoracic segment 2 (mesothorax) – (a) *Cantharis lateralis*; (b) *Cantharis livida* (based on C. Eggers, *unpublished*).

1. Suborder BRACHYCERA
   Usually robust flies with stout, 3-segmented antennae, the third segment sometimes ringed (Fig. 13.33a) and often with an apical spur or style; palps 1- or 2-segmented. Larval head incomplete, generally retractile, and with vertically biting mandibles.

   - Family DOLICHOPODIDAE (long-legged flies) – adults typically metallic bluish or metallic greenish, with (in males) often noticeably prominent genitalia (see Fig. 12.6). The wing venation is relatively simple, with just one obvious cross-vein (Fig. 13.34). Parent (1938) and Assis-Fonseca (1978) give keys for the determination of species.
   - Family EMPIDIDAE (dance flies) – adults essentially similar to those of the following family (Hybotidae), but cubital vein often forked before reaching the wing margin (see Fig. 12.5). Empididae may be identified using the key of Collin (1961).
   - Family HYBOTIDAE (dance flies) – adults small, with a rigid proboscis and with a distinctive 'nick' (indentation) on the inner margin of the compound eyes (Fig.

Fig. 13.33 Antenna – (a) suborder BRACHYCERA, with compound, annulated third segment (arrowed); (b) suborder CYCLORRHAPHA, with plumose arista (arrowed).

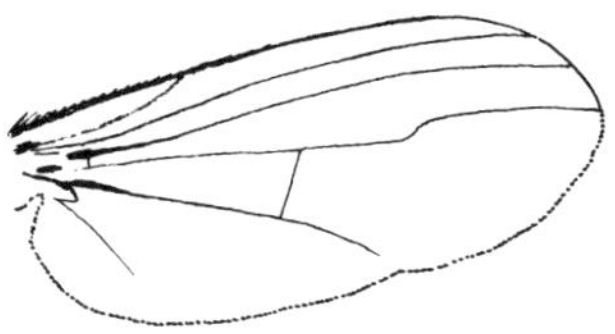

Fig. 13.34 Wing of a long-legged fly – family DOLICHOPODIDAE.

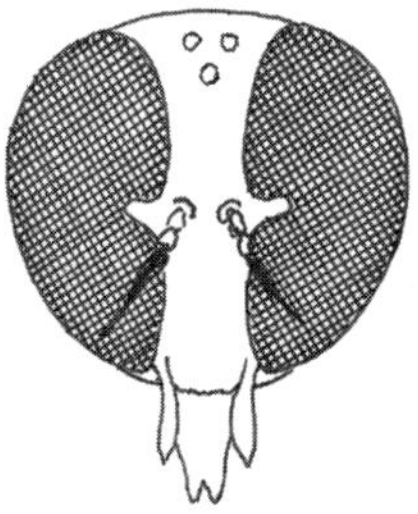

Fig. 13.35 Head of a dance fly – family HYBOTIDAE.

13.35); tibia of each mid-leg often bearing distinct spurs; distinguished from members of the Empididae largely on the basis of wing venation – the veins usually extending to the wing margin without forking (e.g. Fig. 13.36). Owing to the abundance of these flies in oilseed rape, a key to subfamilies and several genera is given here (Box 13.6). All members of the subfamily Tachydromiinae may be identified to species using the key of Chvala (1975) and species in the genus *Platypalpus* only, using the key of Chvala (1989).

- Family RHAGIONIDAE (snipe flies) – mainly brown and yellow, slender-bodied and long-legged flies (see Fig. 12.7); feet with three pads (see Fig. 13.44b). Species of *Rhagio* are often large (10–12 mm long), have patterned wings and can be distinguished from other snipe flies (at least from those with terrestrial larvae) by the open anal cell (Fig. 13.38) and by the two spurs on each hind tibia. Snipe flies can be determined to species using the keys of Roskozny & Spitzer (1965) and Oldroyd (1969).
- Family THEREVIDAE (stiletto flies) – distinctly pubescent, thin-legged flies (see Fig. 12.8); antennae characteristic, with a small style (Fig. 13.39); abdomen slender with a pointed tip; feet with two pads (see Fig. 13.44a); non-predatory as adults. Larvae are white, elongate and snake-like, with the body distinctly seg-

Box 13.6
Key to the subfamilies of the Hybotidae

1. Discal cell and anal cell present (Fig. 13.36a) . . . . Hybotinae and Ocydromiinae
– Discal cell absent; anal cell usually absent but if present then only very small and the anal vein scarcely visible; small species, 1–5 mm in length . . . . Tachydromiinae

Key to abundant genera of Tachydromiinae inhabiting oilseed rape fields

1. Humeri of thorax (see Fig. 13.50) more or less differentiated; eyes bare; head with a distinct neck, not closely set against front of thorax 2
– Humeri of thorax not differentiated; eyes microscopically pubescent; head without a distinct neck, being closely set against front of thorax 5
2. Anal cell present (Fig. 13.36b); humeri rounded or not very long 3
– Anal cell absent; humeri noticeably elongated; wings usually distinctly clouded or banded 4
3. Eyes at least very narrowly separated on frons and on facial part below antennae; mid-femora usually thickened; mid-tibia often with a spur . . . . ***Platypalpus*** Macquart
– Eyes touching on frons or below antennae; legs with all femora equally slender or fore femora the stoutest . . . . other Tachydromiinae
4. Vein usually closing the anal cell present; wings usually uniformly darkly clouded . . . . ***Tachypeza*** Meigen
– Vein usually closing the anal cell absent; wings usually banded . . . . ***Tachydromia*** Meigen
5. First basal cell distinctly shorter than second; wings not conspicuously shortened or narrowed 6
– Both basal cells equal in length or wings short and often narrowed . . . . other Tachydromiinae
6. Anterior pair of ocellar bristles present; wings with the common stem of veins R2 + 3 and R4 + 5 rather long (Fig. 13.37a) 7
– Anterior pair of ocellar bristles absent, the posterior pair very long; wings with the common stem of veins R2 + 3 and R4 + 5 very short (Fig. 13.37b) . . . . ***Crossopalpus*** Bigot
7. Yellow-bodied species . . . . ***Elaphropeza*** Macquart
– Black to blackish-brown species . . . . ***Drapetis*** Meigen

mented. Keys for determination of species are given by Oldroyd (1969) and (for the genus *Thereva* only) Lyneborg & Spitzer (1975).

2. Suborder CYCLORRHAPHA

Adults usually with 3-segmented antennae, the third segment enlarged and bearing a simple or plumose bristle (arista) dorsally (Fig. 13.33b); palps 1-segmented. Larvae are typical maggots, with a vestigial head and the body distinctly tapered anteriorly. The head and mouthparts are retracted into the thorax and form a usually pigmented cephalopharyngeal skeleton, which includes a pair of black, shiny articulating mouth-hooks and is often of characteristic appearance for the family, subfamily or species.

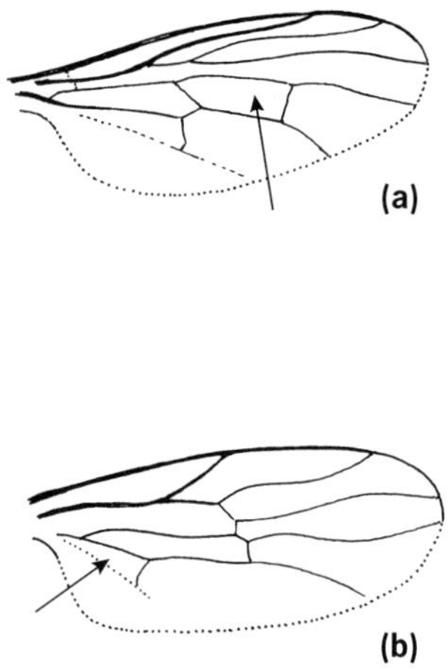

Fig. 13.36 Wing of a dance fly – (a) subfamilies Hybotinae and Ocydromiinae, with discal cell (arrowed) and anal cell; (b) subfamily Tachydromiinae, with anal cell (arrowed).

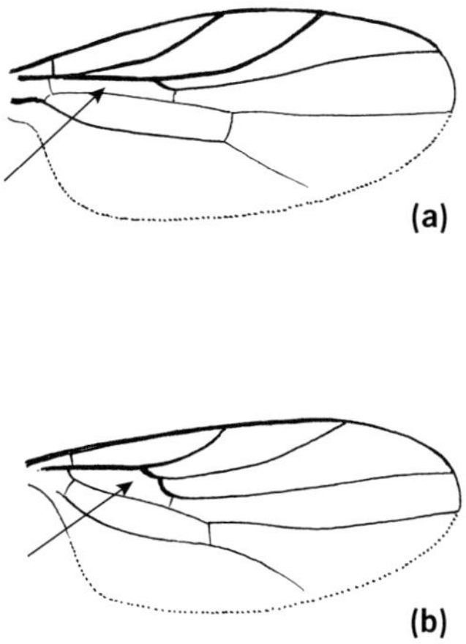

Fig. 13.37 Wing to show common stem of veins R2 + 3 and R4 + 5 (arrowed) – (a) *Drapetis* sp.; (b) *Crossopalpus* sp.

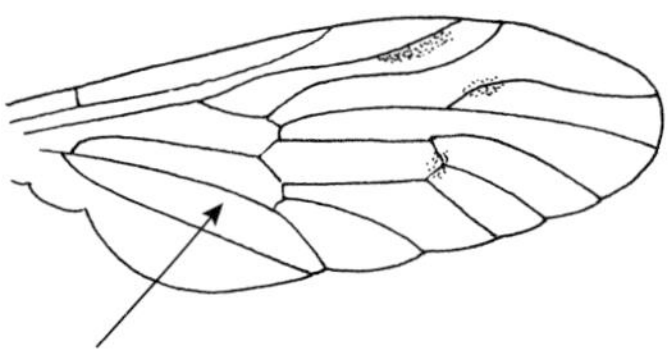

Fig. 13.38 Wing of a snipe fly (*Rhagio* sp.), with open anal cell (arrowed) – family RHAGIONIDAE.

- Family MUSCIDAE – small to relatively large, housefly-like flies, with well-developed squamae covering the halteres (Fig. 13.40), and the second antennal segment distinctly grooved (see Fig. 13.49). Assis-Fonseca (1968) gives a key for determining species of Muscidae.
- Family SYRPHIDAE (hover flies) – medium to large, often brightly coloured and often bee-like or wasp-like flies, with a characteristic wing venation that includes a false vein ('vena spuria') and a false margin (Fig. 13.41). Larvae maggot-like, with

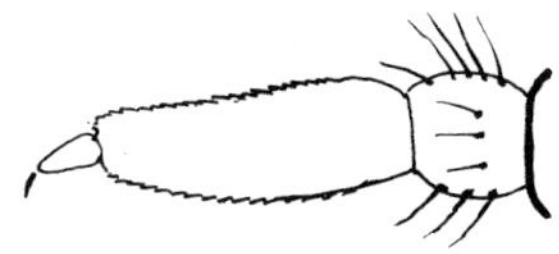

Fig. 13.39 Antenna of a stiletto fly – family THEREVIDAE.

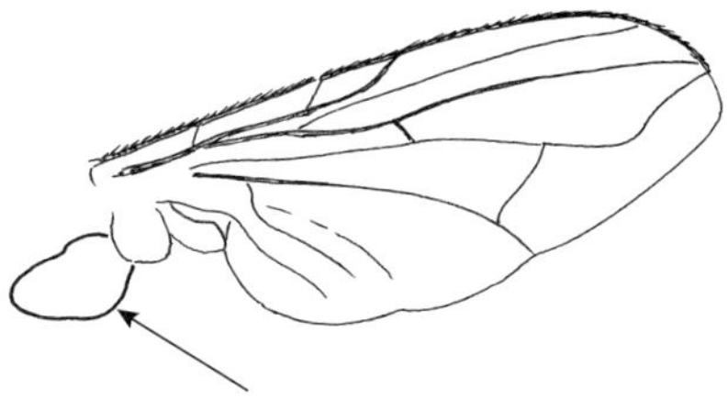

Fig. 13.40 Wing of a calyptrate fly, to show thoracic squama (arrowed) – family MUSCIDAE.

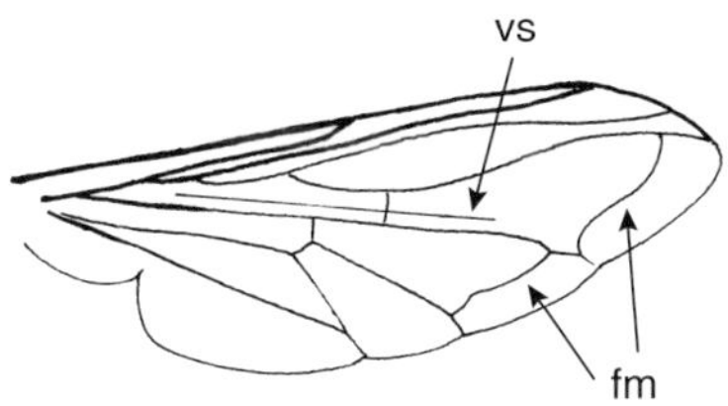

Fig. 13.41 Wing of a hover fly, to show vena spuria (vs) and false margins (fm) – family SYRPHIDAE.

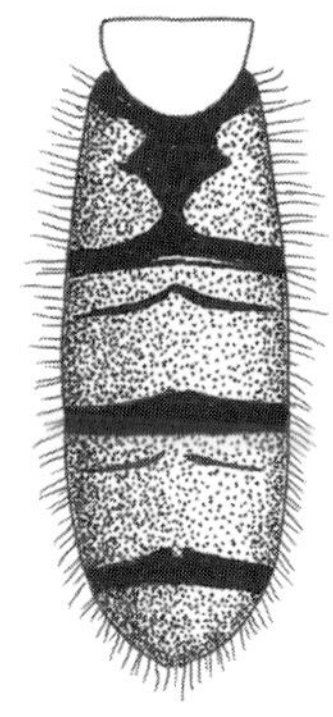

Fig. 13.42 Abdomen of *Episyrphus balteatus* – family SYRPHIDAE.

distinct, often cone-like posterior respiratory processes; body sometimes translucent, exposing the viscera. Hoverflies can be identified using the keys of van der Goot (1981) and Stubbs & Falk (1983). The most frequently occurring species in oilseed rape crops (namely, *Episyrphus balteatus*) can be easily recognized by the distinctive pattern of black and yellow markings on the abdomen (Fig. 13.42).

As a further aid to the identification of flies inhabiting oilseed rape fields to family or generic level see the key in Box 13.7. In view of the frequent presence in oilseed rape of a wide range of flies, the couplets include reference to several non-predatory groups but only to allow their elimination. Family identifications can be checked by the keys of Colyer & Hammond (1968), Oldroyd (1970) and Unwin (1981). A key to midges

Box 13.7

Simplified key to predacious flies (Diptera) inhabiting European oilseed rape fields (with various other families included for elimination purposes)

1. Antennae filiform, with at least six segments .... suborder Nematocera [Note 1]
- Antennae with fewer than six segments 2
2. Wings with a very long and pointed anal cell (Fig. 13.43) 3
- Wings with anal cell short or absent 8
3. Wings with a 'vena spuria' (see Fig. 13.41), reaching into the sub-apical cell, and with false margins; sub-apical cell always closed; abdomen often with yellow and black markings .... family **SYRPHIDAE**
- Wings without a 'vena spuria' 4
4. Feet with two pads (Fig. 13.44a), or with claws in place of pads 5
- Feet with three pads (Fig. 13.44b) 6
5. Wings with five posterior cells between veins R5 and Cu1 (e.g. Fig. 13.45) 7
- Wings with fewer than five posterior cells between veins R5 and Cu1 .... other flies
6. Antennal segment 3 compound, with an apical style or a dorsal arista; wings with the costa continued around the posterior margin, usually with cross-veins in the outer two-thirds; thorax not humpbacked .... family **RHAGIONIDAE**
- Antennal segment 3 annulated or costa of wing discontinued at wing tip or thorax humpbacked .... other flies
7. Vertex distinctly grooved between the eyes (Fig. 13.46a); legs robust (Fig. 13.47a) .... family **ASILIDAE** [Note 2]
- Vertex not grooved (Fig. 13.46b); legs slender (Fig. 13.47b) .... family **THEREVIDAE**
8. Antennae arising from a hollow in the face, surmounted by a ptilinal suture (Fig. 13.48) 9
- Ptilinal suture absent 15
9. Second antennal segment always with a distinct dorsal cleft or seam for nearly its whole length (Fig. 13.49); transverse suture of thorax often entire (Fig. 13.50) or almost so; lower (= thoracic) squamae often large .... (Section CALYPTRATA) – 10
- Transverse suture and lower (= thoracic) squamae poorly developed; second antennal segment without a cleft. .... other flies (Section ACALYPTRATA)
10. Postscutellum strongly developed (Fig. 13.51); thorax with hypopleural bristles ... . family TACHINIDAE [Note 3]
- Postscutellum absent or weakly developed; hypopleural bristles present or absent 11
11. Thorax with hypopleural bristles (Fig. 13.52) .... other flies (families CALLIPHORIDAE and SARCOPHAGIDAE)
- Thorax without hypopleural bristles 12
12. Lower (= thoracic) squamae large, and as long as or longer than the upper (= alar) squamae (see Fig. 13.40) 13
- Lower (= thoracic) squamae short, at most half the length of the upper (= alar) squamae .... other flies (families ANTHOMYIIDAE [in part] and SCATOPHAGIDAE)

13. Anal vein reaching the wing margin (Fig. 13.53) . . . . other flies (ANTHOMYIIDAE [in part])
– Anal vein not reaching wing margin 14
14. Axillary vein longer and usually fainter than anal vein, their paths convergent (Fig. 13.54) . . . . other flies (family FANNIIDAE)
– Axillary vein and anal vein short, their paths not convergent, as in Fig. 13.40 . . . . family **MUSCIDAE**
15. Wing venation reduced, as in Fig. 13.55 . . . . other flies (family PHORIDAE)
– Wing venation otherwise 16
16. Wing tips sharply pointed, with cross-veins at extreme base . . . . other flies (family LONCHOPTERIDAE)
– Wing tips not sharply pointed 17
17. Basal cells minute, with no cross-veins in the outer four-fifths of wing, or anal cell pointed and almost twice as long as the second basal cell . . . . other flies (family PLATYPEZIDAE)
– Wings not as above . . . . subfamily **EMPIDOIDEA** – 18
18. Basal cells very short or absent; 2nd basal cell confluent with discal cell; radial and cubital veins branching from their common stem close to wing base; cubital vein never forked (see Fig. 13.34); body often shiny metallic bluish, greenish or bronze . . . . family **DOLICHOPODIDAE**
– Basal cells sufficiently long to be conspicuous; radial and cubital veins branching from their common stem beyond (or at least only a little before) basal third of wing; 2nd basal cell often separated from discal cell; cubital vein often forked 19
19. Antennal segment 1 bristled (Fig. 13.56), or at least with a few bristly hairs beneath; cubital vein often forked; male hypopygium symmetrical and unrotated, with discrete gonopods (lateral lamellae) . . . . family **EMPIDIDAE**
– Antennal segment 1 very small, without bristles; cubital vein never forked; male hypopygium with gonopods connected dorsally (forming a periandrium) and often asymmetrical and rotated 20
20. Wings with an alula (Fig. 13.57); parts of hind legs laterally compressed and dilated in both sexes . . . . family **ATELESTIDAE** [Note 4]
– Alula not developed 21
21. Basal and anal cells conspicuous but small, the anal cell usually rounded apically; discal cell present, emitting three veins to wing margin; costa running around wing tip (Fig. 13.58) . . . . family **MICROPHORIDAE** [Note 2]
– Basal cells moderately large, the anal cell differently shaped or even absent; discal cell absent or, if present, emitting two or three veins to wing margin (see Figs 13.35 and 13.36); male genitalia rotated through about 90° towards the right . . . . family **HYBOTIDAE**

Note 1. Apart from a few midges (e.g. *Aphidoletes aphidimyza*) (family CECIDOMYIIDAE), this suborder includes no predatory species.
Note 2. No genera or species clearly identified as dominant or subdominant in oilseed rape.
Note 3. A family of parasitoids (see Chapter 9).
Note 4. Predacious flies not yet recorded from oilseed rape.

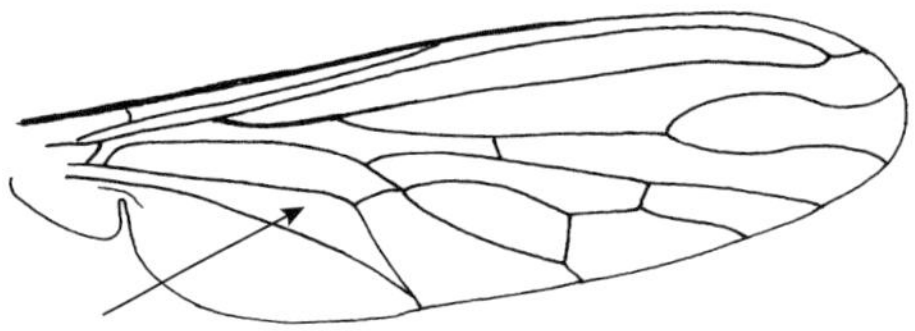

Fig. 13.43 Wing of a robber fly, with pointed anal cell (arrowed) – family ASILIDAE.

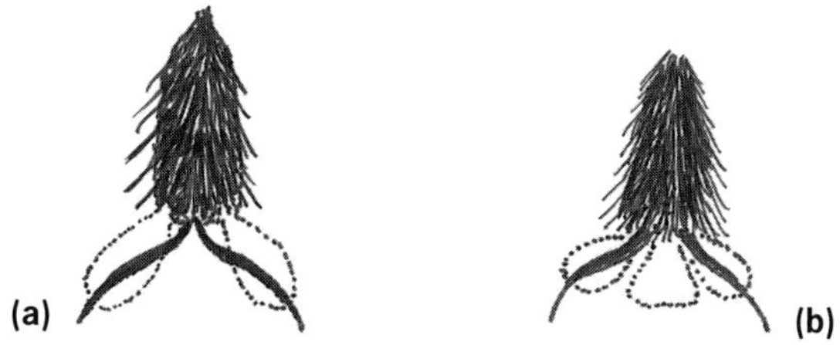

Fig. 13.44 Foot of a fly – (a) with two pads; (b) with three pads.

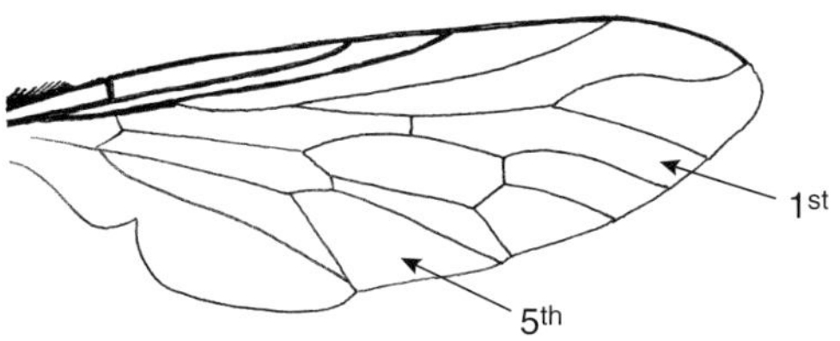

Fig. 13.45 Wing of a stiletto fly, to show five posterior cells (1st and 5th arrowed) between veins R5 and Cu1 – family THEREVIDAE.

Fig. 13.46 Vertex – (a) family ASILIDAE; (b) family THEREVIDAE.

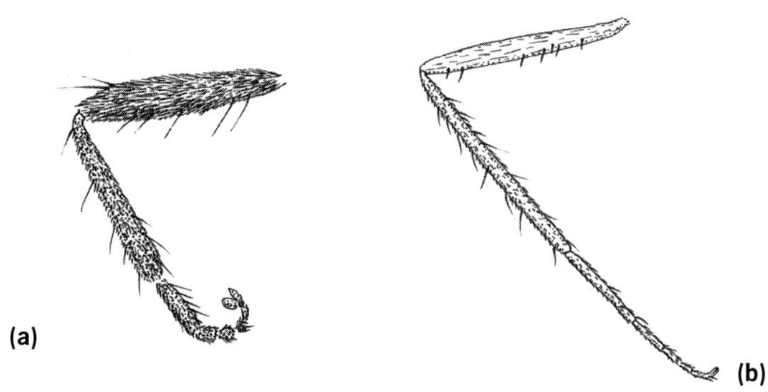

Fig. 13.47 Hind leg – (a) family ASILIDAE; (b) family THEREVIDAE.

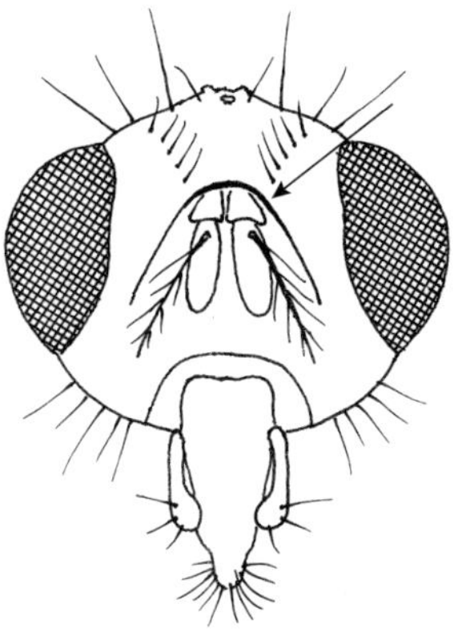

Fig. 13.48 Head of a calyptrate fly, to show ptilinal suture (arrowed).

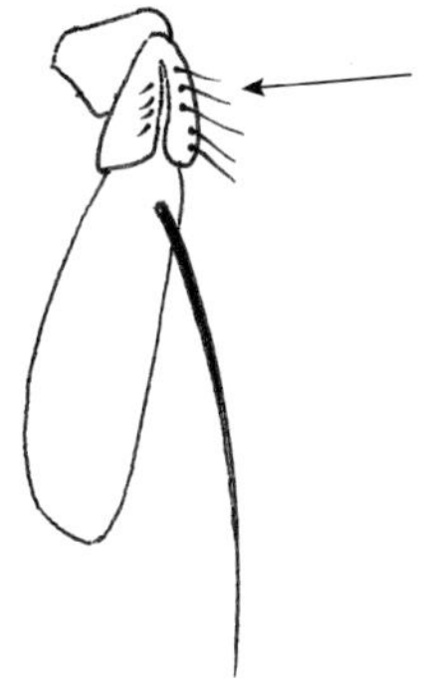

Fig. 13.49 Antenna, to show grooved segment 2 (arrowed) – section CALYPTRATA.

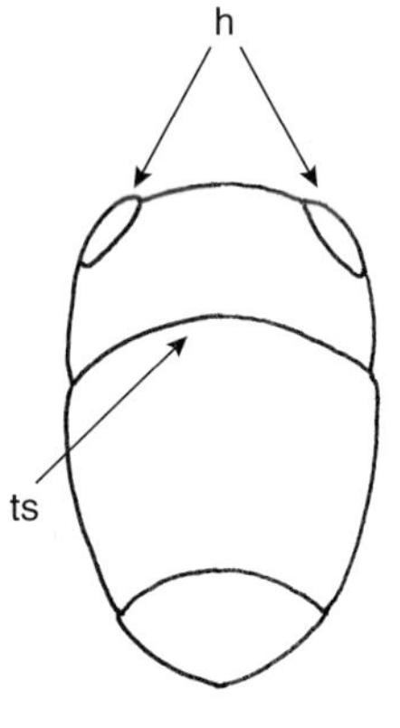

Fig. 13.50 Dorsal view of thorax, to show humeri (h) and transverse suture (ts) – section CALYPTRATA.

Fig. 13.51 Left lateral view of thorax, to show postscutellum (arrowed) – family TACHINIDAE.

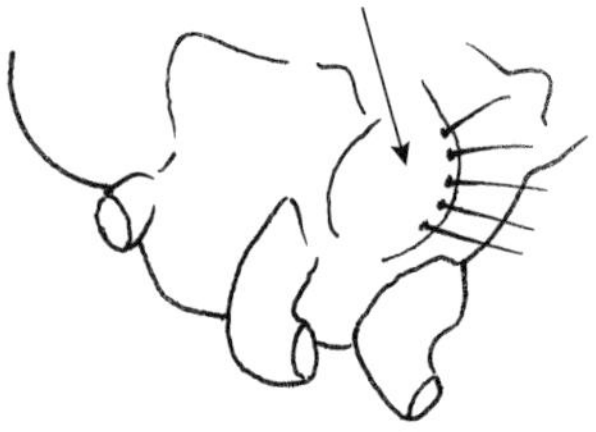

Fig. 13.52 Hypopleural bristles on thorax (arrowed) – family CALLIPHORIDAE.

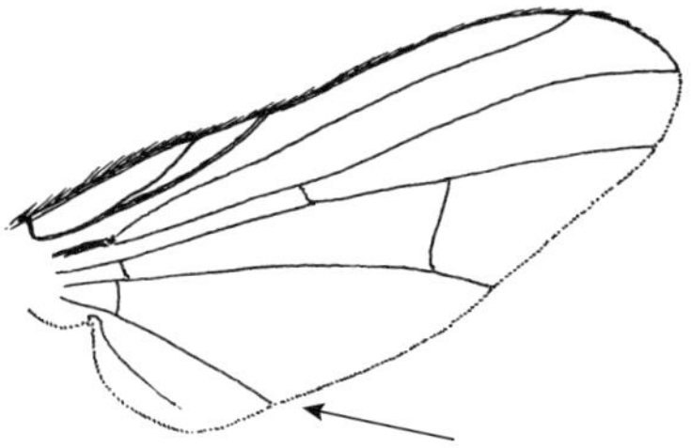

Fig. 13.53 Wing, with anal vein (arrowed) extending to margin – family ANTHOMYIIDAE (in part).

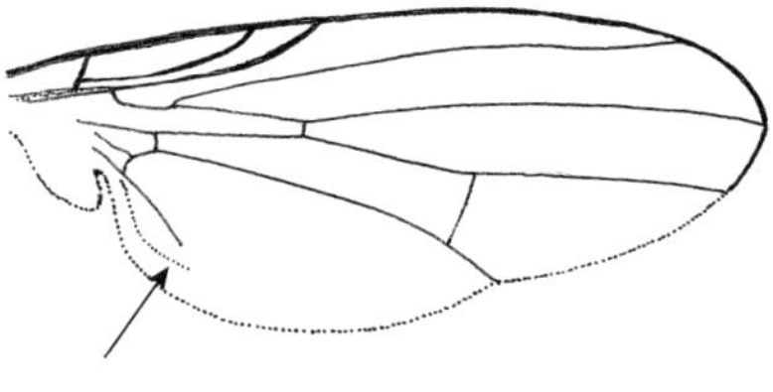

Fig. 13.54 Wing, with axillary vein (arrowed) longer than anal vein – family FANNIDAE.

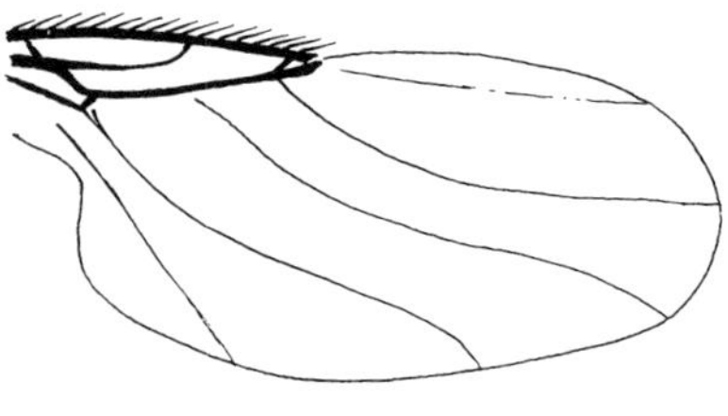

Fig. 13.55 Wing – family PHORIDAE.

Fig. 13.56 Antenna of a dance fly, with bristles on segment 1 (arrowed) – family EMPIDIDAE.

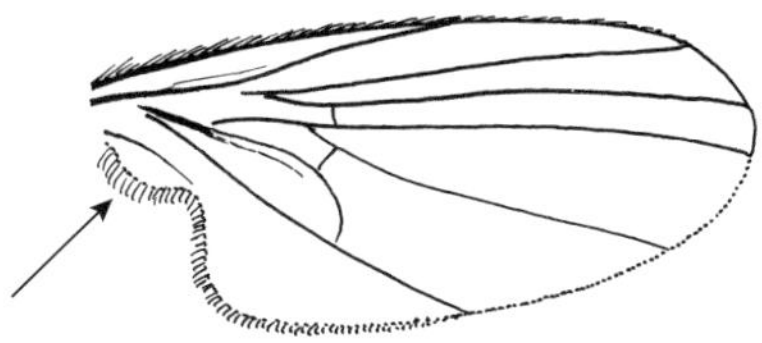

Fig. 13.57 Wing, with an alula (arrowed) – family ATELESTIDAE.

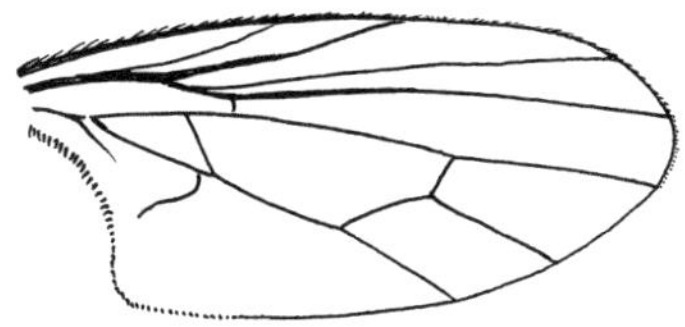

Fig. 13.58 Wing – family MICROPHORIDAE.

and flies particularly associated with oilseed rape (but mainly pests) is given by Kirk (1991).

General keys to dipterous larvae exist (e.g. Brauns, 1954a, 1954b; Smith, 1989), and these works include brief summaries of predatory groups. However, no keys or descriptions are dedicated to the identification to predatory species of significance in oilseed rape fields and, on the basis of currently available knowledge, it would not be feasible to produce one. Nor is it considered that such a key would have any practical application.

Spiders

Unlike insects, spiders (class Arachnida: order Araneae) have the body divided into just two parts (the carapace or cephalothorax and the abdomen or opistosoma); they are also eight-legged and lack antennae and compound eyes. Unlike their close relatives (e.g. harvestmen and mites), the carapace and abdomen are separated by a narrow waist; spiders also possess spinnerets (three pairs) and a pair of palps that are particularly well developed in males. The jaws of a spider are known as chelicerae, each composed of a large basal segment to which a large apical fang is articulated. Most spiders possess three or four pairs of simple eyes, the form and position of which are often characteristic.

Important but specialist keys to spiders include those by Locket & Millage (1951, 1953), Locket *et al.*, (1974), Jones (1984), Roberts (1985, 1987) and Heimer & Nentwig (1991). Their use, however, is unlikely to help the inexperienced to identify species, which usually requires the examination of specimens under a microscope. Recently, a key (which is continuously updated) was published on the Internet (Nentwig *et al.*, 2001). For a more general account of spiders, which includes summaries of family characteristics, see Bristowe (1958).

Spiders from at least 10 families are known to occur in oilseed rape fields (see Table 12.5 for systematic details). Although some individual species are very characteristic, and easily recognized, determination to species is usually a specialist task; this is especially so in the case of small species, such as money spiders (Linyphiidae). To at least aid in the recognition of the main groups associated with oilseed rape, a simplified, illustrated key, based on the above-mentioned standard works, is presented here (Box 13.8).

Box 13.8
Key to the main groups of spiders inhabiting European oilseed rape fields

1. Spiders with eight eyes, of similar or different sizes; body usually longer than 4 mm 2
– Spiders with eight eyes, all of similar size; body usually shorter than 4 mm, if longer then usually with a calamistrum on metatarsus IV (see Fig. 13.65) 8
2. Chelicerae robust and strongly divergent (Fig. 13.59). . . . . family **TETRAGNATHIDAE**
– Chelicerae otherwise 3
3. Legs with two tarsal claws; spiders never constructing webs to catch prey 4
– Legs with three tarsal claws; spiders either free-living or inhabiting orb-webs 7
4. Spinnerets prominent and projecting well beyond tip of abdomen; abdomen long and narrow; spiders often inhabiting silken retreats under stones, etc. 5
– Spinnerets not or only just visible from above; spiders free-living 6
5. Anterior spinnerets (i.e. the lowest pair, viewed from above) cylindrical, clearly separated and slightly longer than posterior pair (Fig. 13.60) . . . . family **GNAPHOSIDAE**
– Anterior spinnerets conical, close-set and shorter than the posterior pair (Fig. 13.61) . . . . family **CLUBIONIDAE**
6. Legs relatively short and robust, all of similar length; middle-anterior eyes very large (Fig. 13.62); carapace very large and square-fronted; body more or less clothed in iridescent (often scale-like) hairs . . . . family **SALTICIDAE**
– General appearance crab-like, with legs I and II noticeably longer than legs I II and IV (see Fig. 12.9); eyes all small, mostly located on small elevations; carapace short and broad; abdomen broader posteriorly . . . . family **THOMISIDAE**
7. Abdomen cylindrical, narrower than carapace (Fig. 13.63a); two large eyes on top and two at front of head (the former more widely spaced), plus an anterior row of four small eyes; legs hairy; ground-dwelling spiders, the female carrying an egg sac attached to the spinnerets . . . . family **LYCOSIDAE**
– Abdomen more or less globular, dorsoventrally flattened, broader than carapace (often markedly so) and often distinctively patterned (Fig. 13.63b); eyes all of similar size; legs with strong bristles; chelicerae strong and vertical, with a lateral condyle (Fig. 13.64); orb-web builders . . . . . family **ARGIOPIDAE**
8. Metatarsus IV with a calamistrum (Fig. 13.65); eyes typically pearly white . . . . family **DICTYNIDAE**
– Metatarsus IV without a calamistrum. 9
9. A row of serrated bristles present on tarsus IV (Fig. 13.66); legs relatively short; abdomen globular; carapace grey or patterned . . . . family **THERIDIIDAE**
– Legs entirely without bristles, but tibia IV with one or two spines; abdomen somewhat flattened; chelicerae with ridges used for stridulation; spiders hang upside down beneath sheet webs. . . . . family **LINYPHIIDAE** – 10
10. Tibia IV with one dorsal spine; female palp without a claw . . . . subfamily **ERIGONINAE**
– Tibia IV with two dorsal spines (Fig. 13.67); female palp with a claw. . . . . subfamily **LINYPHIINAE**

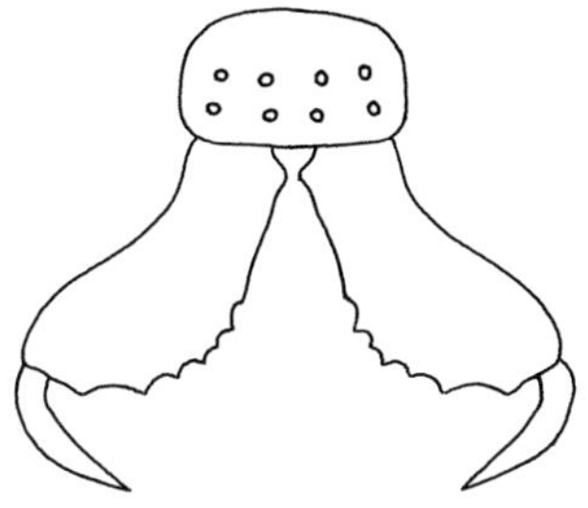

Fig. 13.59 Face and strongly divergent chelicerae – family TETRAGNATHIDAE.

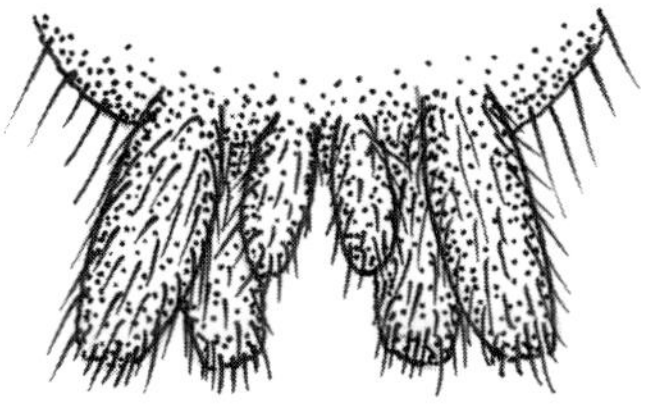

Fig. 13.60 Spinnerets, viewed from below – family GNAPHOSIDAE

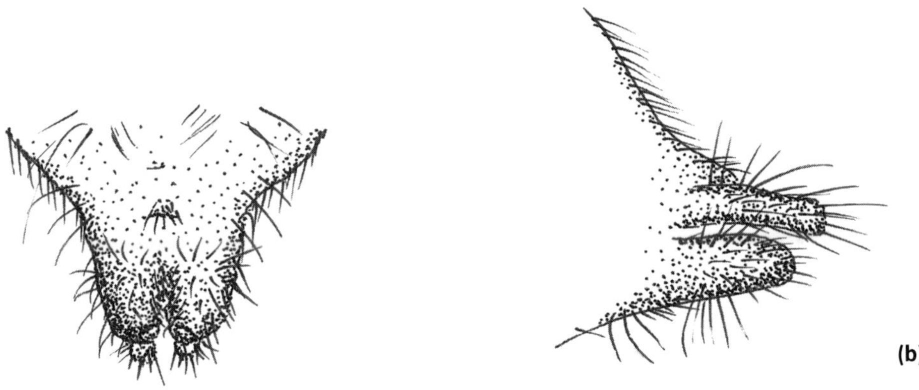

(a)

(b)

Fig. 13.61 Spinnerets – family CLUBIONIDAE. (a) Ventral view; (b) lateral view.

Fig. 13.62 Face and chelicerae of a jumping spider – family SALTICIDAE.

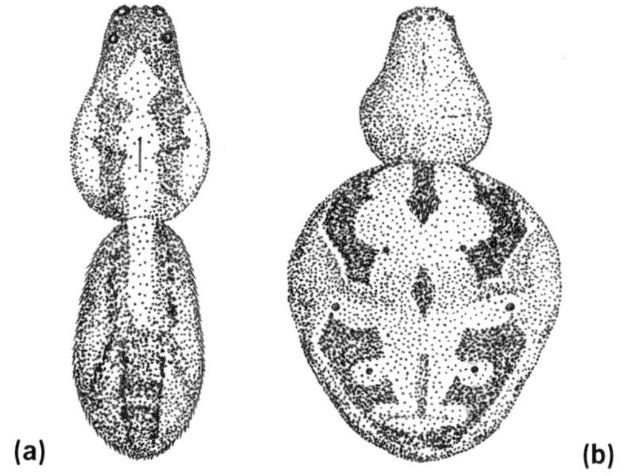

Fig. 13.63 Carapace and abdomen – (a) a wolf spider, family LYCOSIDAE; (b) an orb-web spider, family ARGIOPIDAE.

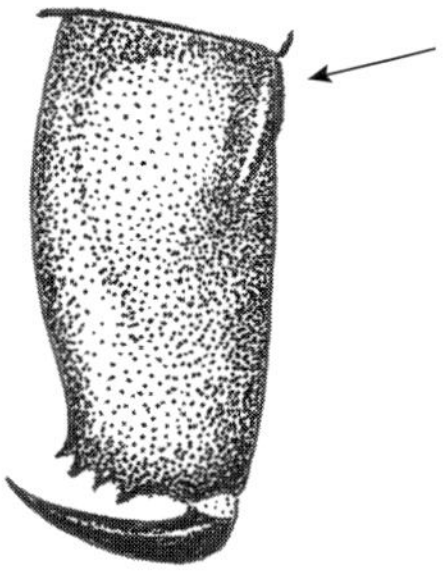

Fig. 13.64 Chelicera of an orb-web spider, to show condyle (arrowed) – family ARGIOPIDAE.

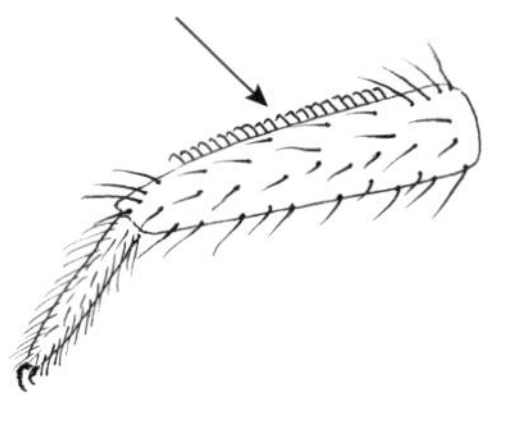

Fig. 13.65 Tarsus and metatarsus IV, to show calamistrum (arrowed) – family DICTYNIDAE.

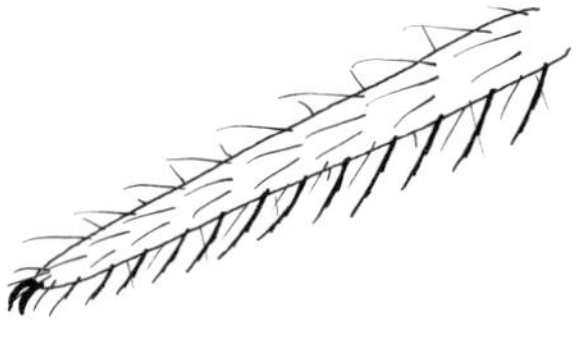

Fig. 13.66 Tarsus IV – family THERIDIIDAE.

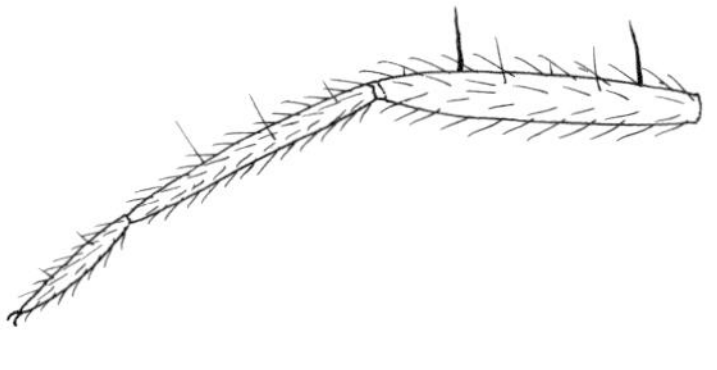

Fig. 13.67 Tip of leg IV of a money spider – subfamily LINYPHIINAE.

Minor Predators

Descriptions and general keys are available for various minor groups of invertebrate predators to be found in arable fields: e.g. for heteropteran bugs see Southwood & Leston (1959); for thrips (Thysanoptera) see Mound *et al.* (1976); for lacewings (Neuroptera) see Killington (1937), Aspöck *et al.* (1980) and McKewan *et al.* (2001); for harvestmen (Arachnida: Opiliones) see Martens (1978); for centipedes (Chilopoda) see Eason (1964). These publications do not relate specifically to enemies of crop pests, such as those to be found in oilseed rape. Again, it was not considered practical or appropriate to present details here, especially as such predators play a relatively minor role as natural enemies of oilseed rape pests (see Chapters 12 and 16).

Acknowledgements

This work was part-funded by the European Commission Concerted Action: *Minimizing pesticide use and environmental impact by the development and promotion of bio-control strategies for oilseed rape pests, FAIR 96-1314.*

References

Alford, D. V. (1999). *A Textbook of Agricultural Entomology*. Blackwell Science: Oxford.

Arndt, E. (1991). *Carabidae.* In: B. Klausnitzer (ed.). *Die Larven der Käfer Mitteleuropas – Adephaga*, Vol. 1. Goecke & Evers: Krefeld.

Arndt, E. (1997). *Carabidae.* In: B. Klausnitzer (ed.). *Die Larven der Käfer Mitteleuropas – Polyphaga Teil 3 sowie Ergänzungen zum 1. bis 3. Band*. Vol. 4. Goecke & Evers: Krefeld.

Aspöck, H., Aspöck, U. & Hölzel, H. (1980). *Die Neuropteren Europas. Vols 1 & 2.* Goecke & Evers: Krefeld.

Assing, V., Frisch, J., Kahlen, M., Löbl, I., Lohse, G. A., Puthz, V., Schülke, M., Terlutter, H., Uhlig, M., Vogel, J., Wunderle, P. & Zerche, L. (1998). *23. Familie Staphylinidae.* In: W. Lucht & B. Klausnitzer (eds). *Die Käfer Mitteleuropas. 4. Supplementband.* Goecke & Evers: Krefeld.

Assis-Fonseca, E. C. M. de (1968). Cyclorrhapha (Muscidae). *Handbooks for the Identification of British Insects* **10** (4b), 1–119.

Assis-Fonseca, E. C. M. de (1978). Diptera: Dolichopodidae. Orthorrapha, Brachycera. *Handbooks for the Identification of British Insects* **9** (5), 1–90.

Assmann, T., Balkenohl, H., Daffner, H., Gebert, J., Hieke, F., Lohse, G. A., Lompe, A., Müller-Motzfeld, G., Schmidt, J., Trautner, J. & Wrase, D. W. (1998). Ergänzungen und Berichtigungen zu 'Die Käfer Mitteleuropas' 1. Familie Carabidae. In: W. Lucht & C. Klausnitzer (eds). *Die Käfer Mitteleuropas. 4. Supplementband.* Goecke & Evers: Krefeld.

Böving, A. G. (1929). On the classification of beetles according to larval characters. *Bulletin of the Brooklyn Entomological Society* **24**, 55–80.

Böving, A. G. & Craighead, F. C. (1931). An illustrated synopsis of the principal larval forms of the order Coleoptera. *Entomologica Americana* **11** (N.S.), 1–351.

Brauns, A. (1954a). *Terricole Dipteren-Larven. Vol. 1.* Musterschmidt, Wissenschaftlicher Verlag: Göttingen.

Brauns, A. (1954b). *Puppen terricoler Dipteren-Larven. Vol. 2.* Musterschmidt, Wissenschaftlicher Verlag: Göttingen.

Bristowe, W. S. (1958). *The World of Spiders.* Collins: London.

Chvala, M. (1975). The Tachydromiinae (Dipt., Empididae) of Fennoscandia and Denmark. *Fauna Entomologica Scandinavica* **3**, 1–336.

Chvala, M. (1989). Monograph of the northern and central European species of *Platypalpus* (Diptera: Hybotidae) with data on the occurrence in Czechoslovakia. *Acta Universitatis Carolinae – Biologia* **32**, 209–376.

Collin, J. E. (1961). *Empididae. British Flies VI.* Cambridge University Press: Cambridge.

Colyer, C. N. & Hammond, C. O. (1968). *Flies of the British Isles.* Warne: London.

Costa, C., Vanin, S. A. & Casari-Chen, S. A. (1988). *Larvas de Coleoptera do Brasil.* Sao Paulo.

Dahlgren, G. & Wittmer, W. (1979). 27. Familie Cantharidae. In: H. Freude, K. W. Harde & G. A. Lohse (eds). *Die Käfer Mitteleuropas. Vol. 6.* Goecke & Evers: Krefeld.

Eason, E. H. (1964). *Centipedes of the British Isles.* London.

Emden, F. I. van (1942). Larvae of British beetles – keys to families. *Entomologist's Monthly Magazine* **78**, 206–26, 253–72.

Fitton, M. G. (1975). The larvae of the British genera of Cantharidae (Col.). *Journal of Entomology* **B 44**, 243–54.

Freude, H., Harde, K. W. & Lohse, G. A. (1965 *et seq.*). *Die Käfer Mitteleuropas. Vols 1–11.* Goecke & Evers: Krefeld.

Freude, H., Harde, K. W. & Lohse, G. A. (1976). *Die Käfer Mitteleuropsas. Vol. 2, Adephaga 1.* Goecke & Evers: Krefeld.

Fürsch, H. (1967). 62. Familie Coccinellidae. In: H. Freude, K. W. Harde & G. A. Lohse (eds). *Die Käfer Mitteleuropas. Vol. 7.* Goecke & Evers: Krefeld.

Fürsch, H. (1992). 62. Familie Coccinellidae. In: G. A. Lohse & W. H. Lucht (eds). *Die Käfer Mitteleuropas. 2. Supplementband. Vol. 13.* Goecke & Evers: Krefeld.

Ghilarov, M. S. (1964). [*Key for Soil Inhabiting Insect Larvae.*] Moscow. [In Russian.]

Goot, V. S. van der (1981). *De zweefvliegen von NW-Europa en Europees Rusland, in het bijzonder von de Benelux. Bibliotheek Koninklijke Nederlandse Natuurhist.* Vereniging, Hougwood: Amsterdam.

Halstead, D. G. H. (1963). Coleoptera: Histeroidea. *Handbooks for the Identification of British Insects* **4** (10), 1–16.

Hayashi, N., Fukuda, A. & Kurosa, K. (1959). *Coleoptera.* In: T. Esaki, N. Yuasa, T. Ishii & T. Motoki (eds). *Illustrated Insect Larvae of Japan.* Tokyo.

Heimer, S. & Nentwig, W. (1991). *Spinnen Mitteleuropas. Ein Bestimmungsbuch.* Parey: Berlin.

Hurka, K. (1996). *Carabidae of the Czech and Slovak Republics.* Kabourek: Zlín.

Jones, D. (1984). *Der Kosmos Spinnenführer.* Kosmos: Stuttgart.

Joy, N. H. (1932). *A Practical Handbook of British Beetles.* 2 vols. Witherby: London.

Killington, F. J. (1937). *A Monograph of the British Neuroptera. Vol. II.* Ray Society: London.

Kirk, W. D. J. (1991). *Insects on Cabbage and Oilseed Rape. Naturalist's Handbook 18.* Richmond Publishing: Slough.

Klausnitzer, B. (1978). *Bestimmungsbücher zur Bodenfauna Europas Ordnung Coleoptera (Larven). 10. Lieferung.* W. Junk: The Hague.

Klausnitzer, B. (ed.) (1991). *Die Larven der Käfer Mitteleuropas – Adephaga. – Vol. 1.* Goecke & Evers: Krefeld.

Klausnitzer, B. (ed.) (1994). *Die Larven der Käfer Mitteleuropas – Myxophage, Polyphaga Teil 1. – Vol. 2.* Goecke & Evers: Krefeld.

Klausnitzer, B. (ed.) (1996). *Die Larven der Käfer Mitteleuropas – Polyphaga Teil 2. – Vol. 3.* Goecke & Evers: Krefeld.

Klausnitzer, B. (1997a). *27. Familie Cantharidae.* In: B. Klausnitzer (ed.). *Die Larven der Käfer Mitteleuropas – Polyphaga Teil 3 sowie Ergänzungen zum 1. bis 3. Band, Vol. 4.* Goecke & Evers: Krefeld.

Klausnitzer, B. (ed.) (1997b). *Die Larven der Käfer Mitteleuropas – Polyphaga Teil 3 sowie Ergänzungen zum 1. bis 3. Band. – Vol. 4.* Goecke & Evers: Krefeld.

Klausnitzer, B. (ed.) (1999). *Die Larven der Käfer Mitteleuropas – Polyphaga Teil 4. – Vol. 5.* Goecke & Evers: Krefeld.

Klausnitzer, B. (ed.) (2001). *Die Larven der Käfer Mitteleuropas – Polyphaga Teil 5. – Vol. 6.* Spektrum Akademischer Verlag: Heidelberg.

Larsson, S. G. (1938). *Cantharidae, Larven.* In: V. Hansen (ed.). *Malacodermata, Fossipedes, Macrodactyla og Brachymera. – Danmarks Fauna* 44. Kopenhagen.

Lawrence, J. F. & Newton, A. F. (1995). *Families and subfamilies of Coleoptera (with selected genera, notes, references and data on family-group names).* In: J. Pakaluk & S. A. Slipinski (eds). *Biology, Phylogeny and Classification of Coleoptera: Papers Celebrating the 80th Birthday of Roy A. Crowson.* Warszawa.

LeSage, L. (1991). *Cantharidae (Cantharidoidea).* In: F. W. Stehr (ed.). *Immature Insects. Vol. 2.* Kendall: Iowa.

Lindroth, C. H. (1974). Coleoptera: Carabidae. *Handbooks for the Identification of British Insects* **4** (2), 1–148.

Locket, G. H. & Millage, A. F. (1951). *British Spiders. Vol. I.* Ray Society: London.

Locket, G. H. & Millage, A. F. (1953). *British Spiders. Vol. II.* Ray Society: London.

Locket, G. H., Millage, A. F. & Merrett, P. (1974). *British Spiders. Vol. III.* Ray Society: London.

Lohse, G. A. (1964). *Staphylinidae I (Micropeplinae bis Tachyporinae).* In: H. Freude, W. Harde & G. A. Lohse (eds). *Die Käfer Mitteleuropas, Vol. 4.* Goecke & Evers: Krefeld.

Lohse, G. A. (1974). *Staphylinidae II (Hypocyphtinae and Aleocharinae).* In: H. Freude, K. W. Harde & G. A. Lohse (eds). *Die Käfer Mitteleuropas, Vol. 5.* Goecke & Evers: Krefeld.

Lohse, G. A. & Lucht, W. H. (1989). *Die Käfer Mitteleuropas. 1. Supplementband mit Katalogteil.* Goecke & Evers: Krefeld.

Lohse, G. A. & Lucht, W. H. (1989–1993). *Die Käfer Mitteleuropas. 1–3. Supplementband, Vols 12–14.* Goecke & Evers: Krefeld.

Lohse, G. A. & Lucht, W. H. (1992). *27. Familie Cantharidae.* In: G. A. Lohse & W. H. Lucht (eds). *Die Käfer Mitteleuropas. 2. Supplementband. Vol. 13.* Goecke & Evers: Krefeld.

Lucht, W. H. (1987). *Die Käfer Mitteleuropas Katalog.* Goecke & Evers: Krefeld.

Lucht, W. H. & Klausnitzer, B. (1998). *Die Käfer Mitteleuropas. 4. Supplementband, Vol. 15.* Goecke & Evers: Krefeld.

Lyneborg, L. & Spitzer, K. (1975). The Czechoslovak species of *Thereva* (Therevidae, Diptera) with the description of a new species from Hungary and Austria. *Acta Scientia Naturalis Musei Bohemica Meridionalis* **14** (1974), 13–42.

Mamaev, B. M. (1972). [*Tables to Determine Insect Larvae.*] Moscow. [In Russian.]

Martens, J. (1978). Weberknechte, Opiliones. In: F. Dahl (ed.). *Die Tierwelt Deutschlands und der angrenzenden Meeresteile, Part 64.* Gustav Fischer: Jena.

McEwan, P., New, T. R. & Whittington, A. E. (eds) (2001). *Lacewings in the Crop Environment.* Cambridge University Press: Cambridge.

Morimoto, K. & Hayashi, N. (eds) (1986 *et seq.*). *The Coleoptera of Japan in Color.* Hoikusha Publishing Company: Osaka.

Mound, L. A., Morison, G. D., Pitkin, B. R. & Palmer, J. M. (1976). Thysanoptera. *Handbooks for the Identification of British Insects* **1** (11), 1–79.

Nentwig, W., Hänggi, A., Kropf, C. & Blick, T. (2001). *Central Europe Spiders – A Determination Key.* (www.araneae.unibe.ch)

Oldroyd, H. (1969). Tabanoidea and Asiloidea. *Handbooks for the Identification of British Insects* **9** (4), 1–132.

Oldroyd, H. (1970). Diptera. Introduction and keys to families. *Handbooks for the Identification of British Insects* **9** (1), 1–104.

Parent, O. (1938). Dipteres Dolichopodidae. *Faune de France* **35**, 1–720.

Paulian, R. (1990). *Atlas des Larves d'Insectes de France.* Paris.

Perris, E. (1877). Larves des Coléoptères. *Annales de la Société Linnéenne de Lyon* **23**, 1–430.

Peterson, A. (1951). *Larvae of Insects. Part II.* Ann Arbor, Michigan.

Pope, R. D. (1953). Coleoptera: Coccinellidae and Sphindidae. *Handbooks for the Identification of British Insects* **5** (7), 1–12.

Pope, R. D. (1977). A check list of British Insects. Coleoptera and Strepsiptera. *Handbooks for the Identification of British Insects* **9** (3), 1–105.

Reitter, E. (1909). *Fauna Germanica. Die Käfer des Deutschen Reiches. II. Band.* K. G. Lutz': Stuttgart.

Reitter, E. (1911). *Fauna Germanica. Die Käfer des Deutschen Reiches. III. Band.* K. G. Lutz': Stuttgart.

Roberts, M. J. (1985). *The Spiders of Great Britain and Ireland. Vol. 1.* Harley Books: Colchester.

Roberts, M. J. (1987). *The Spiders of Great Britain and Ireland. Vol. 2.* Harley Books: Colchester.

Roskozny, R. & Spitzer, K. (1965). Schnepfenfliegen (Diptera, Rhagionidae) in der Tschechoslowakei. *Acta Entomologica Bohemoslovaca* **62**, 340–68.

Schaefer, M. (2000). *Brohmer – Fauna von Deutschland.* 20th edn. Quelle & Meyer: Wiebelsheim.

Smith, K. G. V. (1989). An introduction to the immature stages of British flies. Diptera larvae, with notes on eggs, puparia and pupae. *Handbooks for the Identification of British Insects* **10** (14), 1–280.

Southwood, T. R. E. & Leston, D. (1959). *Land and Water Bugs of the British Isles.* Warne: London.

Stehr, F. W. (ed.) (1991). *Immature Insects.* Kendall: Iowa.

Striganova, B. R. (1964). Cantharidae. In: M. S. Ghilarov (ed.). [*Key for Soil-Inhabiting Insect Larvae.*] Moscow. [In Russian.]

Stubbs, A. E. & Falk, S. J. (1983). *British Hoverflies: An Illustrated Identification Guide.* British Entomological and Natural History Society: London.

Tottenham, C. E. (1954). Coleoptera: Staphylinidae, Piestinae to Euaesthetinae. *Handbooks for the Identification of British Insects* **4** (8a), 1–79.

Trautner, J. & Geigenmüller, K. (1987). *Tiger Beetles. Ground Beetles. Illustrated Key to the Cicindelidae and Carabidae of Europe.* Josef Matgraf: Aichtal.

Trautner, J., Geigenmüller, K. & Diehl, B. (1984). *Laufkäfer.* 2nd edn. Deutscher Jugendring für Naturbeobachtung: Hamburg.

Unwin, D. M. (1981). A key to the families of British Diptera. *Field Studies* **5**, 513–53.

Verhoeff, K. W. (1917). Zur Entwicklung, Morphologie und Biologie der Vorlarven und Larven der Cantharиden. *Archiv für Naturgeschichte* (A) **83**, 102–40.

Verhoeff, K. W. (1923). Beiträge zur Kenntnis der Coleopteren-Larven mit besonderer Berücksichtigung der Clavicornia. *Archiv für Naturgeschichte* (A) **89**, 1–109.

Wachmann, E., Platen, R. & Barndt, D. (1995). *Laufkäfer: Beobachtung, Lebensweise.* Naturbuch: Augsberg.

14

Sampling, Trapping and Rearing of Predators

Wolfgang Büchs
Federal Research Centre for Agriculture and Forestry, Messeweg 11/12, D-38104 Braunschweig, Germany

Introduction

Sampling, trapping and rearing of oilseed rape pests and their parasitoids were reviewed in Chapter 10. Here, information is provided on sampling, trapping and rearing methods for predators, or potential predators, of oilseed rape pests.

Sampling and Trapping Methods

Various techniques for monitoring and sampling predatory arthropods are used in experimental studies in oilseed rape fields. These methods are not unique to studies of predators in such situations; nor are they, in general, novel. Consequently, description of most of the techniques was considered unnecessary here, although an indication of the suitability of each technique for studying target groups of predators in oilseed rape fields was considered of value.

Pitfall traps

Pitfall traps typically consist of a vessel buried in the soil and partly filled with preservative fluid; there are several variations. Epigaeic arthropods fall into the traps when crossing the entrances. Results depend to a great extent on the degree of activity of arthropods running over the soil surface. Therefore, for detailed population assessments they should be used only in combination with an area-related sampling method (e.g. D-vac suction trapping, emergence traps or the flooding method). Pitfall traps were first developed for sampling insects in caves, and they have since become the most widespread sampling method for use in field entomology, including oilseed rape.

Pitfall traps are easy to install and to handle. They give an overview of the species composition of epigaeic arthropods, with a minimum of expenditure and are useful recorders of epigaeic activity. However, results can be difficult to interpret, primarily because the collection data are a mixture of abundance and surface activity (= 'activity-density');

further, results usually cannot be related to a defined area. Thus, a high activity-density can be a result of a large population within the habitat, a requirement to leave the area (e.g. owing to a lack of prey), or movement of organisms through the habitat from one site to another.

Emergence traps

Emergence traps were originally developed to record the appearance of adult aquatic insects. Later, in the 1960s, they were widely used in forest ecosystems to record the emergence of canopy insects overwintering in the soil. In arable fields, emergence traps were first used in the late 1980s (Büchs, 1991) and they have now become a standard method, essential for sampling epigaeic arthropods and for the interpretation of data from pitfall traps (see above).

Emergence traps (round or square-like, covering areas from 0.25 to 1 m^2) are typically tent-like, with a transparent sampling vessel at the top; all arthropods that emerge from the soil surface covered by the trap and that are positively phototactic will be collected. For sampling negatively phototactic arthropods, a pitfall trap may be installed in the soil covered by the emergence trap. If the requirement is to record not only the emergence of predators but also population densities, then location of the traps should be changed regularly (usually every two to four weeks).

Emergence traps have been used in oilseed rape by, for example, Goltermann (1994), Büchs *et al.* (1997, 1999), Harenberg (1997) and Büchs & Nuss (2000). See also Chapter 10 for details of their use in monitoring pest insects in oilseed rape crops.

Usefully, emergence traps are related to a defined area and operate continuously; they can, therefore, provide reliable information on emergence rates and on the point of origin of the arthropods recorded. Further, samples in the collection vessels provide reasonable evidence of the proportion of predators intending to leave the habitat after emergence. However, emergence traps (especially if large) are not easy to handle and, in agro-ecosystems, their position frequently has to be changed owing to the rapid changes in vegetation cover. This can result in some problems regarding interpretation and statistical evaluation of results (see Büchs *et al.*, 1999).

Gutter traps

So-called gutter traps typically consist of four pieces (sections) of plastic guttering, attached together to enclose a defined rectangular area (e.g. 1 m^2), installed with the top of the guttering at soil level. The guttering can be filled with preservation fluid (e.g. natrium benzoic acid), or installed without fluid but with four pitfall-like vessels in the corners and four in the middle of each section.

To prevent arthropods from entering the trap from the inside of the enclosure a metal frame should be mounted along the inner edge of the guttering. Thus, access to the gutter traps is possible only from the outside. Gutter traps are intended to catch all epigaeic predators that try to enter the enclosed area; they do not provide information on how long they might stay and what they might do within the area, but they do allow estimates to be made of the activity of epigaeic arthropods in relation to a defined area.

Several technical problems are posed by gutter traps:

- installing them more or less horizontally is difficult but is necessary in order to maintain the fluid level as nearly equal at each point of the trap as possible;
- rainfall and subsequent drying-out of the soil around the outer edge of the trap may result in the development of a 'step' or gap between the guttering and the surrounding soil surface, restricting access by arthropods;
- removal of trapped specimens is time-consuming, and may require the use of a special collecting implement (e.g. a sweep net), with no guarantee that all arthropods in the guttering will be recorded;
- the area surrounding the traps is inevitably greatly disturbed, owing to the need for access whilst traps are examined and maintained – at the early stages of flowering in oilseed rape crops, for example, plants are often bent over or trampled underfoot.

Funnel traps

Funnel traps are formed from plastic or cloth, typically with a diameter ranging from 13.5 to 21.0 cm or with a square (60 × 60 cm) frame. In oilseed rape, they are used to record the dropping of insect larvae from the flowers or pods of oilseed rape, and so they mainly catch fully fed larvae of pollen beetle (*Meligethes aeneus*), cabbage seed weevil (*Ceutorhynchus assimilis*) and brassica pod midge (*Dasineura brassicae*) (see Chapter 10), but also larvae of predators (such as rove beetles) which drop from flower stands to pupate in the soil. They are installed in an early growth stage of the crop, so that the plants are not destroyed but overgrow the funnel entrance. In combination with data on the emergence rates of the new generations of pests and predators (obtained from emergence traps), funnel traps also allow the success of reproduction to be calculated (Büchs, 1993). Funnel traps have been used to monitor predatory capacity in oilseed rape crops by Büchi (2002). In addition, again using funnel traps, W. Büchs (*unpublished*) recorded considerable numbers of aleocharine (Coleoptera: Staphylinidae, Aleocharinae) larvae (577.2 ± 296.1 individuals/m^2) dropping from oilseed rape flower stands, along with many pollen beetle larvae.

Artificial (aerial) structures

Artificial structures are used to measure the activity and presence of web-building spiders. There are two different groups of web-building spiders: species of families (e.g. Argiopidae) whose members build larger orb-like webs in the vegetation layer; and money spiders (Linyphiidae) which build small, horizontal webs on the soil surface. Sunderland *et al.* (1986) estimated from observations in winter wheat fields that up to 50% of the soil surface could be covered by money spider webs.

Artificial structures were created to standardize assessments of web building by money spiders (W. Büchs & A. Harenberg, *unpublished*). Here, they were made from pieces of wire mesh (49 × 49 cm, with a mesh width of 1 cm). The pieces of wire mesh were placed flat on the soil surface after the vegetation was removed, and the number of webs formed counted and documented photographically once a week. For better detection of webs, wire meshes may be sprayed with a fine mist of water. In these experiments, the density of horizontal webs was significantly greater in oilseed rape than in either winter wheat or set-aside. On the basis of the results obtained, 5.4% of the soil surface in oilseed rape was estimated to be covered by webs from May to mid-July. Money spiders are so-called 'pioneering species' (see Chapter 12) and numbers of their webs per unit area decrease with increasing extensification (see Chapter 15) and are extremely low in set-aside (e.g. Harenberg, 1997).

Soil flooding

Soil flooding records the actual abundance of epigaeic predators and flea beetles, e.g. cabbage flea beetles (*Phyllotreta* spp.) and cabbage stem flea beetle (*Psylliodes chrysocephala*), in oilseed rape fields, especially in autumn and early spring. Unlike other area-related methods (e.g. emergence traps), soil flooding enables distinctions between dead and live insects in the field to be made, and the method can be used (especially) to record the field effects of insecticide seed treatments or spray applications. By way of example, a metal frame (enclosing an area of $0.25\,m^2$) may be pushed into the soil and filled with water (10–40 litres/sample, depending on the soil type). After a consistent film of water has been created for at least two minutes, insects hidden under soil clods, etc. will rise to the surface and may be collected.

Soil flooding was originally developed for recording predators (mainly ground beetles of the tribe Bembidiini) in inundation soils of river banks. It was introduced into agro-ecosystems by Basedow *et al.* (1988) who, according to their comparative assessments with other methods, suggested it would give good results. However, although field experience has since shown the method to be highly efficient at collecting adult flea beetles and larvae of soldier beetles (Cantharidae) in arable situations (e.g. W. Büchs, *unpublished*), it appears not to be so for smaller beetles (e.g. Staphylinidae: Aleocharinae) or, generally, in soils with a tendency for compaction after natural flooding.

Malaise traps

The most standard type is that described by Townes (1982), but other variants are also available (e.g. those distributed by ECOTECH, Bonn, Germany). Malaise traps are most suitable for trapping Diptera (especially Hybotidae and Syrphidae) and Hymenoptera (including bees, parasitoid wasps and sawflies); they have been used to catch parasitoids in oilseed rape fields (see Chapter 10) but not predators. J. Franzen & W. Büchs (*unpublished*) and W. Büchs (*unpublished*) used malaise traps to sample predators in sugar beet, winter wheat and set-aside areas directly adjacent to oilseed rape fields.

Problems associated with malaise traps are:

- their sensitivity to wind, especially in autumn or spring, when the vegetation is low;
- the typical presence of relatively dense spider webs in the entrance of the catching vessel;
- the huge sample size, with masses of specimens trapped over relatively short periods (making it difficult, and sometimes impossible, to segregate completely the important insect taxa);
- their inability to allow catches to be related to defined areas.

Window traps

A window trap consists of a clear surface (window) which is mounted in a frame and ends at the bottom in a pod-like tray (as a window box) filled with water and a small amount of detergent (and, perhaps, preservative). Various sizes are currently in use (e.g. 0.5 × 0.5 m, up to 2 × 1 m). Sometimes the structure is protected by a roof. Insects that fly against the window should fall into the water and may then be collected. In order to allow the flight direction to be assessed, the window should divide the water pod into two sections (Mühlenberg, 1993). Window traps should be examined at least weekly. In oilseed rape, window traps have been used to detect turnip sawfly (*Athalia rosae*), and larger flying predators, but there are very few published accounts.

Immigration and migration of larger (heavier-bodied) flying insects (e.g. Coleoptera: Histeridae) can be recorded continuously, and with good efficiency, using window traps. However, such traps are not efficient for recording smaller flying insects (or for slow-flying larger insects). However, it is very often observed that many insects (e.g. Aleocharinae, Cantharidae, Coccinellidae, Neuroptera and Syrphidae) fly up and down in front of the window before they either fall into the sampling vessel or fly away. Further, the window (mostly hard plastic) is usually reflective, and gets dirty and/or 'blind' the longer it is exposed to the weather. Another disadvantage is that samples collected cannot be related to a defined area; in particular, because flying insects are recorded above the level of the surrounding vegetation it is not known whether captures have flown in from the immediate surroundings or from a much greater distance; nor, unlike some other traps, can emergence sites of the trapped insects be determined.

Yellow water traps

In Germany, the national working group 'Integrated Plant Protection – Subgroup Oil Seed Rape' decided to use the yellow water traps provided by ZENECA (formerly ICI and now Syngenta) as a standard (see Büchs, 1993). These traps are rectangular (33.5 × 26.0 cm) and, as they are provided without charge, widely used in oilseed rape crops.

Yellow water traps are used for monitoring various pests and parasitoids (for details, see Chapter 10); they are also used to monitor hover flies (Syrphidae).

Sticky traps

Sticky traps were mostly used to trap insects (including pests and predators) before yellow water traps (see above) became common: see the historical review in Büchs (1998). In particular, pollen beetle was sometimes 'controlled' from the late 19th century to the 1930s by huge sticky traps that had to be drawn through oilseed rape fields by horses (e.g. Frank, 1897; Blunck & Hähne, 1929; Buhl & Meyer, 1937). Today, sticky traps are more commonly used to trap pests in commercial greenhouses and, sometimes, orchards, and in scientific experiments on crop pests. See Chapter 10 for further details.

Cone traps and omni-directional flight traps

Although cone traps have been used by Walczak *et al.* (1998) to monitor oilseed rape pests (see Chapter 10), they do not appear to have been used to trap predators. Similarly, although omni-directional flight traps have been used in oilseed rape fields to monitor parasitoids (Murchie *et al.*, 1997) (see Chapter 10) there is no information on their use for trapping predators.

Suction traps

The D-vac suction trap is a standard piece of equipment for predator sampling in some regions (e.g. Germany and the UK). It was developed by Dietrick (1961) and is used mainly to record the actual and *ad hoc* abundance of spiders but also of ground beetles and rove beetles. In addition, D-Vac machines have been used to monitor parasitoids (R. Büchi, *pers. comm.*) and also pests in oilseed rape crops, e.g. the spring immigration into oilseed rape crops of adults of cabbage stem weevil (*Ceutorhynchus pallidactylus*), rape winter stem weevil (*C. napi*) and the autumn immigration into oilseed rape crops of adults of cabbage stem flea beetle (Alford, 1979). See Chapter 10 for further details.

Approaches towards standardizing sampling methods using D-vac suction traps were made mainly by Dinter & Poehling (1995) during studies of spiders in winter wheat crops.

Rearing Methods

Mass rearing of predators of value as natural enemies of oilseed rape insect pests can be carried out in order to pursue the following aims:

- to provide standardized insect (predator) material for laboratory experiments on the biology of key predators (e.g. to study their food preferences, food intake, preferred soil and/or micro-climatic conditions);
- to provide standardized insect (predator) material for laboratory/semi-field and field experiments on side-effects of pesticides for use in oilseed crops (e.g. effects on non-target organisms) – a survey of rearing methods for key species used for standard tests to assess side-effects of pesticides was compiled by Samsoe-Petersen *et al.* (1989);
- to provide material (predators) for release into field crops as biocontrol agents (BCAs), in order to enhance population densities in relation to those of pest organisms.

Because rearing of predators is not specific to oilseed rape crops the presentation of rearing methods, procedures and approaches is restricted here to a very brief description of basic principles, illustrated by a few examples. Also, with few exceptions, in the following listings of rearing procedures and field-release experiments the only species mentioned are those known to be, or that might be expected to be, members of the natural fauna of oilseed rape fields.

With regards to future applications of predators as BCAs in practical farming, experiences with the releasing of natural enemies into field crops are of particular interest. Therefore, here, experiences are reported that relate to species of potential significance as BCAs in oilseed rape fields. Basic demands and problems of rearing and subsequently releasing entomophagous insects (including predators) are summarized by Bigler (1989).

Regarding epigaeic predators, the best knowledge in rearing exists for ground beetles (Carabidae), whereas successful procedures for rove beetles (Staphylinidae) are very few. A number of approaches are known for rearing spiders, but none has yet resulted in a procedure that gives satisfactory results.

Ground beetles (Coleoptera: Carabidae)

Various rearing procedures for ground beetles have been published. The following are examples:

- *Bembidion tetracolum* Theiss & Heimbach (1993)
- *Poecilus cupreus* Kegel (1989)
- *Poecilus lepidus* Kegel (1989)
- *Poecilus versicolor* Kegel (1989)
- *Pterostichus melanarius* Johnson & Cameron (1969).

Methods for rearing of ground beetles may be exemplified by summarizing the methods described for *Bembidion tetracolum* and *Poecilus cupreus* by Heimbach (1992) and Theiss & Heimbach (1993). To induce egg laying, seasons must be simulated by changing both the photoperiod and temperature. Therefore, the beetles have to be kept for two months under artificial 'short day' conditions (8h light; 8–10°C), subsequently followed by 'summer' conditions (16h light; 20°C). After an initial two-month period of egg laying (*B. tetracolum* producing 30–35 eggs/female; *P. cupreus* producing *c.* 75–250 eggs/female), a second period of egg laying can be induced by repeating the same procedure, but the number of eggs subsequently produced is reduced by up to 50%.

Among the developmental stages, eggs usually show the greatest mortality (e.g. for *Bembidion tetracolum* and *Poecilus cupreus*, respectively: eggs, 18% and 34%; larvae, 19.5% and 15% – in both cases, pupal mortality was *c.* 3% and adult emergence success 45–50%). Egg production is highly influenced by conditions, and different species have different requirements. For example, whereas for *P. cupreus* a vessel filled with swollen clay pellets and with a mesh gauze bottom was suitable for inducing egg laying (Heimbach, 1989) (which subsequently allowed easy washing-out of the eggs), *B. tetracolum* laid eggs only on a plaster-charcoal layer (Theiss & Heimbach, 1993).

Whereas adults can be kept in groups of ten, or even more, individual (equal sex ratio) larvae have to be kept separately, because of their cannibalistic behaviour. Even the pupae should be kept individually in glass tubes filled with peat, where they are able to form their own pupal chambers.

Fly pupae (*Calliphora* spp.) or mealworms (*Tenebrio molitor*) are suitable as food for larvae and adults. However, such food items should be sliced up, to make them more readily available.

Rove beetles (Coleoptera: Staphylinidae)

With regards rove beetles, most rearing procedures focus on parasitic members of the genus *Aleochara* (e.g. *A. bilineata*) which are used as test species within ecotoxicological standard tests (Naton, 1989; Samsoe-Petersen *et al.*, 1989); they are also mass-reared as natural enemies of cabbage root fly (*Delia radicum*) for subsequent release in horticultural brassica crops (see Chapter 9). For further details, see the following:

- *Aleochara bilineata* Bromand (1980); Hertveldt *et al.* (1984); Whistlecraft *et al.*, 1985); Naton (1989); Samsoe-Petersen *et al.* (1989);
- *Tachyporus hypnorum* Förster (1991).

Details of rearing methods for other predatory species are sparse.

Because of difficulties attributable to their special biological requirements, some species (e.g. *Tachyporus hypnorum*) do not progress beyond the egg stage.

Spiders (Arachnida: Araneae)

Rearing methods for spiders are sparse because the species (especially Lycosidae) react very sensitively to the composition of food and also to slight variations (oscillations) of abiotic factors (e.g. temperature, air and soil humidity). Specific approaches, however, have been developed for some members of the families Linyphiidae and Lycosidae as follows:

- Linyphiidae: *Erigone atra* — Dinter & Poehling (1994);
- Linyphiidae: *Oedothorax apicatus* — Aukema *et al.* (1990); Dinter & Poehling (1994);
- Lycosidae: *Pardosa amentata* — Hof (1993).

Erigone atra and *Oedothorax apicatus* are amongst the most abundant species in arable fields, including oilseed rape (Chapter 12). Following the emergence of spiders from maternal egg cocoons, Dinter & Poehling (1994) kept individual juveniles of both species separately in glass tubes containing a layer of soil pre-inoculated with large numbers of the springtail *Lepidocyrus lanuginosus* as potential prey. Additionally, the juvenile spiders were fed periodically with wingless fruit flies (*Drosophila melanogaster*). Rearing success in these experiments ranged from 73% to 77%.

Bugs (Hemiptera)

Rearing methods have been developed mainly for anthocorid bugs (Anthocoridae) as natural enemies of aphids and thrips. Other approaches have focused mainly on predacious shield bugs (Pentatomidae), nabid bugs (Nabidae) and ground bugs (Lygaeidae). However, most efforts are recorded from North America and East Asia (Japan) and have no relevance to oilseed rape.

Acknowledgements

This work was part-funded by the European Commission Concerted Action: *Minimizing pesticide use and environmental impact by the development and promotion of bio-control strategies for oilseed rape pests, FAIR 96–1314.* Thanks are also due to Dr. Bernhard Kromp for providing data for this chapter.

References

Alford, D. V. (1979). Observations on the cabbage stem flea beetle, *Psylliodes chrysocephala*, on winter oil-seed rape in Cambridgeshire. *Annals of Applied Biology* **93**, 117–23.

Aukema, B., van den Berg, J. H. J., Leopold, A., Jagers, G. A. J. M. & Everts, J. W. (1990). A method for testing the toxicity of residues of pesticides on a standardized substrate of erigonid and linyphiid spiders. *Journal of Applied Entomology* **109**, 76–80.

Basedow, T., Klinger, K., Froese, A. & Yanes, G. (1988). Aufschwemmung mit Wasser zur Schnellbestimmung der Abundanz epigäischer Raubarthropoden auf Äckern. *Pedobiologia* **32**, 317–22.

Bigler, F. (1989). Quality assessment and control in entomophagous insects used for biological control. *Journal of Applied Entomology* **108**, 390–400.

Blunck, H. & Hähne, H. (1929). Fortschritte in der Bekämpfung von Rapskäfern mittels Fangmaschinen. *Fortschritte der Landwirtscahft* **4**, 193–212.

Bromand, B. (1980). Investigations on the biological control of the cabbage root fly (*Hylemya brassicae*) with *Aleochara bilineata*. *Bulletin OILB/SROP* **3** (1), 49–62.

Büchi, R. (2002). Mortality of pollen beetle (*Meligethes* spp.) larvae due to parasitoids and predators in rape fields and the effect of conservation strips. *Agriculture, Ecosystems and Environment* **90**, 255–63.

Büchs, W. (1991). Einfluss verschiedener landwirtschaftlicher Produktionsintensitäten auf die Abundanz von Arthropoden in Zuckerrübenfeldern. *Verhandlungen der Gesellschaft für Ökologie* **20**, 1–12.

Büchs, W. (1993). Investigations on the occurrence of pest insects in oil seed rape as a basis for the development of action thresholds, concepts for prognosis and strategies for the reduction of the input of insecticides. *IOBC/wprs Bulletin* **16** (9), 216–34.

Büchs, W. (1998). Tierische Schädlinge und ihre Antagonisten in Rapskulturen – Arbeiten zu Biologie, Epidemiologie, natürlicher Regulation und chemischer Bekämpfung in der 100-jährigen Geschichte der Biologischen Bundesanstalt für Land- und Forstwirtschaft. *Mitteilungen aus der Biologischen Bundesanstalt für Land- und Forstwirtschaft Berlin-Dahlem* **340**, 86–125.

Büchs, W. & Nuss, H. (2000). First steps to assess the importance of epigaeic active polyphagous predators as natural enemies of oilseed rape insect pests with soil pupating larvae. *IOBC/wprs Bulletin, Integrated Control in Oilseed Crops* **23** (6), 151–63.

Büchs, W., Harenberg, A. & Zimmermann, J. (1997). The invertebrate ecology of farmland as a mirror of the intensity of the impact of man? – An approach to interpreting results of field experiments carried out in different crop management intensities of sugar beet and an oil seed rape rotation including set-aside. *Biological Agriculture and Horticulture* **15**, 83–107.

Büchs, W., Harenberg, A., Prescher, S., Weber, G. & Hattwig, F. (1999). Entwicklung von Evertebratenzönosen bei verschiedenen Formen der Flächenstilllegung und Extensivierung. *Mitteilungen aus der Biologischen Bundesanstalt für Land- und Forstwirtschaft Berlin-Dahlem* **368**, 9–38.

Buhl, C. & Meyer, E. (1937). Ein neues Gerät zum Rapskäferfang. *Zeitschrift für Pflanzenkrankheiten und Pflanzenschutz* **47**, 34–8.

Dietrick, E. J. (1961). An improved backpack motor fan for suction sampling of insect populations. *Journal of Economic Entomology* **54**, 394–5.

Dinter, A. & Poehling, H.-M. (1994). Laborzucht von *Erigone atra* (Blackwall) und *Oedothorax apicatus* (Backwall) (Araneae, Linyphiidae) und Prüfung der Nebenwirkungen verschiedener Insektizide im Labor. *Mitteilungen der Deutschen Gesellschaft für Allgemeine und Angewandte Entomologie* **9**, 419–26.

Dinter, A. & Poehling, H.-M. (1995). Analyse der Populationsstruktur und -dynamik von Spinnen (Araneae) in Winterweizen – Vergleich von Intensiv-D-vac-Methode und Bodenfallentechnik. *Mitteilungen der Deutschen Gesellschaft für Allgemeine und Angewandte Entomologie* **10**, 533–6.

Förster, P. (1991). *Einflüsse von Pflanzenschutzmitteln auf Larven und Adulte von* Platynus dorsalis *(Pont.) (Coleoptera: Carabidae) und auf Adulte von* Tachyporus hypnorum *(L.) (Coleoptera: Staphylinidae) in Labor-, Halbfreiland- und Freilandversuchen.* PhD thesis, Justus-Liebig-Universität, Gießen, Germany.

Frank, A. B. (1897). *Kampfbuch gegen die Schädlinge unserer Feldfrüchte.* Parey: Berlin.

Goltermann, S. (1994). *Das Auftreten von Laufkäfern (Col., Carabidae) auf Winterrapsfeldern und deren Einfluß auf den Massenwechsel von* Meligethes aeneus *F. (Col., Nitidulidae).* PhD thesis, University of Rostock, Germany.

Harenberg, A. (1997). *Auswirkungen extensiv geführter Anbausysteme in verschiedenen Fruchtfolgen (Raps-, Zuckerrübenfruchtfolge) und einer selbstbegrünenden Dauerbrache auf Spinnen (Arachnida: Araneae).* PhD thesis, Technical University of Braunschweig, Germany.

Heimbach, U. (1989). Massenzucht von *Poecilus cupreus* (Col., Carabidae). *Tagungsberichte der Jahrestagung der Gesellschaft für Ökologie* **19**, 228–9.

Heimbach, U. (1992). Laboratory method to test effects of pesticides on *Poecilus cupreus* (Coleoptera: Carabidae). *IOBC/wprs Bulletin* **15**, 103–9.

Hertveldt, L., Keymeulen, M. van & Pelerents, C. (1984). Large scale rearing of the entomophagous rove beetle *Aleochara bilineata* (Coleoptera: Staphylinidae). *Mitteilungen aus der Biologischen Bundesanstalt für Land- und Forstwirtschaft* **218**, 70–5.

Hof, A.-E. (1993). *Weiterentwicklung einer Testmethode zur Prüfung der Auswirkungen von Pflanzenschutzmitteln auf Wolfsspinnen der Gattung* Pardosa. Diploma, University of Frankfurt/Main, Germany.

Johnson, N. E. & Cameron, R. S. (1969). Phytophagous ground beetles. *Annals of the Entomological Society of America* **62**, 909–14.

Kegel, B. (1989). Laboruntersuchungen über die Nebenwirkungen ausgewählter Herbizide und Insektizide auf die Larvenstadien dreier sympatrischer *Poecilus*-Arten (Col., Carabidae). *Journal of Applied Entomology* **108**, 144–55.

Mühlenberg, M. (1993). *Freilandökologie.* 2nd edn. UTB-Taschenbuch: Heidelberg.

Murchie, A. K., Smart, L. E. & Williams, I. H. (1997). Responses of *Dasineura brassicae* Winn. (Diptera: Cecidomyiidae) and its parasitoids *Platygaster subuliformis* Kieffer (Hymenoptera: Platygastridae) and *Omphale clypealis* Thomson (Hymenoptera: Eulophidae) to traps, baited with organic isothiocyanates, in the field. *Journal of Chemical Ecology* **23**, 917–26.

Naton, E. (1989). Die Prüfung der Nebenwirkung von Pflanzenschutzmitteln auf *Aleochara bilineata* Gyll. (Col., Staphylinidae). *Anzeiger für Schädlingskunde Pflanzenschutz Umweltschutz* **62**, 1–6.

Samsoe-Petersen, L., Bigler, F., Bogenschuetz, H., Brun, J., Hassan, S. A., Helyer, N. L., Kuehner, C., Mansour, F., Naton, E., Oomen, P. A., Overmeer, W. P. J., Polgar, L., Rieckmann, W. & Stäubli, A. (1989). Laboratory rearing techniques for 16 beneficial arthropod species and their prey/hosts. *Zeitschrift für Pflanzenkrankheiten und Pflanzenschutz* **96**, 289–316.

Sunderland, K., Fraser, A. M. & Dixon, A. F. G. (1986). Field and laboratory studies on money spiders (Linyphiidae) as predators of cereal aphids. *Journal of Applied Ecology* **23**, 433–47.

Theiss, S. & Heimbach, U. (1993). Fütterungsversuche an Carabidenlarven als Beitrag zur Klärung ihrer Biologie. *Mitteilung der Deutschen Gesellschaft für Allgemeine und Angewandte Entomologie* **8**, 841–7.

Townes, H. (1982). A light weight Malaise trap. *Proceedings of the Entomological Society of Washington* **83**, 225–9.

Walczak, B., Kelm, M., Klukowski, Z., Smart, L. E., Ferguson, A. W. & Williams, I. H. (1998). Effect of trap design and 2-phenylethyl isothiocyanate on catches of stem weevils (*Ceutorhynchus pallidactylus* Marsh. and *C. napi* Gyll.) in winter oilseed rape. *IOBC/wprs Bulletin* **21**, 141–6.

Whistlecraft, J. W., Harris, C. R., Tolman, J. H. & Tomlin, A. D. (1985). Mass-rearing technique for *Aleochara bilineata* (Coleoptera: Staphylinidae). *Journal of Economic Entomology* **78**, 995–7.

15

Impact of On-Farm Landscape Structures and Farming Systems on Predators

Wolfgang Büchs
Federal Research Centre for Agriculture and Forestry, Messeweg 11/12, D-38104 Braunschweig, Germany

Introduction

This chapter reviews the most important factors that impinge upon predatory arthropods in agricultural areas, with particular reference to oilseed rape. Such factors have to be considered in land that is cultivated or more or less uncultivated. Within rural landscapes, the latter includes field margins, hedgerows, ditches, forest and woodland edges, and also set-aside areas that might once again be ploughed and put into production. Within cultivated areas (e.g. arable fields, pastures, orchards and vineyards), the effects of husbandry practices on the predator community are considered, particularly as they relate to oilseed rape production.

As pesticide use is a highly topical issue, agrochemicals (especially insecticides) used in oilseed rape, and their known effects on relevant predator species, are highlighted. Although the main thrust is on insecticides and molluscicides, herbicides are also considered because they have, at least potentially, both direct and indirect effects on predator communities. Many different pesticides are used in oilseed rape in Europe (and also elsewhere in the world), but the legal status regarding registration varies from country to country. Because, in Europe, the most consistent and 'standardized' information on pesticides used in oilseed rape is available in Germany, the situation has been analysed from a German perspective.

The effects of crop husbandry practices on predator communities are derived primarily from comparisons of different farming systems, especially since for the evaluation of positive or negative effects there need to be suitable baselines. Consideration of farming systems with different levels of input or extensification allow this to be done and, in this chapter, the terms 'intensification' and 'extensification' are adopted for high-input farming and low-input farming, respectively. Further, in order to best validate and assess on-farm data, areas of undisturbed land are also needed as a reference. These could be, for example, set-aside land (if at least two years old) with a natural succession of flora, or extremely extensively managed fields, such as those used for feeding game (in

Germany known as 'Wildacker') and introduced for hunting purposes. As one of the main and most long-term projects on different levels of extensification in oilseed rape rotations, the interdisciplinary INTEX project, co-ordinated by the Centre for Agriculture and the Environment of the University of Göttingen, Germany, will be mentioned frequently in this chapter. Details of this project are summarized in Box 15.1.

The following review is restricted mainly to predators in oilseed rape crops. However, the function of uncultivated areas and other landscape structures as overwintering sites, refuges, recreation areas or breeding sites for predators (and the related immigration and emigration processes to and from oilseed rape fields) are considered, as is their importance in contributing to the enhancement, abundance and effectiveness of predation and species richness (i.e. biodiversity).

On-Farm Landscape Structures

On-farm landscape structures can be assumed to influence predator biodiversity and their effectiveness as biological control agents (BCAs) (see Chapter 16). Gärtner (1980), for example, who compared the 'activity-density' (i.e. the measured amount of surface activity) of ground beetles (Carabidae) – which Palosz (1995a) termed 'carabidocoenosis' – on field areas with and without soil consolidation, showed that in fields that were consolidated 10 years before, only 45% of the species numbers of non-consolidated fields could be recorded.

Different crops and different landscape structures result in differences in microclimates (e.g. temperatures) during the season. These differences and the degree of 'patchiness' of the landscape (i.e. its variability within an area) influence species composition of the predator community.

With regards crop microclimate, Pauer (1975) confirmed that, in oilseed rape, temperatures are lower than in wheat and humidity is higher, as had previously been assumed by Tischler (1958). Furthermore, he showed that the temperature decreased gradually from the crop margins to the within-crop areas, and that a bare, uncropped border around a field (if virtually without vegetation) can have a temperature 20°C above that within the adjacent crop. Clearly, with regards epigaeic predators, such factors will have a profound influence on species composition, population development and phenology.

As stated by Tischler (1958), leaf cover within an oilseed rape crop lessens the direct effect of the sun's rays and the drying effect of the wind; microclimatic conditions in oilseed rape, therefore, are more humid than those in most other field crops. In the case of root and tuber crops (such as potatoes and sugar beet), for example, soil is disturbed by tillage in the spring, the vegetation period is rather short and plant cover within the crop is 'closed' from about mid-July only. Even in winter cereals, compared with oilseed rape, plant cover in April/May is relatively poorly developed, by which time winter oilseed rape will have started to flower and the plants will already provide a relatively high humidity below the leaf canopy.

In agricultural landscapes, much if not all of the area can be covered by fields or other cultivated land (e.g. orchards and vineyards). Therefore, cultivated areas are one of the most important of all landscape structures, and it is easy to understand that crop diversity and the make-up of cultivated areas greatly determine the species composition of the existing fauna.

Büchi (2002) investigated the influence of adjacent meadows and flower strips on the population dynamics of pollen beetles (*Meligethes* spp.) in oilseed rape in Switzerland. He

Box 15.1
The INTEX project

The INTEX project started in 1990 at three locations in Germany: two near Göttingen (the first at Reinshof – a lowland type of landscape on loamy soils near the River Leine; the second at Marienstein – a more colline type of landscape, on limestone) and one near Braunschweig (Eickhorst – a soil with greater percentage of sand). At all three locations an oilseed crop rotation was established in four farming systems with different levels of extensification plus a set-aside area as follows:

farming system I = 'conventional'	A three-course crop rotation (oilseed rape, winter wheat and winter rye) that was managed in accordance with official local extension services. A high-input level of pesticides and fertilizers and intensive tillage was defined at the beginning of the project and remained constant over the first six years of the experiment. In special cases management practices were adapted to meet current situations.
farming system II = 'integrated'	A four-course crop rotation (oilseed rape, winter wheat, winter rye and yellow peas), with reductions of nitrogen fertilizers of up to 30% (according to soil and plant analysis) and pesticides of up to 50% (according to damage thresholds), use of resistant cultivars (varieties) and/or varietal mixtures and (especially in oilseed rape) adaptation of sowing time, plant density and drilling techniques and adoption of mechanical weed control.
farming system III = 'reduced'	Rotation and management according to system I, except for a reduction of nitrogen fertilizer input of up to 50% and no use of insecticides.
farming system IV = 'extensive'	Rotation and management according to system II, but no pesticide application (except seed treatments) and no input of mineral nitrogen fertilizers.
'system' V = 'fallow'	A long-term set-aside area, with natural succession and no management activities.

Field sizes ranged from 1 to 2 ha; all crops of the rotations were grown in parallel, although at Eickhorst the crops of systems III and IV were grown sequentially.
After six growing seasons (1990–1995) the concept of the project was changed and restricted to the two locations near Göttingen; systems I and II were then subdivided.

showed that, in both cases, the abundance of pollen beetle larvae within the crop increased with increasing distance from the field borders. However, the border effect was less marked in fields with adjacent strips of wild flowers. This is, no doubt, at least in part influenced by the impact of BCAs on the pollen beetle larvae – the effects of field margins on parasitoid populations is discussed earlier (Chapter 4) – but there have been no studies illustrating the effects on adjacent meadows and flower strips on in-field populations of predators. Frank (1996), again in Switzerland, also showed the significant effects of sown weed strips of different age on the diversity of epigaeic predators (ground beetles and spiders) as well as of hover flies (Syrphidae) and solitary wasps (Sphecidae) as flower-visiting predators, and their tendency to disperse within adjacent fields (maize, oilseed rape and wheat). However, with regards predators inhabiting the vegetation layer, Hausamman (1996) recorded very few by sweep netting in oilseed rape flower stands (at the pod stage), either within the field or close to the sown weed strips. According to Nilsson & Andreasson (1987), in oilseed rape, predators on the soil surface are far more numerous than those actually on the plants. For example, by sweep-net catching, very few predators were caught – no more than 0.75 individuals/m^2. These authors explain how to transform sweep-net catches to a defined area.

Pauer (1975) attributed differences in the species community of ground beetles to the individual phenology of a crop, which is particularly exemplified by differences in the height of oilseed rape plants in summer compared with those of cereals. Differences in plant height are amplified by the relative densities of the crop canopies. Thus, oilseed rape 'covers' much of the soil from a relatively early stage of crop development, whereas the sun's rays are able to reach much of the soil surface in the case of cereals, at least until close to harvest.

Ground beetles leave oilseed rape fields in autumn to overwinter in field margins (Goltermann, 1994). He obtained transect samples and, in the autumn, recorded 74% of the activity-density of ground beetles near the field border and only 5.8% within the field (70 m from the border). In spring, the opposite pertained. Initially, i.e. in early spring, considerable activity occurred but only in areas close to the field margin. Later, up to July, more species (on average 2.7) and a larger number of individuals (0.6 beetles/trap/day) occurred in traps directed towards the field margin, compared with traps directed away from the field margin (1.6 species and 0.4 individuals/trap/day). Similar results were obtained for species of *Pterostichus* by W. Büchs (*unpublished*) in capture–recapture experiments.

Generally, as was expected, species density was clearly higher at the field margin than within the field (averages of 12.9, 5.4 and 2.8 species at distances from the field margin of 2, 35 and 70 m, respectively). In winter, the activity of ground beetles ceased for a longer period with increasing distance from the field margin:

- at the crop margin (if adjacent to grass) – no activity in January only;
- at the crop margin (if adjacent to forest) – no activity;
- 35 m into the crop – no activity from December onwards;
- 70 m into the crop – no activity from mid-November onwards.

In the central part of the field (70 m from the margin) the most frequently occurring species were: *Anchomenus (= Platynus) dorsalis*, *Bembidion femoratum*, *B. lampros*, *B. tetracolum*, *Calathus fuscipes*, *C. melanocephalus* and *Trechus quadristriatus*.

A decrease in species diversity also occurs with increasing distance from the field edge (Knauer & Stachow, 1987).

According to Knauer & Stachow (1987), Stachow (1988), and Goltermann (1989, 1994), the number of species present, activity-density and the number of invading (succession) species in oilseed rape fields are greatest near the field margin, especially if fields are adjacent to uncultivated marginal habitats. With increasing distance from the field margin, within fields a dilution of the ground beetle fauna (namely, a reduction in the number of species and individuals) can be observed. Goltermann (1994) found this phenomenon to be more distinct in oilseed rape than in winter wheat because, as indicated by Pauer (1975), ambient air temperatures near the soil surface are, to some extent, higher in winter wheat than in oilseed rape.

Pollard (1971) showed that populations of hover flies are enhanced by landscape structures such as hedges, and the insects act more efficiently as predators of cabbage aphid (*Brevicoryne brassicae*) if a greater diversity of landscape features is realized.

The effects of on-farm landscape structures on spiders is well documented, but has no special relation to the crop. Harenberg (1997) compared two locations: site A – a sugar beet crop rotation in an intensively used agricultural area virtually devoid of landscape structures; and site B – an oilseed rape rotation in an area with a greater hetereogeneity of habitats (e.g. field margins, hedgerows, trees, fallows and ditches). These differences greatly influenced species numbers during the crop rotation (A: 92 species; B: 131 species) and also affected abundance within species (larger populations in A). With regards the species communities, it was striking that larger species (Lycosidae – e.g. *Alopecosa* spp., *Pardosa* spp., *Trochosa* spp., *Xerolycosa* spp.; Gnaphosidae – *Zelotes* spp.; and Thomisidae – *Xysticus* spp.), which occurred rather frequently at site B, either were absent from site A or were represented by a larger number of species (e.g. Clubionidae). The same tendency (i.e. the absence of larger species in intensively farmed landscapes without any structures) was noted by Zimmermann & Büchs (1999) for rove beetles (Staphylinidae) and by Kokta (1989) for ground beetles.

Further, Vansgaard *et al.* (1990) compared an isolated field surrounded by a forest, peat bogs and moorland with one in an intensively farmed agricultural landscape and recorded a greater abundance of free-living (i.e. non-web-building) spiders in the latter. According to Klinger (1987) the occurrence of wolf spiders (Lycosidae) within fields depends on the existence of suitable field margin habitats. However, soil type may also have a significant influence on the spider community. While the first site was characterized by a heavy loam soil [and dominated by money spiders (Linyphiidae)], the soil at the second was lighter, with a greater percentage of sand. This agrees with Heydemann (1983), who distinguished between the main spider communities in two different fields: one, with a loamy, damp soil with a high clay content, in which money spiders dominated; the other with a drier and sandy soil, in which the spider community was represented by species such as *Alopecosa pulverulenta*, *Pardosa agrestis*, *Trochosa terricola*, *Xerolycosa* spp. (all Lycosidae), *Zelotes* spp. (Gnaphosidae), *Xysticus* spp. (Thomisidae) and *Pachygnatha degeeri* (Tetragnathidae).

Stippich & Krooß (1997) demonstrated the importance of fallow strips at the field margin for overwintering of spiders. With the exception of extensively managed fields (no pesticide and no fertilizer inputs), no spiders could be found in oilseed rape fields in October; however, very large numbers were recorded in the surrounding fallow strips and in a nearby set-aside area. Ground beetles and rove beetles showed similar relationships; amongst the former, *Anchomenus dorsalis* and *Notiophilus* sp. were the most abundant taxa in the fallow strips in winter.

Using so-called 'directional' pitfall traps (i.e. pitfall traps that are set to limit the direction from which target organisms can approach and then fall into them) between a set-aside area and (winter wheat) fields, Stippich & Krooß (1997) (for spiders) and Kopp *et al.* (1997) (for ground beetles) could demonstrate the migration of predators in spring from weedy field margins back into the fields. Stippich & Krooß (1997) assumed that *Carabus nemoralis*, for example (which they recorded regularly in the centre of oilseed rape fields in summer), immigrated from fallow strips adjacent to the fields.

Density and abundance of major predators

Basedow (1973) found that the abundance of ground beetles in oilseed rape fields varied from 20 to 80 individuals/m^2. However, in a comparison of winter wheat, winter barley and winter oilseed rape, Knauer & Stachow (1987) recognized oilseed rape as the crop with the least number of unique species (i.e. species that occurred only in that crop).

In most cases, *Amara similata* is the most dominant ground beetle species in oilseed rape fields (e.g. Pfiffner & Luka, 1996; Büchs & Nuss, 2000; D. Felsmann & W. Büchs, *unpublished*), especially if crops are intensively managed. However, Stippich & Krooß (1997) found that *Amara ovata* was the most abundant species (217.3 individuals/trap/year = 80.0% of all ground beetles) in an oilseed rape field under an integrated management system. The authors relate this phenomenon to the higher weed density in the integrated field, since only mechanical weed control was performed. In France, Houpert (1983) found that the ground beetles *Anchomenus dorsalis*, *Poecilus cupreus* and *Pterostichus melanarius* were most abundant in oilseed rape, with a dominance of about 65%.

Comparing winter wheat and oilseed rape fields, Goltermann (1994) recorded over six years (from 1987 to 1992) 68 ground beetle species in different oilseed rape fields. Fourteen species occurred only in oilseed rape (in spring, for example, *Amara eurynota*, *Carabus cancellatus* and *C. nemoralis*; in autumn, for example, *Badister bipustulatus*, *Demetrias atricapillus*, *Dromius melanocephalus* and *Poecilus cupreus*). In the spring he found from 29 to 47 species per field and season. Seventeen of these species were recorded in oilseed rape in all six years. *Anchomenus dorsalis* and *Pterostichus melanarius* were the dominant species; *Agonum muelleri*, *Amara similata*, *Bembidion tetracolum*, *Loricera pilicornis* and *Nebria brevicollis* were sub-dominant. In the autumn he recorded an average of 27 species (over three years), of which 14 were found in every year; *A. dorsalis*, *Bembidion tetracolum*, *Nebria brevicollis* and *Trechus quadristriatus* were dominant.

Density estimations for spiders in oilseed rape fields are available mainly from Germany and Switzerland (e.g. Büchs *et al.* 1997; Harenberg, 1997; Ulber *et al.*, 1990).

In a comparison of different agricultural crops, Nyffeler (1982) reported 0.1–0.5 individuals/m^2 of foliage-dwelling spiders in intensively grown oilseed rape. This was similar to that found in cereals (0.1–0.6 individuals/m^2) but considerably lower than that found in other intensively grown crops or meadows (e.g. 0.6–7.0 indivduals/m^2 in rye; 2.2 in barley; 2.2–3.6 in alfalfa; 0.6–29.6 in potato; 0.6–4.4 in meadows).

Using pitfall traps, Tischler (1958) recorded 24 species of spider in fields on loamy soils with oilseed rape in the rotation. The most numerous were the money spiders *Erigone atra*, *E. dentipalpis* and *Oedothorax apicatus*, and the wolf spider *Pardosa amentata*. Most other species collected were, again, either money spiders (genera *Araeoncus*, *Bathyphantes*, *Dicymbium*, *Lepthyphantes*, *Meioneta*, *Porrhomma* and *Savignia*) or wolf spiders (genera *Tarentula* and *Trochosa*), although representatives of three other families (Clubionidae – *Clubiona reclusa*; Gnaphosidae – *Zelotes lutetianus*; Tetragnathidae – *Pachygnatha clercki* and *P. degeeri*) were also found.

Weed strips

Frank & Nentwig (1995) and Frank (1997) showed the importance of weed strips for the species richness (spiders as well as ground beetles) of oilseed rape fields. With regards ground beetles, species numbers in older (second-year) weed strips were greater than in adjacent oilseed rape crops. However, compared with other situations (such as maize, wheat and a meadow), species numbers in oilseed rape were the greatest. Nevertheless, species numbers decreased by 15% with increasing distance from the weed strip, and the difference in species numbers at the two within-field sites (13 and 50 m from the weed strip) was significant. However, in two subsequent years, species diversity was greatest in the crop. The kinds of species present in the oilseed rape were similar to those inhabiting the weed strip. The similarity was also closer than between two oilseed rape fields, demonstrating to what extent weed strips can influence populations in fields.

Frank (1997) has shown that weed strips are colonized by ground beetles shortly after sowing, but an increase of the species number in adjacent fields can be observed only in the following year. The results for spiders are rather similar: higher activity-densities in the weed strips in spring but, later in the season, a change to higher activity-densities in the adjacent oilseed rape fields, which clearly indicates that immigration takes place. However, some species that dominate the spider community within the weed strip do not show any exchange with the adjacent fields (Frank & Nentwig, 1995).

Overwintering

Within oilseed rape fields, at a distance of 5 m from the field margin, Goltermann (1994) found very few rove beetles and no adult ground beetles, but numerous ground beetle larvae. In the central area of the field (60 m from the field margin) numbers of predatory beetles (ground beetle adults and larvae, and rove beetles) were considerably greater and small species of adult ground beetles were also found hibernating.

With regards penetration of oilseed rape fields from the margins to the central area, different patterns of immigration were observed:

- *Asaphidion flavipes*, *Bembidion* spp., *Pterostichus nigrita* and *P. oblongopunctatus* remained close to the field margins;
- *Clivina fossor* and *Loricera pilicornis* immediately (and *Agonum muelleri*, *Anchomenus dorsalis* and *Nebria brevicollis* eventually) colonized the whole field;
- *Amara* spp. tended to frequent areas of the field that had a higher density of weed cover.

Further, Goltermann (1994) observed that colonization of arable fields in spring (starting from the field margins) took place more quickly in winter wheat than in oilseed rape. He also showed that biodiversity decreased more rapidly in oilseed rape than in winter wheat with increasing distances from the field margin, although over time there was an increasing tendency in both crops for species numbers, overall, to even out. The indices of similarity – according to Sörensen (1948), for species; according to Renkonen (1938), for the dominance structure of species – provide evidence that distribution of various species of ground beetle happens more slowly within oilseed rape crops than within winter wheat crops, as does an increase in their activity-density.

Goltermann (1994) specifically cited *Anchomenus dorsalis*, *Bembidion tetracolum*, *Clivina fossor*, *Loricera pilicornis*, *Poecilus cupreus*, *Pterostichus melanarius* and *Trechus quadristriatus*, and new-generation members of *Amara familiaris*, *A. similata*, *Harpalus affinis*, *Nebria brevicollis* and *P. melanarius* as typical 'succession' species in oilseed rape (i.e. those subsequently recolonizing the crop in spring).

Goltermann (1994) stated that, in winter oilseed rape fields, the new generation of spring breeders (but only some autumn breeders) were recorded at the field borders; only a small number of beetles (species as well as specimens) were recorded in the central area of fields. In spring, the fields were first colonized by beetles that overwintered as adults; such beetles were, at first, active only in the border area of the fields, so it can be assumed also that eggs would be laid near the field margin. Also, some autumn breeders (e.g. *Pterostichus melanarius*) emigrated from the fields in autumn and immigrated in spring. By using a capture–recapture technique, with marked *Pterostichus* spp. (*P. macer*, *P. melanarius* and *P. niger*), in a winter oilseed rape crop in autumn, Büchs (1991) showed that, up to the end of October, the majority of specimens were still recaptured in the central area of the field. Only 1.9% of all the recaptured *Pterostichus* spp. (578 specimens) left the experimental area (100 m in width) and only 0.7% of them were recaptured at the field border. Also, in October (in several years), very large numbers of *Pterostichus* larvae (and also larvae of *Cantharis lateralis*) were observed right in the centre of oilseed rape fields (Büchs *et al.*, 1989, 1990; W. Büchs, *unpublished*). From these observations it can be assumed, at least for *Pterostichus* spp., that only a small proportion of the ground beetle population vacates oilseed rape fields to overwinter in surrounding, semi-natural, field-margin habitats.

In the case of ground beetles, many authors (e.g. Hoßfeld, 1963; Weber, 1965; Sotherton, 1984; Goltermann, 1994) have recorded a relatively constant species community of up to five species (*Asaphidion flavipes*, *Bembidion tetracolum*, *B. lampros*, *Nebria brevicollis* and *Trechus quadristriatus*) as active in oilseed rape crops (and also other crops) during the winter.

Pauer (1975) was able to identify three different types of activity patterns (termed 'type I', 'type II' and 'type III') in the border region between winter oilseed rape and spring wheat fields.

Type I

Type I species (e.g. *Bembidion lampros*) showed a significant delay of nearly eight weeks in their initial post-hibernation activity between spring wheat and winter oilseed rape fields. Specifically, the activity of *B. lampros* in spring wheat commenced in the first week of April (if not earlier), whereas in oilseed rape significant activity-density was not recorded until the end of May. Further, in spring wheat, the activity of *B. lampros* virtually ceased by the end of June, whereas activity continued in oilseed rape almost until harvest (i.e. until the end of July).

Generally, the period of activity of *B. lampros* in oilseed rape was noticeably shorter than in spring wheat, because temperatures were already higher when the ground beetles entered the crop; thus, they would have required less time for 'maturity feeding' than they would if in a wheat crop.

In Pauer's (1975) study, the field margin between the spring wheat and the oilseed rape field functioned as a true border line, which was crossed only occasionally. When activity commenced in the rape field, activity-density recorded in the field margin area was extremely high, but this decreased very rapidly towards the field centre. *Bembidion lampros* is a thermophilic, sun-loving, oligostenohygric species, well adapted to the field-margin area of crops. In oilseed rape, there is a strong microclimatic gradient away from the field margin to the field centre, with the in-field areas becoming increasingly unsuitable for this species. Not surprisingly, therefore, specimens of *B. lampros* enter the more central areas of fields only occasionally, and then for only short periods.

According to Pauer (1975), type I ground beetle species:

- become active early in the spring, when soil temperatures vary in different crops according to the density of the plant cover – for autumn breeders, these microclimatic differences result in a delay or an acceleration in larval development;
- have a rather limited tolerance to changes in microclimatic conditions (temperature and air humidity).

Type II

Type II is represented by *Anchomenus dorsalis*, which shows a nearly equal spatio-temporal relationship, with regards commencement of activity and spatial distribution in fields and field margins. However, activity-density for this species is generally higher in oilseed rape fields than in spring wheat.

Bembidion tetracolum (unlike *B. lampros*) showed high activity-densities in field centres but extremely low ones in the warmer, drier field margins. The larvae of the new (i.e. summer) generation of *Nebria bevicollis* find especially favourable conditions for their development in oilseed rape fields. However, they are nocturnally active and do not cross the field margin because of the lower night-time temperature (field margins cool down more rapidly after sunset than the field as a whole). Populations, therefore, tend to be static within fields. According to Pauer (1975), *Carabus auratus*, *C. nemoralis*, *Clivina fossor* and *Pseudoophonus (= Harpalus) rufipes* also belong to this type. Presumably, owing to its demand for higher temperatures and low humidity, *Harpalus affinis* was more active in the border area between fields. *Carabus nemoralis* seemed to change between the fields purely by chance; however, few specimens of *C. auratus* were recorded in the margins of fields, suggesting that this species usually does not cross field borders, although it is relatively large bodied and is able to travel considerable distances. Pauer (1975) concludes from these patterns of activity that a proportion of the population hibernates within the fields, as observed also by Scherney (1961); however, Renken (1956), Kirchner (1960) and Novák (1968) found *Anchomenus dorsalis* hibernating in grassy field margins and in hedgerows.

Clivina fossor commenced its activity in both fields at the same time. The degree of activity, however, depends on soil type and on the tillage strategy employed. Thus, for this species, surface activity will be greater in more compacted soils, since the beetles tend to burrow into subsurface layers in sandy soils; also, because the effects of tillage on soil compactness decrease over time, spring tillage (for example) will lead to less compacted 'winter' soils than will autumn tillage.

Type II species generally become active later in the year than type I species, and are able to tolerate a wide range of microclimatic conditions.

Generally, only thermophilic and xerophilic species occur in large numbers near field edges (e.g. *Bembidion lampros*, *Harpalus affinis*, *Pseudoophonus rufipes* and *Poecilus cupreus*), whereas more hygrophilous and oligostenothermic species are more abundant within the fields (Pauer, 1975).

Pauer (1975) concluded from his investigations of spatio-temporal activity patterns in adjacent crops that the majority of ground beetle species (especially types I and II) stay within one crop and that there is very little spontaneous migration between different crops. Only a small number of species (e.g. *Amara* spp., *Bembidion* spp., *Harpalus* spp. and *Loricera pilicornis*) are able to colonize suitable new habitats by flying.

It is known that populations of ground beetle species include macropterous as well as brachypterous individuals (e.g. *Bembidion lampros*, *B. tetracolum* and *Pterostichus melanarius*), or are generally macropterous, e.g. *Anchomenus dorsalis*, *Clivina fossor* and *Pseudoophonus rufipes*. However, many macropterous specimens are unable to fly as their wing muscles are insufficiently developed. These factors, along with the extreme microclimatic conditions in the borders of fields, significantly restrict the exchange of populations between different fields and crops. According to Pauer (1975), most of these species hibernate within the fields, so that the abundance and/or activity-density, respectively, recorded during the vegetation period depend mainly on the number of overwintering larvae and adults. The number of beetles overwintering within fields is especially influenced by the previous crop and by the timing of tillage operations.

Within one and the same field, special environmental (microclimatic) conditions lead to patchy, local concentrations of certain species. Thus, in an oilseed rape field, Pauer (1975) recorded significantly more specimens of *Bembidion lampros* in areas where the crop was relatively sparse than in areas where the crop was more dense.

Type III

Some species show a high activity-density in one kind of crop but are absent in another. These may be considered as representing a third group (type III). Most of these species are phytophagous and prefer certain weed plants as food, often restricting their occurrence in any numbers to one kind of crop. In such cases, crop preference is not governed by microclimate but by the availability of food. Most species of *Amara* (especially *A. familiaris* and *A. similata*) and *Bembidion quadrimaculatum* belong to this type.

Interestingly, *Amara similata*, which is usually one of the most abundant species in oilseed rape fields (see, for example, Nuss & Büchs 2000) and has been recorded in France as a pest (Pfiffner & Luka, 1996; Y. Ballanger, *pers. comm.*), was virtually absent from the oilseed rape fields in Pauer's (1975) investigations. The probable explanation is that oilseed rape cultivars have changed dramatically since the 1970s, when plants were larger and seed rates higher (then nearly 6.0 kg/ha, compared with *c.* 3.0 kg/ha today); further, today's cultivars show greater weed tolerance, and insecticide use is restricted to pyrethroids (see Chapters 1–3). Such reasons make the development of phytophagous ground beetle populations in oilseed rape crops more possible than in former times.

Amongst type III species, microclimatic reasons were probably responsible for the absence of *Bembidion quadrimaculatum* in oilseed rape over the two-years' field observations by Pauer (1975), but its abundance in other crops was low (especially in sugar beet and in the subsequent wheat crop). However, the conclusion that microclimatic effects determine spatio-temporal distribution and crop preferences is valid only for years in which rainfall and temperature (i.e. weather conditions) are about average. Where there are extremes (for example, in fields with a sparse vegetation, combined with very dry, hot conditions), the tolerance thresholds for even thermophilic and xerophilic species may be exceeded. Such species may then find better conditions for their development in, for example, oilseed rape crops that would usually be considered too cold and humid.

Effect of previous and neighbouring crops

Pauer (1975) compared the activity-density of ground beetles in winter oilseed rape with either barley or winter oilseed rape as the previous crop. Compared with the cereal/oilseed rape rotation, growing continuous oilseed rape led to a higher activity-density amongst cold-tolerant, hygrophilic species (e.g. *Anchomenus dorsalis*, +65%; *Carabus granulatus* +133%; *Clivina fossor*, +67%; *Pseudoophonus rufipes*, +157%; *Pterostichus melanarius*, +139%). Conversely, activity-density was reduced for thermophilic, less hygrophilic species (e.g. *Asaphidion flavipes*, –90%; *Bembidion lampros*, –75%; *B. tetracolum*, –99%; *Carabus auratus*, –68%). This suggests that a 'typical' oilseed rape fauna of ground beetles can develop only if the same crop is grown in subsequent years; it will not develop if there is a diverse crop rotation.

A positive influence of rape (and wheat) fields on neighbouring cotton crops is reported. It has been observed, for example, that ladybirds (Coccinellidae) (*Adonia variegata*, *Coccinella septempunctata* and *Propylea japonica*), as well as spiders, move from rape fields in mid- to late May into cotton crops, so that aphids on the latter were nearly completely controlled and insecticide use was unnecessary (Fang *et al.*, 1984).

Effect of landscape structures

Most ground beetle species recorded in cultivated areas have their origin in habitats such as sea-shores (= littoral) and river banks (= riparian), and belong to the so-called 'Littoraea-coenosis' fauna. *Carabus auratus*, as a species that is able to cross small streams by swimming (Tischler, 1958), is an example and has spread from western Europe to central Europe. A relationship with forests also exists, since typical forest species (such as *Anchomenus (= Platynus) assimilis*, *Carabus coriaceus*, *C. convexus*, *Notiophilus biguttatus* and *Pterostichus strenuus*) can spread up to 40 m into nearby fields, and typical field-dwelling species (such as *Crabus auratus*, *C. granulatus*, *Harpalus seladon* and *Loricera pilicornis*) will penetrate into adjacent forest stands.

Some rove beetle species (e.g. *Anotylus inustus* and *Ilyobates subopacus*) nowadays seem to prefer cultivated fields rather than their 'original' littoral and riparian habitats.

As already stated by Zimmermann & Büchs (1999), adult rove beetles (especially members of the subfamily Aleocharinae) change their habitat very quickly after emerging from the pupal stage. Thus, Tischler (1958) found that nearly all the species he recorded in fields during the summer occurred in considerable numbers in hedgerows, forest edges and within forests during the winter.

In the case of aphidophagous hover flies, their early immigration into crops in the spring could be enhanced if suitable overwintering sites, such as agricultural coppices and hedgerows [especially those containing trees and shrubs such as elder (*Sambucus niger*)], were present in the immediate neighbourhood of arable fields (Tenhumberg, 1993; Tenhumberg & Poehling, 1995).

Oilseed rape itself functions as a 'landscape structure'. Especially in the southern states of the USA and in China, for example, it has a strong relationship with cotton. A comparative survey was carried out in Texas by Parajulee & Slosser (1997), to assess whether it was possible to enhance the abundance of predators in cotton fields by planting adjacent strip crops. These strip crops included both spring and winter rape, and also sorghum, *Vicia* beans and wheat. The predators in question included lacewings (Neuroptera: Chrysopidae), ladybirds, soldier beetles (Cantharidae) and spiders, and also various predatory bugs (Heteroptera). The results showed that spring crops generally had smaller numbers of predators compared with autumn crops. However, most strikingly, a rape–sorghum strip not only lead to the greatest numbers of predators in cotton but also resulted in the smallest numbers of melon & cotton aphid (*Aphis gossypii*) on the crop. In China, the sowing of cotton adjacent to rape (and also wheat and clover) fields enabled farmers to reduce the average number of insecticide applications on the cotton from an unbelievable 64.5 to 7.8 (Wang & Wang, 1996). In Taiwan, oilseed rape was planted within citrus orchards, in order to increase aphid predators (Tao, 1973). These examples indicate the considerable potential of oilseed rape as a reservoir of predators.

With regards strip crops, negative interactions between oilseed rape and cotton should also be mentioned. Thus, in the USA, Boyd & Lentz (1999) collected plant bugs (*Lygus* spp., mostly *Lygus lineolaris*) (Lygaeidae) and thrips (Thysanoptera) (mostly *Frankliniella tritici* and *F. fusca* and *Neohydatothrips variabilis*) (Thripidae) from oilseed rape, where they feed on buds, flowers and pods. These bugs and thrips are serious pests of cotton, at both the seedling and fruiting stages. Cotton is one of the most important crops in the southern States, and the authors postulate that an increase of pest problems in cotton could arise if oilseed rape were grown nearby.

Farming Systems

Influence of field management

In general terms, the structural dominance of soil-inhabiting arthropod populations (including predators) appears to be greatly influenced by field management systems, whereas species composition will tend to reflect crop type (e.g. oilseed rape, sugar beet or winter wheat). At least as far as ground beetles are concerned, soil conditions will also have an influence on species composition (see Kromp, 1999). According to Thiele (1977), field populations (as opposed to woodland populations) of ground beetles tend to be dominated by those that 'prefer' warm, dry sites (i.e. thermophilic, xerophilic species); soil moisture and soil particle size can also influence habitat choice (e.g. Meissner, 1984).

Field management practices in arable crops also have a major influence on spider populations. By way of example (see Büchs *et al.*, 1997), populations of some spider species – e.g. *Oedothorax retusus* and *Pelecopsis parallela* (Linyphiidae); *Pachygnatha degeeri* (Tetragnathidae) – show an increase with decreasing management intensity; *Oedothorax apicatus*, however, prefers a high intensity of management and may be absent from set-aside land.

The regime of tillage adopted can also affect insect predators. For example, an ongoing eight-year study in Germany [established in 1989 to observe the agro-ecological effects of integrated farming systems on a field scale; see, for example, Steinmann (1998)] showed that populations of dance flies (Hybotidae) increased significantly with decreasing cultivation intensity (W. Büchs & S. Prescher, *unpublished*).

Palosz (1995a) compared the activity-density of ground beetles on intensively, conventionally and extensively managed oilseed rape fields, which were separated mainly by the range of the crop rotation and by the number of pesticide treatments applied. In this experiment the activity-density of predators during the whole growing season of the conventional treatment was two to three times as high as in the intensive and extensive treatments. However, the number of species present was considerably lower in the intensive treatment (27 species) compared with the others (conventional = 35 species; extensive = 32 species).

Knauer & Stachow (1987) demonstrated that, on arable farms, oilseed rape fields form important refuges for predators in June and July, when there is often widespread use of insecticides on crops such as winter wheat. Thus, unlike the situation in winter wheat, an increasing activity-density could be shown in oilseed rape for the most frequently occurring ground beetle species: namely, *Agonum muelleri*, *Anchomenus assimilis*, *Anchomenus dorsalis*, *Amara similata*, *Bembidion lampros*, *B. tetracolum*, *Carabus granulatus*, *Clivina fossor*, *Loricera pilicornis*, *Nebria brevicollis* and *Pterostichus melanarius*.

The apparent absence of species from fields after, for example, tillage, harvesting, the application of pesticides and mechanical weed control is often interpreted as indicating that they have emigrated (i.e. that they have moved away, following the onset of unsuitable conditions). However, their continued presence in pitfall traps suggests that this may not be the case, and Pauer (1975) concluded that, although ground beetles can emigrate, they will also retreat into the soil.

The occurrence of larger ground beetles is greatly influenced by the farming system. Using pitfall traps set in oilseed rape fields during the period of insecticide use, Büchs (1994a) recorded only 0.125 and 0.5 individuals of *Carabus auratus* per trap under intensive and integrated management, respectively; however, in an extensively managed oilseed rape field (e.g. no pesticides and no fertilizer applied) an activity-density of 15.8 beetles/trap was recorded.

Büchs (1994b) demonstrated the positive effects of long-term set-aside areas with a natural faunal succession. On set-side following winter oilseed rape a considerably higher activity-density was recorded than in set-aside that followed sugar beet. This is because of the more delayed development of a vegetation layer in fallows following sugar beet, owing to the late harvest date (October) in sugar beet compared to that of oilseed rape (July or August). Also, owing to the occurrence of volunteer rape soon after harvest, set-aside areas following oilseed rape provide a more balanced microclimate and have a relatively rich vegetative structure. In contrast, set-aside areas following sugar beet usually develop only a very sparse vegetative structure before winter, because of low autumn temperatures.

Büchs *et al.* (1997) reported that in an oilseed rape rotation without insecticide input (farming system III – see Box 15.1) and in the most extensive farming system (IV) (without pesticide and fertilizer inputs) greatly increased activity-density of the ground beetle *Carabus auratus* occurred within three years (600% and 900%, respectively). For spiders, in oilseed rape, an increase in emergence rates of about 500% was recorded within farming systems I (conventional) and III (reduced, without insecticides) (compared with winter rye, the previous crop), suggesting that oilseed rape is one of the most suitable crops for the reproduction of spiders. The advantages to predators are probably a result of the special conditions afforded by the crop (e.g. rapid and long-term coverage of the soil by vegetation, relatively high humidity, richness of plant structures and sufficiency of prey). Unlike other crops, emergence rates in oilseed rape in the integrated farming system II were the lowest, although comparatively large numbers of spider species were recorded. Of the farming systems considered here, system II was the only one with a four-crop rotation. This suggests that extension of the crop rotation with the introduction of a spring crop (with an additional tillage in spring) is not suitable for spiders.

Predacious Diptera (mainly suborder Brachycera) in oilseed rape crop rotations have been considered by Prescher & Büchs (1997, 1998, 1999). Analysis of different trophic groups for the autumn period of oilseed rape (August–January) showed that the brachyceran community was dominated by saprophagous species, and only a few individuals of species with carnivorous life stages were recorded.

According to Prescher & Büchs (1998), in central Europe, carnivorous flies started to appear in the second half of April and their numbers continuously increased (up to nearly 25 individuals/m^2 within 14 days) through to harvest-time. Peak numbers were positively correlated with the increasing extensification, and most brachycerans with predacious larvae occurred under the reduced farming system III, with no insecticide input. This positive correlation of the emergence rates was caused mainly by the emergence of Hybotidae and by other predacious Brachycera, for example Dolichopodidae and predacious Muscidae. Very few adults of all other brachyceran families with predacious larvae (e.g. Empididae, Rhagionidae, Syrphidae, Tachinidae and Therevidae) emerged in oilseed rape fields. However, adults of some of these families may not emerge within the fields but may breed and subsequently immigrate from elsewhere (e.g. Syrphidae).

Members of many other important predatory taxa (for example, ground beetles of the genera *Bembidion* and *Trechus*) also show increasing activity-densities and emergence rates with decreasing management intensity of the farming system (Büchs *et al.*, 1999; Prescher & Büchs, 1999, 2000). Wishart *et al.* (1956) and Jones (1975) showed the importance of these taxa as natural enemies of anthomyiid pests such as cabbage root fly (*Delia radicum*), the beetles preying on the eggs. Further, Wright *et al.* (1960) and Coaker (1966) found a correlation between ground beetle population densities and the abundance of *Delia* larvae in the soil. As a consequence of increased predator activity, numbers of many (potential) pests show a decrease with decreasing management intensity. However, some of the main pest species on oilseed rape [e.g. brassica pod midge (*Dasineura brassicae*), cabbage seed weevil (*Ceutorhynchus assimilis*), cabbage stem weevil (*C. pallidactylus*) and pollen beetles] apparently do not follow this principle and are recorded with increasing emergence rates when farming systems are extensified.

Within the INTEX project (see Box 15.1), comparison of farming systems I and III showed that the reduced management system (without insecticides and reduction in nitrogen fertilizer of about 50%) resulted in:

- a significantly higher activity-density of predators in farming system III, which corresponded with there also being many more oilseed rape pests in that system (64.6% of all pest insects);
- a larger number of predator species (43 as opposed to 38) and two more dominant species in farming system III.

At the species level, the ground beetle *Amara similata* showed a considerably greater dominance in system I than in system III (I: 44.1%; III: 29.6%). Conversely, *Pterostichus melanarius* (a species typical of intensification and disturbance) was considerably more dominant in the reduced farming system (I: 7.3%; III: 13.6%). With regards *Poecilus cupreus*, which is assumed to be a key predator species, significantly greater activity (60% of all specimens) was recorded in the reduced management farming system III where no insectides were applied, compared with the intensively managed farming system I. This was assumed to be related to the significantly greater abundance of various pests – e.g.

brassica pod midge, cabbage seed weevil, cabbage stem weevil, pollen beetles, rape stem weevil (*Ceutorhynchus napi*), cabbage stem flea beetle (*Psylliodes chrysocephala*) and turnip sawfly (*Athalia rosae*) – as potential prey in the former system.

The nutritional status of ground beetle adults (which is reflected in biomass and, especially, gut filling) should be related to the availability of prey items and so also to management intensity. However, although Hokkanen & Holopainen (1986) recorded a greater biomass on organically managed spring rape fields in Finland, no relationship between gut filling and management intensity could be found within the INTEX data (intensive management I; reduced management III). Also, no relationship between management intensity and sex-specific functions (e.g. development of egg mass by females) was found. However, generally, male beetles did show less filling of the gut compared with females, which was most obvious in the mid-gut.

Most striking were differences in the phenology of *Poecilus cupreus*. Whereas in farming system III (without insecticides) the peak activity-density occurred by mid-May (as is usual), in the intensively managed farming system I the peak of activity was delayed for a whole month, clearly as a result of a pyrethroid spray application against ceutorhynchid weevils and pollen beetles at the end of April. Furthermore, in the reduced system III a second peak was recorded before harvest, possibly indicating the appearance of beetles of the new generation. This second peak of activity was not detected in farming system I, suggesting that significant reproduction can take place only if no insecticides are applied. It was assumed that these phenological differences were caused by a slower development of the rape plants in system III, owing to a 50% reduction of nitrogen fertilizers. The density of rape plant was considerably lower in system III (45.5 plants/m^2 in I, compared with 39.0 in III – at growth stage 51); thus the temperature of the soil surface in early spring could be more suitable for *P. cupreus* in the reduced system III and so allow an earlier start of activity.

In spring rape, Hokkanen & Holopainen (1986) recorded more specimens and species of predators in organically managed fields than conventionally managed ones, which they related to the availability of a higher density of prey.

For spiders, Büchs *et al.* (1997) and Harenberg (1997) showed that the percentage of juveniles was greatest (64% of all spiders) in the extensively managed farming system IV (without pesticide and fertilizer input) and similar to the percentage of the set-aside area with natural succession (70%). However, it was considerably smaller (36%) in the more intensively managed farming systems I and II, even if no insecticides were applied (as in III). This indicates that the percentage of juvenile spiders recorded by trapping can be used to assess the effects of management intensity and husbandry practices.

Stippich & Krooß (1997) investigated the two locations (Marienstein and Reinshof) near Göttingen, Germany (see Box 15.1). The former site, with a range of variable soil types in short distances (Teiwes, 1997), was set in a varied landscape and was also surrounded by forests and hedgerows; the latter, more uniform site, however, was located in an area poor in landscape structures. Not surprisingly, activity-density and number of species (both for ground beetles and spiders) were higher at Marienstein. Also, over four years, the numbers of species were mostly greater and populations reached their peaks in the reduced (no insecticide input; nitrogen fertilizer reduction to 50%) and the extensive (no pesticides and fertilizers) farming systems (III and IV) compared with the other two (I and II).

Unlike, Harenberg (1997), Stippich & Krooß (1997) recorded the highest activity-density of spiders (and also of ground beetles) in the extensively managed oilseed rape fields and the lowest in the conventional system. However, the effects of management intensity were not as strongly developed in oilseed rape as in winter wheat. As stated by the authors, in an assessment of farming systems, distinctions between crops must be made. Thus, whereas in winter wheat species numbers and/or activity-densities of spiders and ground beetles could be enhanced by the integrated farming system II, in oilseed rape such enhancement will be observed more often in the reduced (III) and extensive (IV) farming systems. Tillage (see below) was identified as a major factor in causing such differences. Unlike winter wheat (ploughed only in farming systems I and III), oilseed rape fields in all farming systems had to be ploughed before sowing; therefore, any effect of reduced tillage could not be observed in the latter crop. As a consequence of tillage differences, *Oedothorax apicatus* made up 31% of all spiders in winter wheat but only 22% in oilseed rape.

Comparing the ground beetles of two oilseed rape fields with different pesticide treatment (a two-course rotation with 13 and a four-course rotation with four insecticide applications in four years of the crop rotation), with *Pseudoophonus rufipes* dominating in both fields, Palosz (1995b) found significantly greater abundance and species numbers in the conventionally managed field compared with the intensively managed one (2,186 cf. 777 individuals; 35 cf. 27 species, respectively). Including comparison with an extensively managed area (Palosz, 1995c), activity-density and species number were significantly lower in the least-intensive treatment; however, most strikingly, especially under intensive management, species that were more phytophagous or more saprophagous predominated (62.9% of all ground beetles).

Harvest effects

Harvesting has a major effect on the fauna of field crops in general. The microclimate inevitably changes as the protective cover of vegetation disappears, and sun, rain and wind impact the soil surface more directly than previously. Also, the pressure of heavy harvesting machines affects soil-inhabiting organisms by causing soil compaction. These factors affect predators directly and also have an indirect effect by reducing the number of prey items. According to Rieckert & Lockley (1984), harvest effects have a greater impact on predator communities than does the use of pesticides.

The intensity of these harvesting effects depends on the time that harvest actually occurs. In crops such as maize and sugar beet, large numbers of predators (e.g. ground beetles and, especially, spiders) are observed in September. Harvest time in these crops (October), however, is sufficiently late that most predatory species have reached the end of their seasonal period of activity in the fields. If harvesting of a crop takes place in July or August (as, for example, in oilseed rape and in most cereals) this coincides with the peak abundance of most spider species, but also with that of some ground beetles (e.g. *Calathus* spp., *Pterostichus* spp., *Trechus* spp.) and the emergence of the new generation of adults (e.g. *Carabus auratus* and *C. nemoralis*) or the presence of juvenile spiders and their cocoons (Nyfeller & Benz, 1979). However, Harenberg (1997) showed for winter barley that populations of dominant spider species (such as *Erigone atra*, *E. dentipalpis*, *Meioneta rurestris* and *Oedothorax apicatus*) (all Linyphiidae), although greatly depleted at harvest, reached pre-harvest levels just two to three weeks later.

Kopp *et al.* (1997) showed that *Nebria brevicollis* adults, which often are one of the most abundant ground beetle species in oilseed rape fields (Stippich & Krooß, 1997), left oilseed rape fields in significant numbers following harvest, because [as assumed by Müller (1968)] of the loss of vegetative cover and because hydro-thermal conditions became less suitable.

Tillage effects

A number of investigations have shown that spiders are damaged by tillage and that greater numbers, and greater biodiversity and biomass, occur in unploughed as opposed to ploughed fields (House & Stinner, 1983; Haskins & Shaddy, 1986; Stinner *et al.*, 1986). In particular, tillage operations are particularly disruptive to Lycosidae and Tetragnathidae. The most important impact of ploughing is the complete removal of structures provided by organic material and weeds on the soil surface which remain when tillage without ploughing (such as mulching or direct drilling) is practised.

The timing of tillage is critical for its effects on predator communities, especially if this involves ploughing. As Büchs *et al.* (1999) showed, ploughing in spring (e.g. for spring rape, sugar beet, maize, oats, rotational set-aside and other spring crop) has a more severe effect on the population development of large ground beetles (e.g. *Carabus auratus* and *C. nemoralis*) than does ploughing in late summer or autumn. It is assumed that the effects on predator populations are less harmful the later in the season that the tillage procedures occur (tillage for winter wheat is possible until the end of October) because, at that time, most predators will have moved deeper into the soil or will have emigrated to their overwintering quarters. Tillage for oilseed rape, however, is likely to occur relatively early in the summer (e.g. in northern Germany, for example, by the middle of August). At that time the tillage process, which occurs immediately after harvest, has a significant effect on the emergence of the new generation of spring-breeding adults of ground beetles (e.g. *C. nemoralis*) such that stable populations could become established only on set-aside areas with a natural succession of plants; they could not become established on cultivated areas such as oilseed rape fields (Büchs, 1994a).

Winter crops are more suitable for ground beetles hibernating as adults, and spring crops are more suitable for autumn breeders. Büchs *et al.* (1999) confirmed that the different time of tillage was the main factor. Most authors agree that ploughing in spring is particularly hazardous for larger adults and larvae of ground beetles (e.g. *Carabus* spp. and *Pterostichus melanarius*), whereas smaller beetles and larvae are said to be not endangered (Tischler, 1958; Scherney, 1962).

In the transdanubian region of Hungary, Szel *et al.* (1997) compared the ground beetle fauna in a wide range of winter and spring crops, and also in uncultivated habitats [a field margin and a set-aside conservation area for bustards (*Otis tarda*)]. They identified *Anchomenus dorsalis*, *Brachinus explodens*, *Poecilus cupreus*, *P. sericeus* and *P. punctulatus* as the most frequent ground beetles in all crops, although numbers were greater in winter and early-sown spring crops than in late-sown spring crops. The timing of tillage could be one of the main reasons for this phenomenon.

Effects of crop rotation

Owing to the dense, branchy structure of oilseed rape plants, the soil surface is more shaded and, therefore, less warmed by the sun than is the case in, for example, oat fields (e.g. Harenberg, 1997). Therefore, compared with oats, oilseed rape (and also winter wheat) is less suitable for wolf spiders, which are mostly thermophilic. Favourable conditions for thermophiles are better realized in oats, a crop with a considerably lower biomass of shade-producing leaves than oilseed rape and, even, winter wheat. Ulber *et al*. (1990) found that the spiders *Meioneta rurestris* and *Pachygnatha degeeri* (both typical of open habitats) occurred in large numbers in (and showed a preference for) oilseed rape. Harenberg (1997) confirmed that this was so for *M. rurestris*; however, with regards *P. degeeri*, she recorded smallest numbers in oilseed rape within a three-course crop rotation. This suggests that crop preference of polyphagous predators is very weak and heavily influenced by local conditions (e.g. soil type, plant density, cultivar and seasonal weather conditions).

In order to rank the importance of spiders within three- or four-course crop rotations, Büchs & Harenberg (2003) related the spider communities of each crop to a set-aside area by calculating the Jaccard index of similarity. The set-aside area was defined as a reference area, because natural succession was possible as it was neither managed nor disturbed in any way. From this reference area it was possible to assess just how strongly and how far a species community could be influenced by husbandry practices or crop, respectively. In particular, it was shown that the more extensively that crops were managed, the more similar the range of spider species in the cultivated areas was to that in the set-aside area. This phenomenon held sway for oilseed rape, but, compared with other crops, spider communities in this crop showed a closer affinity with those in the set-aside, even under the more intensively managed farming systems. Thus, at least for spiders, oilseed rape may be considered an 'environmentally friendly' crop, with communities close to those of semi-natural habitats such as a set-aside area with natural succession. Interestingly, Frank & Nentwig (1995) also recorded a larger number of species and individuals in oilseed rape than in maize or winter wheat.

Crop, weed and prey density

Comparing the community of ground beetles in pea, winter oilseed rape and winter wheat fields, Honek & Jarosik (2000) identified the density of the crop stand, the presence of falling seeds (weed or crop) and the presence of prey (aphids) as the most important factors steering species composition. With regards the effect of microclimatic differences (caused by the density and/or presence of crop cover), ground beetles preferred crop-shaded areas if plant density was low, but moved to open areas if crop plant density was high. Numbers of more or less phytophagous species (*Amara*, *Harpalus*, *Ophonus* and *Pseudoophonus*) increased especially in oilseed rape stands, because of the availability of both crop and weed seeds. The availability of aphids [e.g. cabbage aphid and peach/potato aphid (*Myzus persicae*)], however, increased the number of ground beetles such as *Bembidion lampros* and *Trechus quadristriatus*.

Physiological features of crop plants

Eigenbrode *et al.* (2000) (in the USA) showed (for *Brassica napus* and *Brassica rapa* lines) that leaves with genetically reduced crystalline surface waxes were 'preferred' by coccinellids of the dominant species *Hippodamia convergens*, resulting in a lower density of cabbage aphids per plant in the field. However, reduction of crystalline surface waxes following herbicide application, with S-ethyl dipropylthiocarbamate as the active ingredient, did not have a similar effect.

Endangered and rare predatory species

The status of endangered and/or rare species varies from country to country and, within countries, from region to region. Therefore, any assessment relating to crop management practices can be valid only on regional or smaller natural geographical units. Few investigations in oilseed rape, however, have focused on endangered and/or rare species.

With regards spiders, Harenberg (1997) found five species in oilseed rape that are either rare or endangered in Germany: *Argenna subnigra* (Dictynidae), *Erigone longipalpis* and *Savignya frontata* (Linyphiidae), *Xerolycosa miniata* (Lycosidae) and *Zelotes pusillus* (Gnaphosidae); of these, *Z. pusillus* was the most abundant (25 specimens in two seasons) and was also present in other crops. The occurrence of the endangered and/or rare spiders, however, had no observable relationship to crop management practices.

Long-term changes and predator communities

In Germany, over the past half century (Heydemann, 1953; Tischler, 1958; Hoßfeld, 1963; Pauer, 1975; Knauer & Stachow, 1987; Büchs *et al.*, 1999) there has been a continuous change in the species composition in arable fields (including oilseed rape). This is especially so in the case of ground beetles. For example, there has been a tendency for:

- an increase of hygrophilous species, such as *Anchomenus assimilis*, *Bembidion tetracolum*, *Carabus granulatus* and *C. nemoralis*, and a decrease of species that prefer drier and warmer conditions, such as *Bembidion lampros*, *Carabus auratus*, *C. cancellatus* and *Pseudoophonus rufipes*;
- a change from larger to smaller species – in the 1950s larger species (e.g. *Carabus auratus* and other *Carabus* spp.) were abundant in field crops, but they occur today only occasionally; on the other hand, smaller species (e.g. *Bembidion tetracolum*) were rather rare in the 1950s but today are most abundant in field crops;
- pesticides (especially insecticides) to have an increasing influence on the phenology of predators.

As stated for oilseed rape fields by Goltermann (1994), the dominance structure of the species community of ground beetles has clearly changed over the last 60 years. Species such as *Bembidion lampros* and *Carabus cancellatus*, which were found to be abundant by, for example, Gersdorf (1937) and Heydemann (1953), have decreased continuously, whereas more hygrophilous species such as *Bembidion tetracolum* and *Carabus granulatus* have become more dominant; this is also the case for *Pterostichus melanarius*. In the past, *Carabus* spp. (such as *C. auratus* and *C. nemoralis*) would form 40% of pitfall trap catches; however, in more recent investigations (e.g. Goltermann, 1994) they are uncommon. A similar situation is apparent for *Calathus* spp., and also for *Pseudoophonus rufipes*, which were described as the most abundant ground beetles in oilseed rape fields by Gersdorf (1937) – they were found only occassionally in oilseed rape by Goltermann (1994) and by other recent workers. Goltermann (1989) has previously explained variations in dominance positions as long-term cyclic fluctuations, and Basedow (1987) identified abiotic conditions and food supply as being more important. Observations by Stippich & Krooß

(1997), as well as those by W. Büchs & S. Land (*unpublished*), however, suggest that the shifting of dominance is a result of changing agricultural practices. Knauer & Stachow (1987) have previously stated that differences in the dominance structure of ground beetles in several crops are related to factors such as growing season, microclimatic conditions within the crop stand, crop rotation, plant protection measures and tillage.

Changes in agricultural practices have also resulted in declines in other predatory (or partly predatory) taxa in oilseed rape crops, e.g. carrion beetles (Silphidae), histerid beetles (Histeridae), harvestmen (Opiliones) and centipedes (Chilopoda).

Indirect effects of genetically modified oilseed rape on predators

Winter oilseed rape

Genetically modified (GM) winter oilseed rape expresses tolerance to glyphosate- or glufosinate-containing herbicides (see Chapter 1). In conventional winter oilseed rape crops, herbicide application usually takes place in autumn, because it is more difficult to control weeds such as goosegrass (*Galium aparine*) in the spring and also then difficult to avoid collateral herbicide damage to the crop. Where transgenic crops are grown, however, a herbicide may be applied in the spring without detriment to the crop; this timing also enables the farmer to tell whether the treatment threshold for weeds is reached and, therefore, to judge whether spraying is actually justified. Thus, herbicide-resistant oilseed rape could have the advantage of allowing a longer undisturbed period of development for the natural weed flora within the crop, which could be potentially beneficial to populations of predators.

Volkmar *et al.* (1999) compared the effects on spiders, ground beetles and rove beetles of transgenic oilseed rape (with a glufosinate application or with mechanical weed control as an alternative) with those of non-transgenic oilseed rape (with conventional treatments). In the case of spiders, the highest activity-density occurred in the transgenic oilseed rape where glufosinate was applied, and all the most abundant spider species showed their peaks of activity during this treatment. In the transgenic oilseed rape with mechanical weed control, species numbers were considerably larger than in all treatments with a herbicide application (whether transgenic or non-transgenic cultivars). This suggests that it was not the transgenic oilseed rape cultivar itself but the herbicide that affected biodiversity of the spider community – pioneer spider species such as *Erigone atra* and *Oedothorax apicatus*, which are typical of intensively managed fields, would be enhanced following herbicide application.

In the case of ground beetles (unlike spiders), activity-density was considerably lower in all treatments with transgenic oilseed rape, mostly attributable to species in the genera *Amara* (namely, *A. ovata* and *A. similata*) and *Harpalus* (namely, *H. affinis*), whereas some species in the genus *Bembidion* (e.g. *B. obtusum* and *B. quadrimaculatum*), and also *Calathus fuscipes* and *Loricera pilicornis*, were more frequent in the transgenic oilseed rape. Nevertheless, species sometimes showed distinct positive (e.g. *B. quadrimaculatum* and *Trechus quadristriatus*) or negative (e.g. *Pseudoophonus rufipes* and *Poecilus cupreus*) reactions to the herbicide treatment. In all cases, fewest species were recorded in the non-transgenic, conventionally grown oilseed rape.

With regards rove beetles, the transgenic and non-transgenic cultivars (with their different regimes of weed control) showed no significant differences in the total number of individuals caught in pitfall traps. The numbers of species in non-transgenic and transgenic cultivars with mechanical weed control were similar (*c.* 35 species), but considerably fewer species were present in the transgenic oilseed rape treated with the herbicide glufosinate. At the species level, considerable differences between transgenic and non-transgenic oilseed rape could be observed. For example, *Lesteva longoelytrata* had a much greater activity-density in the glufosinate-treated transgenic oilseed rape (25 individuals/traps/season), whereas in the non-transgenic cultivar individuals occurred only singly. Similar pictures emerged, although at a lower level (maximum of 2.5 individuals/trap/season) for *Anotylus rugosus* and *Gabrius nigritulus*. Of these rove beetle species, *Gabrius nigritulus* and *Lesteva longoelytrata* were far less active in the transgenic oilseed rape with mechanical weed control where glufosinate was not applied, so these two species obviously benefited from the herbicide application. However, *L. longoelytrata* is active mainly in winter (from October to December) and cannot profit, therefore, from the application of glufosinate if this takes place only in the spring. Unfortunately, the time of herbicide application was not stated by Volkmar *et al.* (1999). A conspicuously low predator activity-density in the transgenic oilseed rape with mechanical weed control was also registered for species such as *Olophrum assimile* and *Xantholinus linearis*, suggesting that, at least for these species, there is a detrimental effect as a result of the regular disturbance of soil microhabitats.

Comparing the effects on predators of a glufosinate-ammonium treatment with that of metazachlor + quinmerac in GM winter oilseed rape tolerant to the former herbicide, the largest numbers of individuals and species (mainly ground beetles and rove beetles) were recorded in the plots treated with glufosinate-ammonium. Spiders 'preferred' untreated plots. A single application of glufosinate-ammonium resulted in a high diversity of species and considerable biological activity of epigeal arthropods. Generally, differences in predatory arthropod activity were related to different weed densities (Volkmar *et al.*, 2000).

Spring oilseed rape

In spring oilseed rape, shortly before harvest (22 July 1996), Volkmar *et al.* (1999) recorded the peak of activity of predators in both trangenic and conventional crops. In transgenic spring rape (cv. 'PHY/22') (included for one year in their investigations), as in winter rape (see above), the highest activity-density of spiders was recorded in the treatment where the herbicide glufosinate was applied. The average activity (expressed as the number of spiders/trap/day) in the non-transgenic, conventionally grown spring was 4.89 individuals/trap/day, compared with from 6.11 to 7.50 individuals/trap/day in all treatments with transgenic cultivars; the species *Erigone atra* was particularly responsible for these differences. Owing to considerable variation in the results, no differences were observed in populations of ground beetles and rove beetles.

Endangered predatory species

Volkmar *et al.* (1999) recorded five species of spider in oilseed rape (*Argenna subnigra* – Dictynidae; *Mioxena blanda* and *Walckenaeria acuminata* – Linyphiidae; *Pardosa agricola* and *Xerolycosa miniata* – Lycosidae) and 11 species of ground beetle (*Amara consularis*, *A. curta*, *A. eurynota*, *A. littorea*, *Brachinus explodens*, *Calathus ambiguus*, *Harpalus zabroides*, *Microlestes minutulus*, *Notiophilus aestuans*, *Pterostichus quadrifoveolatus* and *Synuchus vivalis*) that –

according to the national German Red List or to the Red List of the federal state in which the experimental site was located – are said to be endangered. However, most of these organisms belong to categories that indicate only potential endangerment. The records include more than 100 specimens of *Mioxena blanda* (a money spider), 19 of *Notiophilus aesthuans*, 12 of *Calathus ambiguus* and 8 of *Amara curta* (all ground beetles). Also, more than 70% of the 'endangered' spiders and about half of the 'endangered' ground beetles were recorded in the fields growing transgenic oilseed rape. Similar data from elsewhere in Europe are not available.

Side-effects of pesticides on predators

The effects of pesticides on target and non-target organisms have been addressed in Chapter 2. Here, additional information is given that focuses on the side-effects of pesticides on predators, with particular reference to their relevance within pesticide registration processes. At an international level, these refer mainly to testing guidelines developed by the OECD and to collective ring testings promulgated by the IOBC Working Group 'Pesticides and Beneficial Organisms' (e.g. Hassan *et al.*, 1994); the latter include the testing of pesticides registered for use in oilseed rape, mostly based on methods developed by the IOBC Working Group.

Palosz (1997) showed that the abundance of epigaeic predators such as ground beetles and rove beetles (and also necrophagous carrion beetles) was significantly reduced in fields where intensive chemical crop protection occurred. Field contamination of predators by pesticides is possible by direct contact during the time of application. Subsequently, predators may come into contact with pesticides by running over sprayed soil and/or parts of plants; they may also imbibe pesticides or their metabolites within the food chain (Sunderland, 1992). Usually, species that inhabit higher (i.e. more aerial) parts of the field vegetation are more exposed to (and, potentially, at greater risk from) the direct effects of pesticides than those that live on the ground (Jepson, 1989; Vickerman, 1992). Spiders that are active at night are less exposed than those with diurnal activity periods, and the juvenile stages are more susceptible than adults. Also, males are more affected than females, owing to their greater activity range (Dinter, 1995).

Nowadays, chemical control of insect pests in oilseed rape relies on the application of pyrethroid sprays. Pyrethroid insecticides act as either 'contact' or 'stomach' poisons. In the case of wolf spiders (*Pardosa* spp.), so-called 'knockdown' effects can occur, with disruptions in motor co-ordination and also reductions in food intake. Recovery is possible, but, during the 'knockdown' period, affected indivdiuals will be unable to escape from other predators.

Most insecticide applications (sprays) in oilseed rape occur in early spring (to control, for example, pollen beetles and ceutorhynchid weevils) and in autumn (to control cabbage stem flea beetle), at comparatively low temperatures. However, the effects of pyrethroids depend on temperature and soil humidity, and their efficacy is enhanced when they are applied at low temperatures (Harris & Turnbull, 1978). Frampton & Çilgi (1993) recorded the strongest adverse effects of pyrethroids on money spiders after an application in autumn. Also, Everts *et al.* (1989) showed that the effects of pyrethroids are stronger if the humidity of the air is low, because diuretical processes (loss of water) caused by the insecticide lead to a lower resistance against desiccation. In oilseed rape, humidity of soil and air is comparatively high, compared (for example) with cereal crops.

Potentially, high humidity of the soil can cause more severe effects, because this increases the probability of epigaeic predators becoming contaminated whilst running over the soil surface (Harris & Turnbull, 1978). Mullié & Everts (1991) showed that poisoning of *Oedothorax apicatus*, the most abundant field-inhabiting money spider, results mainly from contamination on the soil surface and less from food intake or direct spray contact. In oilseed rape, the direct exposure of epigaeic predators to insecticides is relatively low, owing to the dense layers of vegetation that (to a greater or lesser extent) protect against contamination; this is especially so in the later stages of crop growth when insecticides are applied to combat pests such as brassica pod midge and cabbage seed weevil.

According to Brasse (1990), among the typical predator taxa, the ground beetles *Bembidion lampros* and *Poecilus cupreus* and the rove beetle *Aleochara bilineata* (which is more parasitic than predatory – see Chapter 9) are proposed as standard test organisms for oilseed rape; however, typical aphid predators such as the bug *Anthocoris nemorum* (Anthocoridae), the hover fly *Metasyrphus corollae*, the lacewing *Chrysoperla carnea* (Chrysopidae), the ladybird *Coccinella septempunctata* and the spider *Lepthyphantes tenuis* (Linyphiidae), which are recommended for other crops, are not.

Typically, a testing scheme includes several steps, starting with a rather artificial 'worst case' laboratory test; this is continued in the event of definite effects being observed (assessment based mainly on percentage mortality) with a semi-field test and (in exceptional cases) with a full-field test. The degree of harmfulness in laboratory or semi-field tests is classified as follows:

- 'harmless' <50% (laboratory), <25% (semi-field);
- 'slightly harmful' 50–79% (laboratory), 25–50% (semi-field);
- 'harmful' 80–99% (laboratory), 51–75% (semi-field);
- 'extremely harmful' >99% (laboratory), >75% (semi-field).

Further information on side-effects on predators, derived from laboratory, semi-field and field experiments, is also available. Büchs *et al.* (1991), for example, showed that insecticide seed treatments (of oilseed rape) may have severe adverse effects on epigaeic predators such as ground beetles, because even predatory species (e.g. *Poecilus cupreus*) are partly phytophagous and will also attack the seeds. In their experiments, germination of untreated rape seed was reduced, on average, by up to 10% when adults of *P. cupreus* were present. However, in the field the effect on ground beetle populations depends considerably on the structure of the soil. Thus, Büchs *et al.* (1991) recorded mortality of 36% in a laboratory-like, semi-field test (in which *P. cupreus* adults were released into enclosures with isofenphos-treated seeds and sieved, finely structured, artificial soil), whereas in the true field situation (in which the soil was cloddy) mortality of the beetles reached 80%. A cloddy soil has many crevices, and these allow beetles to penetrate below the surface and to contact the treated seeds. Schmuck (1992) also showed that, in the field, the toxicity of isofenphos-treated rape seed is considerably lessened if the soil structure is more fine. However, the toxicity of insecticide seed treatments depends also on the active ingredient. For example, under field conditions, a carbosulfan seed treatment causes far lower mortality to *P. cupreus* than isofenphos. Moreover, laboratory experiments have shown that carbosulfan seed treatments are slightly repellent to ground beetles, although this is merely a short-lasting effect.

According to Goltermann (1994), ground beetles (especially spring breeders such as *Anchomenus dorsalis* and *Bembidion tetracolum*) will be affected by insecticide seed treatments because, at the time of sowing, the very active final-instar larvae and the young, new-generation adult beetles are particularly at risk. In detail, however, the hazardous effects of oilseed rape seed treatments depend on:

- the active ingredient of the insecticide (e.g. isofenphos has a strong effect whereas carbosulfan does not; the strong effect of isofenphos is also applicable to rove beetles);
- mode of drilling (toxicity of seeds incorporated into the soil is lower and more delayed, compared with seed left on the soil surface);
- soil type (risks on sandy or well-prepared soils are considerably lessened, compared with those on cloddy, clay soils with many fissures and crevices, which make rape seed more readily accessible to epigaeic predators).

Similar results have been obtained for molluscicides. For example, in semi-field tests in oilseed rape crops, whereas severe detrimental effects on several ground beetle species (*Broscus cephalotes*, *Carabus granulatus*, *Pseudoophonus rufipes* and *Poecilus cupreus*) were observed following an application of methiocarb, for *Pterostichus melanarius* the mortality was considerably lower (Büchs *et al*., 1991). However, in a methiocarb treatment, emergence rates of smaller species in the genera *Amara*, *Bembidion*, *Clivina*, *Lasiotrechus* and *Trechus* were reduced by up to 87% compared with a metaldehyde treatment and an untreated control. Virtually no mortality occurred after an application of metaldehyde. In oilseed rape, molluscicides can be applied either in autumn or in spring. However, owing to their lifecycles, non-target organisms (ground beetles, for example) are more affected the earlier molluscicides are applied in autumn and the later they are applied in spring (Büchs *et al*., 1989, 1990). Samsoe-Petersen *et al*. (1992) showed in laboratory experiments that methiocarb pellets also have detrimental effects on the rove beetle *Aleochara bilineata*, reducing survival to at best 78%; metaldehyde pellets, however, cause no mortality, but they do lead to a significant increase in the hatching of eggs.

Using pitfall traps, Rzehak & Basedow (1982) compared the side-effects of three insecticides (deltamethrin, methoxychlor and parathion – applied separately) on selected species of predatory beetles (carabids and rove beetles) and 'small' spiders (mostly Linyphiidae) in oilseed rape two weeks before and four weeks after application. In each case the insecticide was applied shortly before the flowering of the crop, in order to control adult pollen beetles. Especially in the case of the carabid *Loricera pilicornis*, the rove beetles *Lathrobium fulvipenne* and *Tachyporus hypnorum* and 'small' spiders, a decrease in activity post-application was observed over the whole trapping period. With regards deltamethrin, the reduction was significant for rove beetles and especially noticeable and long-lasting in the case of spiders. The negative effects of parathion on the activity of the various taxa were also significant throughout. After the application of metoxychlor, considerable reduction in activity was observed for *Loricera pilicornis*; in the case of *T. hypnorum*, however, an enormous increase in activity was observed, interpreted by the authors as behavioural disturbance, resulting from sub-lethal effects of the pesticide.

According to Goltermann (1994) the intensity of the side-effects of pesticides depends on the time of application and on the lifecycles of the predators in question. Early-season application of a pesticide, e.g. an insecticide applied to combat rape stem weevil (see Chapter 2), may not have strong side-effects on predators, because this will occur when the predators are still within their hibernation quarters. Later applications of insecticides, however, are at least potentially harmful, especially if these coincide with the reproductive period of the predators (when they are then highly active in the field); such applications may have long-lasting effects on predator population development because reproduction, egg laying and so on are disrupted. Further, in the event of subsequent spraying, successful colonization of fields by predators can also be delayed. Goltermann (1989) showed that, at the time of application, plant density within the crop can also influence the intensity of side-effects. Following the application of deltamethrin at 0.2 litres/ha (against pollen beetle, *Meligethes aeneus*), he collected 50% more dead rove beetles and 160% more dead ground beetles in a sparse stand of oilseed rape than in a more dense one.

Effects on predators within non-oilseed rape crops have also been demonstrated. Kaczmarek (1992), for example, observed in potato fields strong and long-lasting effects (up to 1.5 months) of a pyrethroid application (deltamethrin at 0.3 litres/ha) on the activity of ground beetles.

Everts *et al.* (1991) investigated effects of deltamethrin on the predominant linyphiid spider *Oedothorax apicatus* in an oilseed rape field in the Netherlands. They considered residues and contaminated food items, i.e. springtails (Collembola), by using fluorescent tracers. They showed that toxicity to this spider (which is a typical inhabitant of intensively managed fields) was greatest at field water capacity, and that uptake of pesticide residues from the soil surface was nearly twice as great as from the topical treatment; additional uptake from contaminated prey, however, was low.

Re-establishment of predator populations in fields following pesticide application is clearly an important issue, which may be assumed to be achieved by re-invasion from adjacent areas and subsequent repopulation by in-field survivors. In the case of ground beetles, Pauer (1975) concluded that the re-establishment of populations following application of a pesticide to the growing crop depends more on the number of eggs and larvae that are actually in the soil than on immigration from outside the field. Re-invasion by spiders (for example, where in-field populations have been depleted by a pyrethroid spray), however, is most likely to depend on subsequent immigration, particularly in the case of non-epigaeic species.

Concerning the effects of farming systems with different levels of extensification, Büchs *et al.* (1999) separated predators into 'indicators for extensification' and 'indicators for intensification'. The former included the following predators:

- rove beetle larvae (e.g. *Philonthus* spp.), ground beetles [e.g. *Bembidion* spp., *Carabus auratus*, *Loricera pilicornis*, *Poecilus* spp., *Pterostichus* spp. (*sensu lato*)], soldier beetle larvae, harvestmen, dance flies (Empidoidea) and lacewings, all of which showed an increase in their abundance and activity-density with increasing extensification;

and the latter included:

- the ground beetle *Calathus melanocephalus*, ladybirds (adults and larvae), ants (Formicidae) and the spider *Oedothorax apicatus*, all of which showed an increase in abundance with increasing intensification.

Oedothorax apicatus probably remains unaffected by early pyrethroid applications on oilseed rape (such as those applied against cabbage stem weevil and pollen beetle) because its activity in the field commences rather late in the season. The same is true of *Erigone atra* and *Meioneta rurestris*. However, if such insecticides are applied later in the season (as is usual in winter wheat against aphids), then these species can be affected adversely, as may be shown by a decline in numbers emerging in the next generation. Similar observations have been made by Basedow *et al.* (1985). Thus, for some predators, insecticide applications in oilseed rape may have fewer adverse effects than in other arable crops, because applications occur relatively early in the spring (owing to the early immigration into fields of spring pests). However, in central Europe the insecticide application period on oilseed rape can vary from year to year by up to one and a half months (from mid-March to the end of April), according to weather conditions; this suggests that there could be differences in the effects of insecticides on predators from year to year.

As shown by Büchs *et al.* (1999), effects of the farming system (see Box 15.1) and insecticide input in oilseed rape were more obvious for predators that start their activity in the fields relatively early in the season. The spider *Pachygnatha degeeri* is an example. Its phenology showed a continuously increasing activity-density and a significantly higher activity-density in farming system III which was not treated with insecticides. However, with regards farming systems I and II, adverse effects were noted that were related directly or indirectly to the application of insecticides. All the above-mentioned species (and also *Oedothorax retusus* and *Pelecopsis parallela*) occurred in significantly greater numbers in farming system III. It can be assumed that these species are sensitive to insecticide applications. However, the use of herbicides and the lower input of fertilizers could explain the preference for system III. Both factors lead to a lower density of vegetation (crops or weeds), so the spiders would benefit from the more 'open' conditions in farming system III. Although no insecticides (and no other agrochemicals) were applied in the set-aside area with natural succession, the above-mentioned spider species were present in very low numbers, presumably because of plant density. This indicates that they are typically field species and adapted to areas subject to periodic tillage.

An interesting phenological pattern was observed by Harenberg (1997) for the money spider *Araeoncus humilis* in relation to the different crops or farming systems and the set-aside area with natural succession. In oilseed rape fields this spider showed a noticeable increase in activity-density only in the first half of June, whereas in winter rye (for example) this species was active throughout most of whole vegetative period and activity commenced one month ahead of that in oilseed rape. The highest activity-density was recorded in system II with mechanical weed control and, thus, no input of herbicides. Further, data from emergence traps and pitfall traps suggest that *A. humilis* inhabits the vegetation layers in set-aside areas (owing to plant density), whereas in oilseed rape it remains on the surface. The data also suggest that slight population exchange between uncultivated (set-aside) areas and field (especially oilseed rape) may occur.

Stippich & Krooß (1997) were able to identify short-term effects of pyrethroid applications on the rove beetle *Tachyporus hypnorum*, which is said to be an effective aphid predator (e.g. Dennis & Wratten, 1991). An application of deltamethrin at the end of March reduced the abundance of *T. hypnorum* in the treated system from 20 to 3 individuals/m^2 (emergence trap samples) shortly afterwards. Whereas, in the untreated system, a continuous increase in numbers of *T. hypnorum* was observed during the whole sampling period, in the treated system, numbers of the predator recovered only at the end of May, probably due to immigration from elsewhere. Their pitfall trap catches, however, show a significant 'peak' shortly after the insecticide application, interpreted by the authors as hyperactivity owing to direct effects of the pyrethroid (see above for 'knockdown' effects, etc.). In developing countries, pesticides such as endosulfan, phosalone and phosphamidon are commonly used in oilseed rape, but their use is no longer permitted in Europe or other developed countries, because of their side-effects on non-target organisms (including man). Singh *et al.* (1999) have stated that, on oilseed rape crops in India, endosulfan and phosalone are 'safe' to predatory ladybirds (*Coccinella septempunctata*) and pollinating honey bees (*Apis cerana*), but the long-term, sub-lethal effects were not considered.

Uncultivated areas

Nyffeler (1982) compared the spider fauna of uncultivated and cultivated areas in Switzerland. Out of eleven spider families in uncultivated areas he recorded only five (Argiopidae, Linyphiidae, Tetragnathidae, Theridiidae and Thomisidae) in the field vegetation. He concluded that in cultivated areas, only those that can tolerate the extreme environmental stress afforded by pesticides, tillage, harvesting and other cultivation means are able to recolonize the fields every year. Thus, the species community of spiders in cropped fields can be seen as a selection of spider populations in surrounding, undisturbed (i.e. uncultivated) areas.

The biomass per square metre of spiders in crops is significantly lower than that in uncultivated land or forests. This is because, in fields, the majority of spiders are very small compared with those in surrounding, undisturbed areas. For example, the average individual fresh body weight for an orb-web spider (e.g. *Araneus* sp.) typical of undisturbed areas may exceed 200 or 300 mg, whereas aeronautic linyphiids (typical for fields) have an average fresh body weight of less than 1 mg.

As stated by Knauer & Stachow (1987) and Büchs *et al.* (1999), there is an increasing need to conserve uncultivated areas within farm landscapes, because some of the most abundant ground beetle predators in field crops depend on the presence of such areas as overwintering and aestivation sites and as refuges during harvest and tillage. Also, their value for other predatory taxa cannot be overemphasized, especially if predators are to function as effective biocontrol agents in arable cropping systems.

Acknowledgements

This work was part-funded by the European Commission Concerted Action: *Minimizing pesticide use and environmental impact by the development and promotion of bio-control strategies for oilseed rape pests, FAIR 96-1314.* Thanks are also due to Dr. Bernhard Kromp for providing data for this chapter.

References

Basedow, T. (1973). Der Einfluß epigäischer Raubarthropoden auf die Abundanz phytophager Insekten in der Agrarlandschaft. *Pedobiologia* **13**, 410–22.

Basedow, T. (1987). Der Einfluß gesteigerter Bewirtschaftungsintensität im Getreidebau auf die Laufkäfer (Coleoptera, Carabidae). *Mitteilungen avs der Biologischen Bundesanstalt für Land- und Forstwirtschaft Berlin-Dahlem* **235**, 1–123.

Basedow, T., Rzehak, H. & Voß, K. (1985). Studies on the effects of deltamethrin sprays on the numbers of epigeal predatory arthropods occurring in arable fields. *Pest Science* **16**, 325–31.

Boyd, M. L. & Lentz, G. L. (1999). Seasonal occurrence and abundance of the tarnished plant bug (Hemiptera: Miridae) and thrips (Thysanoptera: Thripidae) on rapeseed in West-Tennessee. *Journal of Agriculture and Urban Entomology* **16**, 171–8.

Brasse, D. (1990). Einführung der obligatorischen Prüfung der Auswirkung von Pflanzenschutzmitteln auf Nutzorganismen in das Zulassungsverfahren. *Nachrichtenblatt für den Deutschen Pflanzenschutzdienst* **42**, 81–6.

Büchi, R. (2002). Mortality of pollen beetle (*Meligethes* spp.) larvae due to parasitoids and predators in rape fields and the effect of conservation strips. *Agriculture, Ecosystems and Environment* **90**, 255–63.

Büchs, W. (1991). Ergebnisse der Fang-Wiederfangmethode bei *Pterostichus* spp. mit einer Anmerkung zur erforderlichen Größe von Versuchsparzellen. *Nachrichtenblatt Deutsche Gesellschaft für Allgemeine und Angewandte Entomologie* **5** (2), 38.

Büchs, W. (1994a). Förderung von großen Laufkäferarten (Gattungen *Carabus*, *Calosoma*, *Cychrus*) durch selbstbegrünende Dauerbrache und abgestufte Extensivierungsmaßnahmen. *Mitteilungen aus der Biologischen Bundesanstalt für Land- und Forstwirtschaft Berlin-Dahlem* **301**, 47.

Büchs, W. (1994b). Auswirkungen der Brache auf die Fauna. *Mitteilungen aus der Biologischen Bundesanstalt für Land- und Forstwirtschaft Berlin-Dahlem* **301**, 41–57.

Büchs, W. & Harenberg, A. (2003). Biodiversity, the ultimate agroenvironmental indicator? Potential and limits for the application of faunisitic elements as gradual indicators in agroecosystems. *Agriculture, Ecosystems and Environment, Supplement (in press)*

Büchs, W., Harenberg, A., Prescher, S., Weber, G. & Hattwig, F. (1999). Entwicklung von Evertebratenzönosen bei verschiedenen Formen der Flächenstillegung und Extensivierung. *Mitteilungen aus der Biologischen Bundesanstalt für Land- und Forstwirtschaft Berlin-Dahlem* **368**, 9–38.

Büchs, W., Harenberg, A. & Zimmermann, J. (1997). The invertebrate ecology of farmland as a mirror of the intensity of the impact of man? An approach to interpreting results of field experiments carried out in different crop management intensities of a sugar beet and an oil seed rape rotation including set-aside. *Biological Agriculture and Horticulture* **15**, 83–107.

Büchs, W., Heimbach, U. & Czarnecki, E. (1989). Effects of snail baits on non-target carabid beetles. *BCPC Monograph No. 41 (Slugs and Snails in World Agriculture)*, pp. 245–52.

Büchs W., Heimbach, U. & Czarnecki, E. (1990). Untersuchungen zu Auswirkungen von Schneckenbekämpfungsmitteln auf einige Laufkäferarten (Coleoptera: Carabidae) bei Anwendung verschiedener Testverfahren im Labor und Halbfreiland. *Zeitschrift für Angewandte Zoologie* **77**, 479–500.

Büchs, W., Heimbach, U. & Czarnecki, E. (1991). Auswirkungen von inkrustiertem Rapssaatgut auf Laufkäfer: Labor- und Halbfreilandversuche. *Gesunde Pflanzen* **43**, 299–306.

Büchs, W. & Nuss, H. (2000). First steps to assess the importance of epigaeic active polyphagous predators as natural enemies of oilseed rape insect pests with soil pupating larvae. *IOBC/wprs Bulletin, Integrated Control in Oilseed Crops* **23** (6), 151–63.

Coaker, T. H. (1966). The effect of soil insecticides on the predators and parasites of the cabbage root fly (*Erioischia brassicae* (Bouché)) and the subsequent damage caused by the pest. *Annals of Applied Biology* **57**, 397–407.

Dennis, P. & Wratten, S. D. (1991). Field manipulation of populations of individual staphylinid species in cereals and their impact on aphid populations. *Ecological Entomology* **16**, 17–24.

Dinter, A. (1995). *Untersuchungen zur Populationsdynamik von Spinnen (Arachnida: Araneae) in Winterweizen und deren Beeinflussung durch insektizide Wirkstoffe.* PhD thesis, University of Göttingen, Germany.

Eigenbrode, S. D., Kabalo, N. N. & Rutledge, C. E. (2000). Potential of reduced waxbloom oilseed *Brassica* for insect pest resistance. *Journal of Agricultural and Urban Entomology* **17**, 53–63.

Everts, J. W., Aukema, B., Hengeveld, R. & Koeman, J. H. (1989). Side effects of pesticides on ground-dwelling predatory arthropods in arable ecosystems. *Environmental Pollution* **59**, 203–25.

Everts, J. W., Aukema, B., Mullié, W. C., van Gemerden, A., Rottier, A., van Katz, R. & Gestel, C. A. M. van (1991). Exposure of the ground dwelling spider *Oedothorax apicatus* (Blackwall) (Erigonidae) to spray and residues of deltamethrin. *Archives of Environmental Contamination and Toxicology* **20**, 13–19.

Fang, C. Y., Wen, S. G., Cui, S. Z. & Wang, Y. H. (1984). [The role of natural enemies in the integrated control of insect pests in cotton.] *China Cotton* **2**, 42–3. [In Chinese.]

Frampton, G. K. & Çilgi, T. (1993). Boxworth and SCARAB: approaches to the long-term study of side effects of pesticide use on invertebrates. *ANPP-BCPC 2nd International Symposium on Pesticide Application Techniques, Strasbourg, 1993*, pp. 399–406.

Frank, T. (1996). Species diversity and activity densities of epigaeic and flower visiting arthropods in sown weed strips and adjacent fields. *Bulletin OILB/SROP* **19** (3), 101–5.

Frank, T. (1997). Species diversity of ground beetles (Carabidae) in sown weed strips and adjacent fields. *Biological Agriculture and Horticulture* **15**, 297–307.

Frank, T. & Nentwig, W. (1995). Ground dwelling spiders (Araneae) in sown weed strips and adjacent fields. *Acta Oecologia* **16**, 179–93.

Gärtner, G. (1980). *Ökologisch-faunistische Veränderungen durch Flurbereinigungsmaßnahmen. Dargestellt am Beispiel der Carabidenfauna von Zuckerrübenkulturen in ausgewählten Kraichgau-Gemeinden.* PhD thesis, University of Heidelberg, Germany.

Gersdorf, E. (1937). Ökologisch-faunistische Untersuchungen über die Carabiden der mecklenburgischen Landschaft. *Zoologische Jahrbücher für Systematik* **70**, 17–86.

Goltermann, S. (1989). *Das Auftreten von Carabiden auf Winterrapsschlägen.* Diploma, University of Rostock, Germany.

Goltermann, S. (1994). *Das Auftreten von Laufkäfern (Col., Carabidae) auf Winterrapsfeldern und deren Einfluß auf den Massenwechsel von* Meligethes aeneus *F. (Col., Nitidulidae).* Dissertation, University of Rostock, Germany.

Harenberg, A. (1997). *Auswirkungen abgestuft extensiv geführter Anbausysteme in verschiedenen Fruchtfolgen (Raps-, Zuckerrübenfruchtfolge) und einer selbstbegrünenden Dauerbrache auf Spinnen (Arachnida: Araneae).* PhD Thesis, Technical University of Braunschweig, Germany.

Harris, C. R. & Turnbull, S. A. (1978). Laboratory studies on the contact toxicity and activity in soil of four pyrethroid insecticides. *Canadian Entomologist* **110**, 285–8.

Haskins, M. F. & Shaddy, J. H. (1986). The ecological effects of burning, mowing and ploughing on ground inhabiting spiders (Araneae) in an old-field ecosystem. *Journal of Arachnology* **14**, 1–14.

Hassan, S. A., Bigler, F., Bogenschütz, H., Boller, E., Brun, J., Calis, J. N. M., Coremans-Pelseneer, J., Duso, C., Grove, A., Heimbach, U., Helyer, N., Hokkanen, H., Samsoe-Petersen, L., Sauphanor, B., Stäubli, A., Sterk, G., Vainio, A., Veire, M. van de, Viggiani, G. & Vogt, H. (1994). Results of the sixth joint pesticide testing programme of the IOBC/WPRS working group 'Pesticides and beneficial organisms'. *Entomophaga* **39**, 107–19.

Hausamman, A. (1996). Strip management in rape crop: is winter rape endangered by negative impacts of sown weed strips? *Journal of Applied Entomology* **120**, 505–12.

Heydemann, B. (1953). *Agrarökologische Problematik, dargetan an Untersuchungen über die Tierwelt der Oberfläche der Kulturfelder.* PhD thesis, University of Kiel, Germany.

Heydemann, B. (1983). Aufbau von Ökosystemen im Agrarbereich und ihre langfristigen Veränderungen. *Daten und Dokumente zum Umweltschutz, Sonderreihe Umwelttagung* **35**, 53–83.

Hokkanen, H. & Holopainen, J. K. (1986). Carabid species and activity densities in biologically and conventionally managed cabbage fields. *Journal of Applied Entomology* **102**, 353–63.

Honek, A. & Jarosik, V. (2000). The role of crop density, seed and aphid presence in diversification of field communities of Carabidae (Coleoptera). *European Journal of Entomology* **97**, 517–25.

Hoßfeld, R. (1963). Synökologischer Vergleich der Fauna von Winter- und Sommerrapsfeldern. *Zeitschrift für Angewandte Zoologie* **52**, 209–54.

Houpert, G. (1983). Les carabiques et la culture du colza en Lorraine. *6ème Congrès International sur le Colza, Paris, 17–19 mai*, **2**, 1178–83.

House, G. J. & Stinner, B. R. (1983). Arthropods in no tillage soybean agroecosystems community composition and ecosystem interactions. *Environmental Management* **6**, 23–8.

Jepson, P. C. (1989). *The temporal and spatial dynamics of pesticide side-effects on non-target invertebrates.* In: P. C. Jepson (ed.). *Pesticides and Non-Target Invertebrates.* Intercept: Andover.

Jones, M. G. (1975). The predators of the wheat bulb fly. *Annals of Applied Biology* **80**, 128–30.

Kaczmarek, S. (1992). Wplyw preparatu "Decis 2,5 EC" na Carabidae w uprawie ziemniaka. *Polskie Pismo Enomologiczne* **61**, 125–9.

Kirchner, H.-A. (1960). *Untersuchungen zur Ökologie feldbewohnender Carabiden.* Dissertation, University of Cologne, Germany.

Klinger, K. (1987). Auswirkungen eingesäter Randstreifen an einem Winterweizen-Feld auf die Raubarthropodenfauna und den Getreideblattlausbefall. *Journal of Applied Entomology* **104**, 47–58.

Knauer, N. & Stachow, U. (1987). Aktivitäten von Laufkäfern (Carabidae, Coleoptera) in einem intensiv wirtschaftenden Ackerbaubetrieb. Ein Beitrag zur Agrarökosystemanalyse. *Journal of Agronomy and Crop Science* **159**, 131–45.

Kokta, C. (1989). *Auswirkungen abgestufter Intensität der Pflanzenproduktion auf epigäische Raubarthropoden, insbesondere Laufkäfer (Coleoptera, Carabidae) in einer dreigliedrigen Fruchtfolge.* PhD thesis, University of Darmstadt, Germany.

Kopp, A., Gerowitt, B. & Poehling, H. M. (1997). *Auswirkungen verschiedener Brachestreifentypen in der Agrarlandschaft auf Vegetation und Laufkäfer*. In: B. Gerowitt & M. Wildenhayn (eds). *Ökologische und ökonomische Auswirkungen von Extensivierungsmaßnahmen im Ackerbau. Ergebnisse des Göttinger INTEX-Projektes 1990–4.* University of Göttingen, Germany.

Kromp, B. (1999). Carabid beetles in sustainable agriculture: a review on pest control efficacy, cultivation impacts and enhancement. *Agriculture, Ecosystems and Environment* **74**, 187–228.

Meissner, R.-L. (1984). Zur Biologie und Ökologie der ripicolen Carabiden *Bembidion femoratum* Sturm und *B. punctulatum* Drap. II. Die Substratbindung. *Zoologisches Jahrbuch für Systematik* **111**, 369–83.

Müller, G. (1968). Faunistisch-ökologische Untersuchungen der Coleopterenfauna der küstennahen Kulturlandschaften bei Greifswald. Teil 1: Die Carabidenfauna benachbarter Acker- und Weidenflächen mit dazwischen liegendem Feldrain. *Pedobiologia* **8**, 313–39.

Mullié, W. C. & Everts, J. W. (1991). Uptake and elimination of [14C] deltamethrin by *Oedothorax apicatus* (Arachnida: Erigonidae) with respect to bioavailability. *Pesticide, Biochemistry and Physiology* **39**, 27–34.

Nilsson, C. & Andreasson, B. (1987). Parasitoids and predators attacking pollen beetles (*Meligethes aeneus* F.) in spring and winter rape in southern Sweden. *IOBC/wprs Bulletin* **10**, 64–73.

Novák, B. (1968). Bedingungsgrad der Imagines einiger Feldcarabiden an die Lebensbedingungen in einem Winterweizenbestand (Col. Carabidae). *Acta Universitatis Palackianae Olomucensis* **28**, *Biologie* **10**, 99–131.

Nuss, H. & Büchs, W. (2000). Zur Bedeutung epigäischer Raubarthropoden als Antagonisten von Blüten- und Schotenschädlingen des Winterrapses mit Verpuppung im Boden. *Mitteilungen der Deutschen Gesellschaft für Allgemeine und Angewandte Entomologie* **12**, 141–7.

Nyfeller, M. & Benz, G. (1979). Zur ökologischen Bedeutung der Spinnen der Vegetationsschicht von Getreide- und Rapsfeldern. *Zeitschrift für Angewandte Entomologie* **87**, 348–76.

Nyffeler, M. (1982). *Field Studies on the Ecological Role of the Spiders as Insect Predators in Agroecosystems (Abandoned Grassland, Meadows, and Cereal Fields).* Unpublished PhD thesis, Swiss Federal Institute of Technology (ETH), Zürich, Switzerland.

Palosz, T. (1995a). Ground beetles on winter rape crops with different plant protection, rotation and soil characteristics. *Rapeseed Today and Tomorrow. Proceedings of the 9th International Rapeseed Congress, Cambridge, 4–7 July 1995*, pp. 1321–3.

Palosz, T. (1995b). Intensywne technologie w rolnictwie a fauna biegazowatych. *Ochrona-Roslin* **39**, 8.

Palosz, T. (1995c). Slad gatunkowy biegaczowatych (Col., Carabidae) na plantacjach rzepaku ozimego o roznej technologii i intensywnosci uprawy. *Materialy Sesji Instytutu Ochrony Roslin* **35**, 108–15.

Palosz, T. (1997). Wystepowanie biegaczowatych (Col., Carabidae) i innych stawonogow epigeicznych na plantacjach rzepaku ozimego o roznej technologii uprawy. *Rosliny Oleiste* **18**, 343–8.

Parajulee, M. N. & Slosser, J. E. (1997). Potential relay strips for predator enhancement in cotton. *Proceedings of the Beltwide Cotton Conference, New Orleans, USA, 6–10 January 1997*, **2**, 1124–7.

Pauer, R. (1975). Zur Ausbreitung der Carabiden in der Agrarlandschaft, unter besonderer Berücksichtigung der Grenzbereiche verschiedener Feldkulturen. *Zeitschrift für Angewandte Zoologie* **62**, 457–89.

Pfiffner, L. & Luka, H. (1996). Laufkäfer-Förderung durch Ausgleichsflächen. Auswirkungen neu angelegter Grünstreifen und einer Hecke im Ackerland. *Naturschutz und Landschaftsplanung* **28**, 145–51.

Pollard, E. (1971). Hedges. VI. Habitat diversity and crop pests: a study of *Brevicoryne brassicae* and its syrphid predators. *Journal of Applied Ecology* **8**, 751–80.

Prescher, S. & Büchs, W. (1997). Zum Einfluß abgestufter Extensivierungsmaßnahmen und selbstbegrünender Dauerbrache im Ackerbau auf funktionelle Gruppen der Brachycera (Diptera). *Verhandlungen der Gesellschaft für Ökologie* **22**, 385–91.

Prescher, S. & Büchs, W. (1998). Der Einfluß abgestufter Extensivierungsmaßnahmen im Raps- und Erbsenanbau auf nützliche Fliegen (Diptera, Brachycera). *Gesunde Pflanzen* **50**, 213–18.

Prescher, S. & Büchs, W. (1999). Fliegenzönosen (Diptera, Brachycera) aus abgestuft extensiv bewirtschafteten Raps- und Erbsenkulturen. *Verhandlungen der Gesellschaft für Ökologie* **29**, 265–9.

Prescher, S. & Büchs, W. (2000). Der Einfluß abgestufter Extensivierungsmaßnahmen im Ackerbau auf die Struktur der phytophagen Brachycerazönose. *Mitteilungen der Deutschen Gesellschaft für Allgemeine und Angewandte Entomologie* **12**, 347–52.

Renken, W. (1956). Untersuchungen über Winterlager der Insekten. *Zeitschrift für Morphologie und Ökologie der Tiere* **45**, 34–106.

Renkonen, O. (1938). Statistisch-ökologische Untersuchungen über die terrestrische Käferwelt der finnischen Bruchmoore. *Annales Zoologici Societatis Zoologico-Botanicae Fennicae* **6**, 1–231.

Rieckert, S. E. & Lockley, T. (1984). Spiders as biological control agents. *Annual Review of Entomology* **29**, 299–320.

Rzehak, H. & Basedow, T. (1982). Die Auswirkungen verschiedener Insektizide auf die epigäischen Raubarthropoden in Winterrapsfeldern. *Anzeiger für Schädlingskunde Pflanzenschutz Umweltschutz* **55**, 71–5.

Samsoe-Petersen, L., Bieri, M. & Büchs, W. (1992). Interpretation of laboratory measured effects of slug pellets on soil dwelling invertebrates. *Aspects of Applied Biology* **31**, 402–9.

Scherney, F. (1961). Beiträge zur Biologie und ökonomischen Bedeutung räuberisch lebender Käferarten. *Zeitschrift für Angewandte Entomologie* **48**, 163–75.

Scherney, F. (1962). Laufkäfer als nützliche Helfer der Schädlingsbekämpfung. *Pflanzenschutzinformationen der Bayerischen Landesanstalt für Pflanzenbau und Pflanzenschutz* **3**.

Schmuck, R. (1992). Auswirkungen einer Rapssaatbehandlung mit dem Saatgutpuder OFTANOL® T auf Laufkäfer. *Gesunde Pflanzen* **44**, 264–8.

Singh, M. P., Singh, K. M. & Singh, K. I. (1999). Efficacy and economics of some insecticides against mustard aphid, *Lipaphis erysimi*, and their safety to predators and foraging bees. *Indian Journal of Entomology* **61**, 11–18.

Sörensen, M. (1948). A method of establishing groups of equal amplitude in plant sociology based on similarity of species content and its application to analyses of the vegetation on Danish commons. *Kongelige Danske Videnskabernes Selskabs Skrifter* **5**, 4.

Sotherton, N. W. (1984). The distribution and abundance of predatory arthropods overwintering on farmland. *Annals of Applied Biology* **105**, 423–9.

Stachow, U. (1988). Zur Bedeutung der Wallhecken in einem Agrarökosystem Schleswig-Holstein. II. Laufkäfer (Carabidae, Col.). *Zeitschrift für Kulturtechnik und Flurbereinigung* **29**, 299–306.

Steinmann, H.-H. (1998). Can integrated arable farming systems achieve long-term stability of crop yield and weed infestation? *Mededelingen van de Faculteit Landbouwwetenchappen Rijksuniversiteit Gent* **63**, 697–703.

Stinner, B. R., Krueger, H. R. & Mc Cartney, D. A. (1986). Insecticide and tillage effects on pest and non-pest arthropods in corn agroecosystems. *Agriculture, Ecosystems and Environment* **15**, 11–21.

Stippich, G. & Krooß, S. (1997). *Auswirkungen von Extensivierungsmaßnahmen auf Spinnen, Laufkäfer und Kurzflügelkäfer.* In: B. Gerowitt & M. Wildenhayn (eds). *Ökologische und ökonomische Auswirkungen von Extensivierungsmaßnahmen im Ackerbau. Ergebnisse des Göttinger INTEX-Projektes 1990–4.* University of Göttingen, Germany.

Sunderland, K. (1992). Effects of pesticides on the population ecology of polyphagous predators. *Aspects of Applied Biology* **31**, 19–28.

Szel, G. Y., Kadar, F. & Farago, S. (1997). Abundance and habitat preference of some adult-overwintering ground beetle species in crops in western Hungary (Coleoptera: Carabidae). *Acta Phytopathologica et Entomologica Hungarica* **32**, 369–76.

Tao, C. C. C. (1973). Citrus insect control in Taiwan and a new integrated control method. *Journal of Taiwan Agricultural Research* **22**, 131–4.

Teiwes, K. (1997). Standortbedingungen. In: B. Gerowitt & M. Wildenhayn (eds). *Ökologische und ökonomische Auswirkungen von Extensivierungsmaßnahmen im Ackerbau. Ergebnisse des Göttinger INTEX-Projektes 1990–4.* University of Göttingen, Germany.

Tenhumberg, B. (1993). *Untersuchungen zur Populationsdynamik von Syrphiden in Winterweizenbeständen und Quantifizierung ihrer Bedeutung als Antagonisten von Getreideblattläusen.* Cuvillier: Göttingen.

Tenhumberg, B. & Poehling, H. M. (1995). Syrphids as natural enemies of cereal aphids in Germany. Aspects of their biology and efficacy in different years and regions. *Agriculture, Ecosystems and Environment* **52**, 39–43.

Thiele, H. U. (1977). *Carabid Beetles in Their Environments.* Springer: Berlin.

Tischler, W. (1958). Synökologische Untersuchungen an der Fauna der Felder und Feldgehölze. *Zeitschrift für Morphologie und Ökologie Tiere* **47**, 54–114.

Ulber, B., Stippich, G. & Wahmhoff, W. (1990). Möglichkeiten, Grenzen und Auswirkungen des gezielten Pflanzenschutzes im Ackerbau: III. Auswirkungen unterschiedlicher Intensitäten des chemischen Pflanzenschutzes auf epigäische Raubarthropoden in Winterweizen, Zuckerrüben und Winterraps. *Zeitschrift für Pflanzenkrankheiten und Pflanzenschutz* **97**, 263–83.

Vansgaard, C., Gavesen, E. & Toft, S. (1990). The spider fauna of a marginal agricultural field (Araneae, Linyphiidae). *Entomologiske Meddelelser* **58**, 47–54.

Vickerman, G. P. (1992). *The effects of different pesticide regimes on the invertebrate fauna in winter wheat.* In: P. W. Greig-Smith, G. K. Frampton & A. R. Hardy (eds). *Pesticides Cereal Farming and Environment*. HMSO: London.

Volkmar, C., Wetzel, T., Lübke-Al Hussein, M., Jany, D. & Richter, L. (1999). Ökologische Begleituntersuchungen beim Anbau transgener Rapspflanzen am Standort Friemar (Freistaat Thüringen) in den Jahren 1994 bis 1996. *Archiv für Phytopathologie und Pflanzenschutz* **32**, 89–130.

Volkmar, C., Kreuter, T., Richter, L., Lübke-Al Hussein, M., Jany, D., Schmutzler, K. & Wetzel, T. (2000). Zur ökologischen Situation in herbizidtorelanten bzw. konventionellen Rapsbeständen im mitteldeutschen Agrarraum in den Jahren 1996 bis 1998. *Zeitschrift für Pflanzenkrankheiten und Pflanzenschutz Sonderheft* **17**, 337–45.

Wang, A. Y. & Wang, L. P. (1996). [An explanatory study on cotton insect control by natural enemies.] *China Cottons* **23**, 31. [In Chinese.]

Weber, F. (1965). Feld- und Laboruntersuchungen zur Winteraktivität der Carabiden auf Kulturfeldern. *Zeitschrift für Morphologie und Ökologie der Tiere* **54**, 551–65.

Wishart, G., Doane, J. F. & Maybee, G. E. (1956). Notes on beetles as predators of eggs of *Hylemya brassicae* (Bouché) (Diptera: Anthomyiidae). *Canadian Entomologist* **138**, 634–9.

Wright, D., Hughes, R. D. & Worral, J. (1960). The effect of certain predators on the numbers of cabbage root fly (*Erioischia brassicae* Bouché) and on the subsequent damage caused by the pest. *Journal of Animal Ecology* **28**, 343–57.

Zimmermann, J. & Büchs, W. (1999). Kurzflügelkäfer (Coleoptera: Staphylinidae) in unterschiedlich intensiv bewirtschafteten Ackerflächen. *Agrarökologie* **32**, 1–154.

16

Predators as Biocontrol Agents of Oilseed Rape Pests

Wolfgang Büchs
Federal Research Centre for Agriculture and Forestry, Messeweg 11/12, D-38104 Braunschweig, Germany

Introduction

Predatory arthropods are important biocontrol agents (BCAs) in arable cropping systems. Considerable attention has been focused on ground beetles (e.g. Kromp, 1999), but members of many other invertebrate groups are also of importance (see Chapter 12). Within an arable context, most investigations have been centred on cereal crops, but studies on oilseed rape also feature. In this chapter the role of various invertebrates as predators of oilseed rape pests is reviewed, with particular attention to ground beetles and spiders. Details of sampling, trapping and monitoring methods used to assess the presence and abundance of predators are given in Chapter 14.

Methods for Assessing Predatory Capacity

Non-field (confined) assessment methods

Goltermann (1994) tested the predatory capacity (i.e. the effectiveness as predators) of adult ground beetles (Carabidae) on pollen beetle (*Meligethes aeneus*) larvae by using Mitscherlich vessels filled with soil sterilized by hot steam. A prepared winter oilseed rape terminal shoot (BBCH growth-stage code 69) with 10 pollen beetle larvae (or a summer oilseed rape shoot with 50 larvae) was pinned upright in the centre of each vessel. One to two male predators (kept for three days without food before release) were then placed in each vessel. During the experiment, sliced meat was offered as alternative food every five days, and the soil was sprayed with 200 ml of water every two days. The vessels were kept outside in the field and covered by emergence traps. In these experiments, six species were tested: *Amara similata*, *Anchomenus (= Platynus) dorsalis*, *Bembidion tetracolum*, *Carabus auratus*, *C. granulatus* and *Pterostichus melanarius*.

Field assessment methods

Field experiments on the assessment of predatory capacity always suffer from a range of uncontrollable factors, such as weather conditions, soil type, population density of naturally occurring predators, parasitoids, plant diseases and insect diseases. However, only field-based studies with naturally occurring populations of BCAs will provide truly reliable data. In laboratory-based experiments, predators are more or less forced to feed on what is on offer and, in many instances, this leads to unrealistically high and over-estimated consumption rates. Under field conditions, predators usually have a wide choice of food items (animal prey as well as parts of plants), so that their effect on pest populations can be considerably lower. In reality, matters are even more complex because predators may, for example, develop a temporary preference for particular pest insects when such prey is available in abundance.

Basically, there are two methodological approaches for assessing the predatory capacity of epigaeic predators in the field:

- the use of enclosures, with the release of a defined number of predators of one or more selected species (T. Palosz, *pers. comm.*);
- the use of enclosures (or exclosures) that lead to different population densities of naturally occurring epigaeic predators within the available species community (Nilsson & Andreasson, 1987; Goltermann, 1994; Büchs & Nuss, 2000).

Practical examples of field assessment methods

Following an experimental design of Gravesen & Toft (1987) in winter barley, Goltermann (1994) initiated field experiments to assess the influence of epigaeic predators on pollen beetle, using three different treatment regimes:

1. Non-exclusion of epigaeic predators – here, following the 'dropping' of fully fed larvae from plants (= the 'dropping' period, i.e. the period when larvae vacate plants in order to pupate in the soil), emergence traps were installed immediately before the emergence of the new generation of adult pollen beetles. The flight period of adult pollen beetles and the 'dropping' period of the larvae were monitored by water traps placed on the soil surface.
2. Exclusion of epigaeic predators when the 'dropping' period of pollen beetle larvae was at an end – here, up to four emergence traps were installed immediately the 'dropping' period had ended; in a second year (when the experiment was repeated) the area was also then covered by gauze.
3. Total exclusion of epigaeic predators during the period between larval 'drop' and the emergence of the new generation of adults – here, plastic frames enclosing an area of 0.5 or 1 m^2 (within which were set one or two pitfall traps) were dug into the soil at the beginning of April. Following the peak larval 'dropping' period, one or two emergence traps were also installed within the enclosed area. When the experiment was repeated in the following year the plots were also sprayed at intervals with an insecticide (dichlorvos), from the beginning of flowering onwards.

Nilsson & Andreasson (1987) used a simpler experimental design in winter oilseed rape and spring-sown turnip rape, installing frames (each enclosing an area of 0.11 m^2) sunk completely into the soil ('low' frame) so that epigaeic predators had free access, or as 'high frames', sunk to a depth of only 5 cm, thereby leaving 10 cm above ground to exclude such predators.

A slightly different experimental design was used by Büchi (2002) to assess the effects of predators on pollen beetles where, unlike Goltermann (1994), all sample equipment was installed at the beginning of April. In the treatment allowing full access of predators, Büchi dug a basket with an open-mesh structure into the soil, to a depth of 10 cm. When emergence of adults of the new generation of pollen beetle was expected, the baskets were covered by gauze and the emerged pollen beetles monitored by using yellow sticky traps. In the exclusion treatment, cylinders (each *c.* 30 cm in diameter) were dug into the soil to a depth of 10 cm and then filled with field soil that had been previously desiccated in the laboratory to ensure freedom from predators. Unlike Goltemann's (1994) method, the cylinders were covered by gauze with a wider mesh size (1.2×1.8 mm) to allow entry of 'dropping' pollen beetle larvae but to prevent access by predators. When the 'dropping' period was finished, this cover was replaced by one with a finer mesh, to prevent adult pollen beetles (of the new generation) from escaping. In both treatments, emerging pollen beetles were monitored by yellow sticky traps placed within the basket (Büchi) or cylinders (Goltemann). 'Dropping' pollen beetle larvae were monitored by funnel traps.

Büchs & Nuss (2000) extended the above-mentioned basic design of field experiments for the assessment of predatory capacity by using barrier systems of three types, to create different and clearly separated density levels of epigaeic predators. These densities were based on observations of naturally occurring predator populations in oilseed rape fields managed under current farming practice but without the use of insecticides.

To obtain **total exclusion** of predators, a soil core (to a depth of 20 cm) was steam-sterilized at 110°C before metal frames (each 1 m^2) were installed; five pitfall traps were set within each frame and oilseed rape was sown in rows by hand. To allow the infestation of buds, flowers and pods by pest insects, but to prevent access by predators, gauze strips were placed between the plant rows, until the petals (and along with them pollen beetle larvae) dropped to the ground (BBCH growth-stage code 68).

A treatment called **reduced exclusion** was similar to treatment 1 of Goltermann (1994). This allowed mainly the recording of predator capacity of those epigaeic predators emerging within the enclosed area. However, in addition, predators adept at climbing (e.g. some spiders) or flying (e.g. anthocorid bugs, lacewings and ladybirds) were, potentially, able to enter the area.

An artificial **enhancement** of predator abundance was achieved in the field by using metal frame barriers, as described above, but with the addition of outer ramps to allow mainly epigaeic predators to enter the enclosures but not to leave them.

When the new generation of pollen beetles was expected to emerge, all barriers (metal frames) were replaced by emergence traps covering an identical area. Earlier, the 'dropping' of pest insect larvae had been monitored with funnel traps, as mentioned above.

Another variation of experimental design (with a spatio-temporal approach, based on naturally occurring predator populations in the field) was introduced by I. H. Williams & D. J. Warner (*pers. comm.*) in the UK, by D. Felsmann & W. Büchs (*unpublished*) in central Europe and by Gabrys *et al.* (1999) in continental Poland. Here, a grid of sample points was spread over a distinct field area. At each point, various assessment methods and observations were employed, as follows: (a) a pitfall trap; (b) a single barrier system, with integrated pitfall traps installed at the beginning of the vegetation period to exclude predators; (c) two emergence traps installed after the peak of larval 'dropping' (one of which replaces the barrier system); and (d) a funnel trap to detect larval 'dropping'. In addition, plant samples and beating-tray samples were taken regularly, to measure pest infestations on the crop. Subsequently, data from each point were calculated [using the so-called SADIE analysis (Spatial Analysis by Distance Indices) (Perry, 1995)], to verify whether differences between neighbouring sampling points were significant and whether, for example, any relationships between the 'activity-density' (i.e. the measured amount of surface activity) of predators and emergence of the new generation of pest insects could be established.

Inevitably, there are some general problems associated with assessing predator capacity in field experiments, especially as the conclusions based on the observed effects on pest insects are to some extent circumstantial, as indicated below.

- In the experimental design of Büchs & Nuss (2000), larger numbers of predators, combined with a reduced emergence rate of the new generation of pest insects, was judged to be due to the effects of the predators. Büchi (2002), however, had no evidence for the presence or absence of predators. Further, with regards the results of Goltermann (1994) and the SADIE experiments, predator abundance was recorded only outside the enclosed plots, supplemented by activity data from the surrounding field. Such data are of different quality and, therefore, difficult to compare.
- Effects of different density levels, as measured by Büchs & Nuss (2000), have to be assessed indirectly, because it is not possible to measure the influence of predators if they are constantly removed by pitfall traps, etc. On the other hand, it is necessary to at least obtain an indication of the numbers present in the different treatment regimes. Therefore, with regards the 'reduced exclusion' and 'enhancement' regimes, two parallel treatments had to be installed: one with pitfall traps (to gain an impression of the density level of predators in each treatment), and one without pitfall traps (where the number of predators present can operate as natural enemies). The effects on pest insects are calculated on the basis of the results from the treatment without pitfall traps.
- In the SADIE experiment, spatio-temporal differences and correlations are based on the number of insects removed from the system by pitfall traps (predators) and collected in funnel traps (pest insect larvae).
- Although most of the experiments that deal with predatory capacity in the field focus on ground beetles (e.g. Goltermann, 1994; Büchs & Nuss, 2000), effects on the emergence success of pest insects cannot be related clearly to one particular predator taxon. The observed effects are also the result of predatory activity by other taxa, such as nabid bugs (Nabidae), rove beetles (Staphylinidae) and spiders, but especially beetle larvae.
- As shown by Büchs & Nuss (2000), total exclusion of predators is not possible. Therefore, it can be assumed that, to some extent, predation also occurs in treatments that aim to exclude (epigaeic) predators.
- Parasitization, and also predation of (flying) carnivorous insects with free access to the treatment areas, cannot be avoided – nor can other biotic and abiotic mortality factors such as bacteria, fungi, drought or wetness and frost or excessive heat. These influences cause the so-called 'basic mortality', which has to be judged as a constant factor in each treatment.

Thus, the huge number of influences and imponderables, which really cannot be controlled (especially under field conditions), may lead to the conviction that viable field experimentation is an impossibility. This may explain why there is little knowledge in this area and why so few field-based investigations have been done. However, results from laboratory-based studies are also far from clear cut and their interpretation needs to be done with great care.

Assessment of predation rates

There are two main approaches for the direct assessment of the quality and quantity of predation rates, prey selection and food intake: microscopical gut dissection and biochemical analysis.

Microscopical gut dissection

This method is commonly used (see, for example, Hengeveld, 1980) but suffers from difficulties associated with identifying clearly the remnants of the consumed prey. It appears to be suitable mainly for stenophagous predators (see Chapter 12), but it is usually difficult to achieve clear-cut results in the case of polyphagous species, except when they have developed a temporary preference for specific prey items. It is assumed that non-sclerotized parts of prey are unlikely to be recorded, as they have probably been digested. Therefore, quantification of the gut contents is possible only very approximately. Nevertheless, the advantage of gut dissections (compared with biochemical methods) is that they can be done relatively easily and with little equipment, and the method is suitable for 'older' (i.e. not fresh) insect material stored in alcohol or any other preservative.

Biochemical methods

There are many relevant publications dealing with the well-known ELISA technique but very few concerning the use of isoenzymes. The general problem associated with these 'advanced' techniques concerns the availability of antibodies (for ELISA) or primer (for PCR, RAPDs isoenzyme techniques). Also, their elaboration is rather laborious, and they are almost impractical for polyphagous predators (especially for those with a wide range of animal, and also vegetable, food). With regards oilseed rape pests, antibodies exist only for aphids and slugs (see, for example, Symondson & Liddell, 1993).

Biochemical methods can be used only if 'fresh' insect material is available. Nevertheless, the advantages of such approaches are clear:

- they allow identity of the consumed prey to be indisputable;
- results can be produced relatively quickly (so long as appropriate primers and/or antibodies are available);
- the methods are suitable also for predators with extra-intestinal digestion, as in the case of spiders, rove beetles and some ground beetles (e.g. *Carabus* spp.).

Currently, most results and experiences using biochemical techniques are derived from investigations in horticultural crops or cereals, and not in oilseed rape.

Key Predators in Oilseed Rape Fields

Unlike parasitoids, polyphagous predators and their larvae are present in oilseed rape fields in considerable numbers for most of the year (except, in more continental regions, for one or two months' absence in the winter). Therefore, they are present not only during the main period when fully fed larvae of, for example, brassica pod midge (*Dasineura brassicae*), cabbage seed weevil (*Ceutorhynchus assimilis*) and pollen beetle drop to the soil (from flowers or pods) in order to pupate (herein termed the 'dropping' period) but also when the crop is infested by (for example) cabbage stem flea beetle (*Psylliodes chrysocephala*), peach/potato aphid (*Myzus persicae*) and turnip sawfly (*Athalia rosae*) in the autumn or by (for example) cabbage stem weevil (*Ceutorhynchus pallidactylus*) and rape stem weevil (*Ceutorhynchus napi*) in the early spring.

The most important predators in oilseed rape fields are ground beetles, rove beetles, various flies and spiders (see Chapter 12 for details). Predation rates, prey selection and biocontrol effects of these, and a few other groups of predators, are discussed below.

Ground beetles

Ground beetles are often abundant in oilseed rape fields. Basedow (1973), for example, recorded from 20 to 80 individuals/m^2. According to Scherney (1958), the daily intake of prey by ground beetles as epigaeic, polyphagous predators can greatly exceed their own body weight.

The oilseed rape pests or pest stages most likely to be affected by ground beetles are aphids and the various pre-adult stages of pests associated with the soil environment. The latter include:

- eggs
  - ➢ cabbage stem flea beetle
  - ➢ cabbage flea beetles (*Phyllotreta* spp.)
  - ➢ cabbage root fly (*Delia radicum*)
- fully grown larvae
  - ➢ brassica pod midge
  - ➢ cabbage stem flea beetle
  - ➢ ceutorhynchid weevils
  - ➢ pollen beetle
- pupae
  - ➢ brassica pod midge
  - ➢ cabbage stem flea beetle
  - ➢ ceutorhynchid weevils
  - ➢ pollen beetle

Quantified data to indicate the impact of such predation, however, are wanting. Although much predation is likely to occur on the soil surface [e.g. by species of the genera *Agonum*, *Bembidion*, *Nebria*, *Notiophilus* and *Pterostichus* (in part)], some species (namely, *Clivina fossor*, *Trechus quadristriatus* and, to some extent, *Calathus* spp.) are adept at seeking prey within the soil surface layers.

Whereas some publications give evidence of ground beetles as effective antagonists of insect pests in cereals (Basedow, 1973; Sunderland & Vickerman, 1980; Griffiths, 1982; Chambers *et al.*, 1983; Scheller, 1984; Poehling *et al.*, 1985; Chiverton, 1988; Ekbom *et al.*, 1992), there are very few, mostly old, publications of predator–prey relationships in oilseed rape. However, these assessments do not correlate with current agricultural practice of oilseed rape growing (e.g. cultivars, tillage and plant-protection methods are now different); the accounts are also, to some extent, contradictory.

According to Scherney (1959), the ground beetle *Clivina fossor* caused 60–65% mortality in pollen beetle larvae; similar results are stated by Bonnemaison (1957) for cabbage seed weevil larvae. Basedow (1973) showed a difference between fully fed pollen beetle and brassica pod midge larvae that dropped from the plant stands to the soil and those that were found in the upper soil layer of from 39% to 65%, respectively; he concluded that predators might be, at least in part, responsible for mortality of the larvae. In cabbage crops, however, Kirchner (1961) did not observe any losses amongst fully fed cabbage seed weevil larvae during the period they were actively burrowing into the soil. Finally, Goltermann (1994) attributed 20% (the average value in a three-year experiment) of the total mortality of pollen beetle larvae to a direct effect of epigaeic predators; a basic mortality (due to other factors, as explained above) of about 60% was also registered.

In contrast to the above-mentioned results, Hokkanen *et al.* (1988) (in Finland) could attribute only a 3% reduction of pollen beetle larvae and pupae to ground beetles and other epigaeic predators. However, Varis *et al.* (1984) attribute the negligible influence of such predators on pollen beetle population in Finland to the generally low level of ground beetle activity during the three to four weeks' mid-summer period, when pollen beetle larvae pupate and new adults emerge. A major reason for the poor activity of predacious ground beetles could also be that in Finland almost exclusively summer (i.e. spring-sown) oilseed rape is grown, so that tillage is carried out in late spring rather than in late summer or early autumn. Büchs *et al.* (1999) showed that tillage activities in spring have considerable negative effects on populations of ground beetles and other epigaeic predators (see Chapter 15).

Predatory efficacy is greatly influenced by the time period over which ground beetles can prey upon pollen beetle larvae and, hence, pest availability. Thus, according to Goltermann (1994), if ground beetles have unlimited access to pollen beetle larvae during the whole 'dropping' period then mortality of the larvae can be considerable (on average, 78.5%); however, if access is possible only after this period (i.e. when larvae have already burrowed into the soil to pupate), mortality resulting from epigaeic predators is far less (on average, 25.8%). However, feeding on pollen beetle pupae within the upper soil layers leads to an additional increase in pollen beetle mortality of, on average, 18.8%.

Predatory efficacy is also influenced by the phenological patterns of crop, pests (including 'dropping' of their larvae) and predators. For example, in years when development of the oilseed rape crop (notably, the flowering stage) and, hence, its infestation by pests is delayed, but the phenological occurrence of predators in fields is normal, the mortality effects by predators on pest larvae will be much greater than in years when predator abundance and activity-densities are similar but crop development and pest invasion are advanced. Three factors, therefore, will enhance the positive effects of polyphagous epigaeic predators (i.e. their 'success' as naturally occurring BCAs) in oilseed rape crops:

- delayed flowering of the crop;
- high density (i.e. availability) of pest larvae;
- longer duration of the larval 'dropping' period.

In cage-release (confined) experiments with ground beetles (Goltermann, 1994; see above), somewhat different effects on the mortality of pollen beetles have been demonstrate for several species. Thus, whereas *Amara similata* induced no increased mortality, species such as *Anchomenus dorsalis*, *Carabus auratus*, *Carabus granulatus*, *Clivina fossor* and *Pterostichus melanarius* led to an increase in pollen beetle mortality of about 80% or more (compared with the control treatment without predators). However, when interpreting these results it has to be considered that densities of 32 individuals/m^2, which were used by Goltermann, are to some extent unrealistic, especially for *Carabus* spp.

A comparison of adult emergence rates for pollen beetle if predators are either present during the whole period of larval 'dropping' or present only after this period shows that most beetles feed mainly on the larvae immediately after 'dropping', i.e. whilst the prey is still lying on the soil surface. This is especially true for *Carabus* spp. but also for *Pterostichus melanarius*, the new generation of which occurs in oilseed rape fields in most regions only after the period of larval 'dropping' (see Büchs & Nuss, 2000); therefore, under natural field conditions, the predatory efficacy of the latter species is probably very low. Unlike other ground beetles to be found in oilseed rape fields, *Clivina fossor* feeds almost exclusively on pollen beetle larvae within the upper layers of the soil. Goltermann (1994) showed that 'dropping' of pollen beetle larvae is not restricted to any particular time of day, so they are available as prey items to both diurnal and nocturnal surface-feeding predators.

Goltermann (1994) assumed that the feeding rates of ground beetles increased with the body size of the predator, as long as prey larvae (e.g. fully fed pollen beetle larvae dropping from plants in order to pupate in the soil) are freely accessible; however, this relation becomes inverted once the larvae have burrowed into the soil, because smaller predatory beetles (e.g. Bembidiinae) tend to crawl into soil crevices and thus will have advantages in finding such prey.

Additionally, it has to be considered that the cultivation technique in these experiments was still slightly influenced by the then current farming practice of the former German Democratic Republic. This may explain why Nilsson & Andreasson (1987) obtained contradictory results (no effects of polyphagous epigaeic predators on pollen beetle larvae) in Sweden. Nevertheless, in winter rape, these authors did observe a 25% higher rate of emergence for adult pollen beetles within 'fenced' plots (from which predators were excluded) compared with that from the surrounding soil.

Studies of predation by ground beetles and other polyphagous epigaeic predators were carried out by Büchs & Nuss (2000), using different barrier systems to create near to zero, natural and artificially increased population densities. Successful adult emergence of pest insects was generally reduced according to the densities of the predators, and reached values of up to 51% for cabbage seed weevil, 56% for pollen beetle, 58% for brassica pod midge and 82% for cabbage stem weevil. Polyphagous epigaeic predators caused a significant increase in the total mortality of up to 4% in pollen beetle populations, up to 10% in the case of brassica pod midge populations and up to 52% in the case of cabbage seed weevil populations. In some cases the effects of predators on the mortality of pest larvae seem to be rather low and may be overlooked. However, as shown by Büchs & Nuss (2000), 'natural' mortality resulting from non-predatory factors (e.g. fungal infections, dryness of the soil, soil humidity and parasitoids) also have to be considered. In the case of brassica pod midge these non-predatory factors may reach 80–90%; therefore, a slight additional increase in mortality (such as that introduced by predators) may tip the balance and result in breakdown of the pest population or at least prevention of a pest outbreak in the following season.

As demonstrated by Büchs & Nuss (2000), most fully grown pollen beetle drop to the ground (mostly within the falling petals of flowers), when the ground beetles *Amara similata* and *Poecilus cupreus* are especially active in the field (with maximum predator densities at that time of *c.* 50 and 20 individuals/m^2). In fact, the appearance of young adults of the new generation of *Amara similata* coincided with the main period of 'dropping' of fully grown larvae of brassica pod midge and cabbage seed weevil. Activity of *Poecilus cupreus* adults, however, is restricted to the time when fully grown pollen beetle larvae drop to the ground. Therefore, this ground beetle species will have an influence on the mortality of this pest species only. In north-eastern Germany, Goltermann (1994) found that *Anchomenus dorsalis* and *Bembidion tetracolum* were the most active ground beetles during the 'dropping' period of pollen beetle larvae. In the case of *Pterostichus melanarius* adults, noticeable emergence rates have been recorded only at the end of June (Goltermann, 1994) or from mid- to late July (Büchs & Nuss, 2000). At that time, 'dropping' of insect pest larvae in oilseed rape was at an end and the emergence of adults of the new generation of pest insects was in progress. Therefore, with regards larvae of these pests, it may be assumed that this species is of lesser importance as a BCA.

Compared with *Poecilus cupreus* and *Pterostichus melanarius*, the species *Amara similata* in general shows the best phenological coincidence with the 'dropping' period of oilseed rape insect pest larvae from both flowers and pods. This could indicate a key role for this species as a natural enemy of pests such as pollen beetle, brassica pod midge and cabbage seed weevil. However, generally, species of *Amara* are said to be phytophagous; thus, Luka *et al.* (1998) observed *Amara ovata* and *A. similata* feeding on oilseed rape flowers, pods and seeds. Goltermann (1994), however, found *A. similata* mainly in areas with a higher weed density, where they fed preferentially on the seeds of shepherd's purse (*Capsella bursa-pastoris*), and he could show no regulatory effects on oilseed rape pest populations. Further, Hartke *et al.* (1998) showed the feeding preference of *Pseudoophonus (= Harpalus) rufipes* larvae for weed seeds. Among seven plant species tested [namely, wild mustard (*Brassica kaber*), birds rape mustard (*Brassica rapa*), dandelion (*Taraxacum officinale*), common hemp-nettle (*Galeopsis tetrahit*), fat hen (*Chenopodium album*), pigweed (*Amaranthus retroflexus*) and perennial ryegrass (*Lolium perenne*)], the weeds fat hen and pigweed were mostly attacked. Also, although the presence of *P. rufipes* larvae reduced the percentage emergence of pigweed seeds, no effect on emergence could be recorded in the case of *Brassica kaber*. Optionally, *Amara* spp., *Harpalus* spp. and *Pseudoophonus* spp. can also be carnivorous, feeding (for example) on the eggs, larvae and pupae of Diptera (Lindroth, 1945; Dobson 1961).

The very instructive results of Büchi (2002) allow the importance of predators as natural enemies of pollen beetle to be estimated. Over an average of three years, about 20% of the total pollen beetle mortality was caused by predators. This is nearly three times lower than mortality due to unspecific factors (as mentioned above), but approximately ten times higher than mortality resulting from parasitization. These relationships explain the very low (but significant) increases in total mortality (calculated as a quotient of the number of 'dropped' larvae and the emergence rate of adults of the new generation), even under the influence of an extremely great abundance of predators recorded by Nuss & Büchs (2000). Also, Nilsson (1989) (who investigated several mortality factors between the egg and the adult stage of pollen beetles in southern Sweden) showed that from 5.3% to 7.7% of larvae were killed whilst on the soil surface.

The importance of predation depends on various factors, such as soil type and soil structure. For example, Leuchs (1956) found a positive correlation between the clay content of soil and the mortality of pollen beetle. Also, Nilsson & Andreasson (1987) conclude from their results that the importance of predation is increased in soils that do not crack; this is because the pest larvae in such conditions spend more time on or close to the soil surface, where they are exposed to the predators.

The following examples of the most important pests (fully fed larvae) 'dropping' to the ground in oilseed rape flower stands provide an example of the potential food source for epigaeic predators:

• brassica pod midge	35,000 individuals/m^2	(Nuss & Büchs, 2000)
	21,000	(Basedow, 1973)
	7,000	(Büchs, 1993)
	1,800	(Axelsen, 1992)
• cabbage seed weevil	820 individuals/m^2	(Nuss & Büchs, 2000)
	165	(Büchs, 1993)
• pollen beetle	2,000 indivduals/m^2	(Basedow, 1973)
	2,000	(Nuss & Büchs, 2000)

Pest–predator spatio-temporal relationships

One of the first attempts to examine the pest–predator spatio-temporal relationships in winter oilseed rape was that of Warner *et al.* (2000) for the brassica pod midge and ground beetles in the UK. Over two years, brassica pod midge larvae were collected from a grid of sampling points across oilseed rape stands and related to the activity-density of ground beetles during the peak of larval 'dropping'. In both years, *Amara similata*, *Anchomenus dorsalis*, *Nebria brevicollis* and *Pseudoophonus rufipes* were the most active and abundant species during this critical period. Their results established a positive, and significant, correlation between the occurrence and spatial distribution of the midge larvae with *A. dorsalis* and *P. rufipes*, but not with *Amara similata* or *N. brevicollis*. Although the general phenology of *N. brevicollis* did not dovetail completely with that of brassica pod midge, it did show (in one year) a spatial correlation, being strongly restricted to areas where the midge larvae occurred in greatest numbers; the authors concluded, therefore, that there could be a well-directed movement of this predator into areas where prey items are abundant. This might reflect a temporal preference to this prey item, triggered by a 'threshold' with regards the numbers of 'dropping' midge larvae. In contrast, the more or less phytophagous (weed-seed-feeding) species *Amara similata* showed no positive spatial correlation to the 'dropping' of brassica pod midge larvae, although the results were not as clear as might have been expected. *Anchomenus dorsalis* showed a positive spatial correlation with the spatial distribution of the 'dropped' midge larvae, but only in the second year of the study.

In the study of Gabrys *et al.* (1999), *Poecilus cupreus* and (the mostly phytophagous) *Amara similata* were the most abundant ground beetle species. *Poecilus cupreus*, which was present over the whole study period (8 April–15 June), occurred in aggregations that varied over time.

The interpretation of the spatio-temporal distribution of pests and predators is rather difficult, owing to some methodological problems and to limiting factors that cannot be avoided. For example:

- a high activity-density of ground beetles has been said to be caused by a 'lack of prey' [e.g. because of an insecticide treatment; see Basedow (1987)] but not by an abundance of prey items;
- from studies of single-plant infestations, it is evident that numbers of brassica pod midge larvae dropping from oilseed rape flower stands might change over very short distances – the activity-density of ground beetles, however, has to be measured over greater distances, so it is unclear whether the two values may be truly related to one another;
- a positive correlation between ground beetle activity-density and brassica pod midge larvae gives no direct evidence of a host–prey relationship;

- when using pitfall traps, ground beetles (which might prey upon the pest larvae) are removed from the system and directly from the area where they should be acting as predators.

Prey consumption

From laboratory feeding experiments, Basedow (1973) calculated that, for winter wheat fields, ground beetles (*Anchomenus dorsalis* and *Pterostichus melanarius*) and spiders (Linyphiidae and Lycosidae) could consume nearly 686,000 yellow wheat blossom midge (*Contarinia tritici*) larvae per hectare daily, and recorded a range of consumption rates for different species of the above-mentioned predatory taxa. However, he discovered that prey consumption rates (especially of ground beetles) were considerably lower if rotting carcasses were available. Owing to that, and since the spectrum of prey is much greater in the field, in practice the feeding rates mentioned above would be considerably lower. Furthermore, it is questionable whether feeding rates on *Contarinia* larvae are representative for dipterous larval prey in general and whether they can be applied to *Dasineura* larvae in oilseed rape.

Rove beetles

Most field-based investigations on the predatory activity of rove beetles (Staphylinidae) refer either to cereal crops or to grasses, with particular reference to the impact on aphids (e.g. Bryan & Wratten, 1984; Dennis & Wratten, 1991). However, the overall diets of rove beetles have also been reviewed (see Good & Giller, 1991; Andersen, 1992; Mamedov, 1995; Krooß, 1996). These publications indicate that rove beetles, especially members of the subfamily Staphylininae, will feed on a wide range of hosts.

As regards oilseed rape pests, apart from aphids, the ground fauna (e.g. fully fed larvae and pupae of brassica pod midge, pollen beetles and various other species – such as ceutorhynchid weevils – that pupate in the soil, and eggs (e.g. of *Delia*) if laid in the soil) is most likely to be preyed upon. Details of the true impact of rove beetles on such pests, however, is wanting. Other gaps in knowledge associated with rove beetles indicate (in common with ground beetles) a need for future work on activity-density, abundance, phenology and prey selection by the larvae.

Other groups of predatory beetles

Ladybirds

Ladybirds (Coccinellidae) are well-known predators of aphids and they are often regarded as important natural enemies of aphid pests of field crops. In one of the first records on host–prey relationships in oilseed rape, Friedrichs (1920) recorded ladybirds (as well as *Malachius* spp. – Melyridae) as predators of pollen beetle larvae within the flower stands. With regards their activity as predators of aphids in oilseed rape crops, their impact is probably marginal because cabbage aphid (*Brevicoryne brassicae*) (the main aphid species on oilseed rape, see Chapter 2) is considered 'inadequate' prey (Blackman, 1965, 1967). Field studies have shown that the coccinellid *Harmonia axyridis* avoids cabbage aphid (George, 1957), although *Coccinella septempunctata* will eat them occasionally. *Coccinella septempunctata* may also act as a predator of pollen beetle larvae (Kurppa, 1989). Food preferences of the British species are summarized by Majerus (1994).

Peach/potato aphid (*Myzus persicae*), a major virus vector in oilseed rape, especially during the autumn, is of rising importance as a pest in oilseed rape crops (see Chapter 2). However, although this pest is frequently eaten by ladybirds (e.g. by *Coccinella septempunctata* and various other species), such predation on oilseed rape in autumn, when the virus infection is set, is probably very low. This underlines the importance of non-specific, polyphagous predators (e.g. ground beetles, rove beetles and spiders) which are present and active in the fields throughout much of the year, including the autumn.

Soldier beetles

The impact of soldier beetles (Cantharidae) as predators of oilseed rape pests has not yet been assessed. There are hints from winter wheat that the larvae feed on aphids, and adults of the generally common species *Rhagonycha fulva* have been found feeding on rosy apple aphids (*Dysaphis plantaginea*) on apple trees (D. V. Alford, *pers. comm.*). In laboratory experiments, Langenstück *et al.* (1998) observed a significant reduction of bird-cherry aphid (*Rhopalosiphum padi*) but not of grain aphid (*Sitobion avenae*), a species that is more able to resist cold temperatures. According to Janßen (1963), cantharid larvae are also natural enemies of (young) field slugs (*Deroceras reticulatum*), one of the recently most serious pest problems in oilseed rape crops. However, feeding of soldier beetle larvae on slugs (e.g. *Deroceras* spp.) could not be confirmed by Langenstück *et al.* (1998) in their laboratory experiments. The abundance of soldier beetle larvae measured in oilseed rape fields by using emergence traps and the flooding method (see Chapter 14) was 5.2 and 22.4 individuals/m^2, respectively (W. Büchs, *unpublished*). Unlike most other predators, soldier beetle larvae are active in winter, especially at temperatures ranging from 0 to 10°C. In oilseed rape crops, they are of particular importance in late autumn and winter as predators of peach/potato aphids.

Miscellaneous beetles

Small numbers of predatory carrion beetles (Silphidae) and histerid beetles (Histeridae) can occur in oilseed rape fields, especially the latter (Büchs *et al.*, 1999), and the relatively high emergence rates for histerids in oilseed rape (2.9–9.3 individuals/m^2) suggest that the crop provides suitable conditions for these insects to complete their lifecycle. Details of the food consumed by the larvae in oilseed rape fields, however, are wanting.

Predatory flies

Members of various families of flies occur as predators in arable crops (see Chapter 12), but there are few reports of their impact on oilseed rape pests. Some groups, however, would appear at least to be of potential value as BCAs. These are discussed briefly below.

Dance flies

Owing to their abundance in oilseed rape fields (up to 59 individuals/m^2) and their feeding rate (3 midges/day over a period of three to four weeks), hybotid dance flies

(Hybotidae) have considerable antagonistic potential as predators of brassica pod midge (Prescher & Büchs, 1998). Hybotids are certainly efficient predators of midges (Cecidomyiidae) in cereal crops (Stark & Wetzel 1987; Stark, 1988, 1994; Klumpp, 1990). Their abundance in oilseed rape was also demonstrated by Brunel *et al.* (1989) who caught 3,100 hybotids (belonging to 10 species) in 26 yellow water traps situated in flowering stands from 22 June to 3 July. They also recorded 97 empids (Empididae) (belonging to 12 species), with members of the genus *Hilara* being most numerous.

Hover flies

The overall effectiveness of aphidophagous hover fly larvae (Syrphidae) as regulators of aphid infestations on crops is reduced because the adults usually appear in crops and lay eggs only after aphid populations have reached critical levels and are too great for predation to halt their progress. This is also true of other aphidophagous insects, such as ladybirds (see above) and lacewings (see below). In the case of hover flies, one reason is that some of the most important aphidophagous species in north-western Europe (e.g. *Episyrphus balteatus*) partly undertake long-distance migrations (down to southern France) where they overwinter and do not return until later in the following spring. Larvae of hover flies are not of relevance as predators of aphids in oilseed rape crops in the autumn.

Long-legged flies

Adult long-legged flies (Dolichopodidae) (e.g. *Dolichopus* spp.) are potentially useful predators of brassica pod midge (S. Prescher, *pers. comm.*), and populations of adults emerging from the soil in oilseed rape fields have been recorded in northern Germany at levels of 11–17 adults/m^2 (Prescher & Büchs, 1999). Brunel *et al.* (1989) caught 2,380 individuals (belonging to 35 species) in the water traps described under dance flies (above).

Midges

Although predatory midges (Cecidomyiidae) (e.g. *Aphidoletes aphidimyza*) often feed on aphids in field crops (Harris, 1973), they are not considered of importance as potential BCAs in oilseed rape. However, Kelm *et al.* (1994) reported an unexpected predatory behaviour on a typical pest midge species. They installed cages that enclosed a defined number of cabbage seed weevils, one treatment with and one without the presence of brassica pod midge. In cages with midge larvae, the mortality of cabbage seed weevil larvae in the pods was 32%, but no such mortality could be observed in cages without midges. These observations suggest that the brassica pod midge larvae may have fed on seed weevil eggs in the pods, and thus behaved as egg predators.

Muscid flies

Muscid flies (Muscidae) are of limited value as predators in oilseed rape fields (see Chapter 12) and few data are available. However, with regards *Phaonia trimaculata*, a predator of larvae of cabbage root fly (*Delia radicum*) (Shaw, 1970; Stubbs & Chandler, 1978) – a pest of rising importance in oilseed rape – adult emergence rates under oilseed rape have been recorded at 8–15 individuals/m^2 (Prescher & Büchs, 1999); larvae of *P.*

trimaculata also prey upon those of cabbage stem weevil and rape stem weevil (Fritzsche, 1955; Jourdheuil, 1960). Larvae of the false stable fly (*Muscina stabulans*) occur within the stems of plants, where they are predators of cabbage stem weevil larvae, but no quantitative data are available. Individual adults of *Coenosia tigrina*, which occur in oilseed rape as a predator of cabbage root fly, have been observed to kill up to five adults per day (Perron *et al.*, 1956).

Snipe flies

In oilseed rape fields, emergence rates of adult snipe flies (Rhagionidae) (whose feeding habits are unknown) have been recorded in northern Germany at 0.5–4.0 adults/m^2 (Prescher & Büchs, 1999). Larvae are predatory, but no other details in relation to oilseed rape are available.

Stiletto flies

Few details are available of the effects of stiletto flies (Therevidae) on oilseed rape pests, although adult emergence rates from the soil (again in oilseed rape fields in northern Germany) have been recorded at 1 adult/m^2 (Prescher & Büchs, 1999).

Lacewings

Lacewings (Neuroptera) are well-known predators of insects, including aphids (McEwan *et al.*, 2001). There is also evidence of predation by larvae of *Chrysoperla carnea* (Chrysopidae) on larvae of the noctuid moth *Mamestra configurata* in oilseed rape fields in Canada. Laboratory studies showed that a density of 10 chrysopid larvae/m^2 could reduce the noctuid population significantly up to the third larval instar (Bucher & Bracken, 1978). As mentioned in Chapter 2, however, noctuid larvae are not important pests of oilseed rape crops in Europe. Sengonça *et al.* (1995) showed in olfactometric studies on the reaction of *C. carnea* that the predator was attracted by kairomones produced by several species of aphid, including peach/potato aphid. Notably, larvae (especially third instars) also responded positively to the odours of the host plants of the prey species tested which, in the case of peach/potato aphid, was *Brassica oleracea* (i.e. oilseed rape).

Spiders

Web-forming spiders

The prey composition of spiders' webs varies considerably from season to season and from location to location, and according to the time assessment is carried out. For example, in a period of two years the number of aphids per web was three to four times greater in the second year in comparison with the first; also, an assessment in June in white mustard (*Sinapis alba*) recorded mainly adult pollen beetles as prey (Harenberg, 1997; B. Weiß & W. Büchs, *unpublished*). Further, the web density of *Theridion impressum* (Theridiidae) in oilseed rape fields varied with different levels of agronomic input, resulting in a maximum density (albeit not very high) where inputs were greatest:

- 2.8 webs/m^2 under conventional farming system, managed according to the advice of local extension services;
- 2.1 webs/m^2 under an integrated farming system (i.e. with lower inputs);
- 0.8 webs/m^2 under a reduced-input farming system (without insecticides).

The decreasing density with greater extensification (i.e. with lower inputs) is unexplained. In rotational set-aside areas with seven different seed mixes and a natural succession, similar densities of *Theridion* webs were counted (W. Büchs & B. Weiß, *unpublished*). However, web density varies greatly from year to year, depending upon weather conditions and the amount of prey available (W. Büchs & A. Harenberg, *unpublished*). Furthermore, the web density of *T. impressum* is strongly influenced by the structure of the available vegetation. Thus, as stated by Hatley & MacMahon (1980), only 'open' plants with vertical branches (e.g. white mustard) are invaded, whereas others affording a denser structure (such as grasses) are not.

It is often overlooked that money spiders (Linyphiidae), the small but most abundant group of pioneer spiders in cultivated areas, are also web builders, constructing small horizontal webs on the soil surface. As Sunderland *et al.* (1986) stated for winter wheat fields, nearly 50% of the soil surface can be covered by their horizontal webs. Comparing oilseed rape with mulch-drilling and conventional tillage (ploughing), W. Büchs (*unpublished*) observed a striking density of linyphiid webs in the mulch-drilling treatment that was significantly greater than in the conventionally ploughed system. Comparing different levels of extensification (see above), web density was measured with mesh wire screens of 49 × 49 cm (mesh width of 1 cm), following the newly developed standardized method described in Chapter 14. Using these artificial structures in oilseed rape, a significantly greater web density could be observed, comparable with those in winter wheat and in a five-year-old set-aside area with natural succession. However, there was no correlation (either positive or negative) with the intensity of cultivation, although, again, web density tended to decrease with increasing extensification.

An analysis of prey trapped in *Theridion impressum* webs was carried out by Harenberg (1997) on the basis of 40 webs collected in two years in mid-July (*c.* 7–10 days before harvesting). In both years, at that time, prey consisted mostly of aphids (50–90%, mainly cabbage aphid), Diptera (mainly *Dasineura* spp., 5–30%) and Thysanoptera (up to 10%). Dingler (1935), however, who investigated the fauna of asparagus fields, afforded *T. impressum* no important role as a natural enemy because he found that, to a large extent, the webs captured more beneficial insects than pests. Nentwig (1980) stated that web-forming spiders mainly catch small, slow-flying insects with long wings (alate aphids and adult midges are common examples). Nyffeler (1982) estimated that, in meadows, spiders eat nearly 150 kg of prey (equivalent to 38 million insects) per hectare each year. He also compared the percentage of aphids and dipterans in the food of orb-web spiders (Argiopidae and Tetragnathidae) in a range of other cultivated situations (barley, maize, oilseed rape and wheat). The comparison showed that oilseed rape had the lowest percentage of aphids but the highest percentage of dipterans (obviously, including brassica pod midges).

Non-web-forming spiders

Wolf spiders (Lycosidae) such as *Pardosa palustris* include pollen beetle larvae and brassica pod midge larvae in their diets, preying on those that have dropped to the ground prior to pupation. Crab spiders (Thomisidae) (e.g. *Misumena vatia*) and jumping spiders (Salticidae) (e.g. *Salticus scenicus*) inhabit the inflorescences of oilseed rape plants, where they prey on a range of small insects including pests such as brassica pod midge adults. However, there are no quantitative data on predation rates.

Miscellaneous predators

Representatives of various other predator groups occur in oilseed rape crops (Chapter 12), including harvestmen (Opiliones) and centipedes (Chilopoda). However, little or no information is available on their effect on pests. Tischler (1958) found that harvestmen would invade oilseed rape fields from nearby woodlands, mainly in late summer after harvest and probably temporarily. Numbers present in oilseed rape, however, are usually small (see Büchs *et al.*, 1999). With regards centipedes, Büchs *et al.* (1999) reported low activity-densities in oilseed rape fields (<1 individual/pitfall trap in any one growing season), but recorded populations were small (up to 1.3 individuals/m^2 in emergence traps). However, the selected sampling methods may not have been particularly suitable for catching centipedes.

Conclusions

Predatory arthropods are useful BCAs in oilseed rape crops, but there is a need to qualify and quantify their effects on oilseed rape pests in greater detail. Potentially, predators could form an important component of decision-support systems for oilseed rape growing. These aims might be achieved (at least in part) by:

- the development and implementation of relevant techniques for gut dissection and subsequent serological tests of predators;
- the development of appropriate methods for the precise density estimation of predator populations;
- the enhancement of particularly efficient (i.e. key) predator species;
- the further evaluation of predator groups that, until now, have been largely overlooked in oilseed rape fields, e.g. carnivorous Diptera and larvae of Staphylinidae.

Acknowledgements

This work was part-funded by the European Commission Concerted Action: *Minimizing pesticide use and environmental impact by the development and promotion of bio-control strategies for oilseed rape pests, FAIR 96–1314.*

References

Andersen, A. (1992). Predation by selected carabid and staphylinid species on the aphid *Rhopalosiphum padi* in laboratory and field experiments. *Norwegian Journal of Agricultural Science* **6**, 265–73.

Axelsen, J. (1992). The population dynamics and mortalities of the pod gall midge (*Dasyneura brassicae* Winn.) (Dipt., Cecidomyiidae) in winter rape and spring rape (*Brassica napus* L.) in Denmark. *Journal of Applied Entomology* **114**, 463–71.

Basedow, T. (1973). Der Einfluß epigäischer Raubarthropoden auf die Abundanz phytophager Insekten in der Agrarlandschaft. *Pedobiologia* **13**, 410–22.

Basedow, T. (1987). Der Einfluß gesteigerter Bewirtschaftungsintensität im Getreidebau auf die Laufkäfer (Coleoptera, Carabidae). *Mitteilungen der Biologischen Bundesanstalt für Land- und Forstwirtschaft Berlin-Dahlem* **235**, 1–123.

Blackman, R. L. (1965). Studies on specificity in Coccinellidae. *Annals of Applied Biology* **56**, 336–8.

Blackman, R. L. (1967). The effects of different aphid foods on *Adalia bipunctata* L. and *Cocinella 7-punctata* L. *Annals of Applied Biology* **59**, 207–19.

Bonnemaison, L. (1957). Le charançon des siliques (*Ceuthorrhynchus assimilis* Payk.). *Annales des Épiphyties* **4**, 387–543.

Brunel, E., Grootaert, P. & Medquida, J. (1989). Entomofaune associée à la floraison du colza (*Brassica napus* L.): note preliminaire sur les Dolichopoididae et les Empididae (Insectes: Diptera). *Mededelingen van de Faculteit Landbouwwetenchappen Rijksuniversiteit Gent* **54**, 727–37.

Bryan, K. M. & Wratten, S. D. (1984). The responses of polyphagous predators to prey spatial heterogeneity: aggregation by carabid and staphylinid beetles to their cereal aphid prey. *Ecological Entomology* **9**, 251–9.

Bucher, G. E. & Bracken, G. K. (1978). Chrysopid predation on the bertha armyworm. *Manitoba Entomologist* **10**, 26–30.

Büchi, R. (2002). Mortality of pollen beetle (*Meligethes* spp.) larvae due to parasitoids and predators in rape fields and the effect of conservation strips. *Agriculture, Ecosystems and Environment* **90**, 255–63.

Büchs, W. (1993). Investigations on the occurrence of pest insects in oil seed rape as a basis for the development of action thresholds, concepts for prognosis and strategies for the reduction of the input of insecticides. *IOBC/wprs Bulletin, Integrated Control in Oilseed Crops* **16** (9), 216–34.

Büchs, W. & Nuss, H. (2000). First steps to assess the importance of epigaeic active polyphagous predators as natural enemies of oilseed rape insect pests with soil pupating larvae. *IOBC/wprs Bulletin, Integrated Control in Oilseed Crops* **23** (6), 151–63.

Büchs, W., Harenberg, A., Prescher, S., Weber, G. & Hattwig, F. (1999). Entwicklung von Evertebratenzönosen bei verschiedenen Formen der Flächenstillegung und Extensivierung. *Mitteilungen aus der Biologischen Bundesanstalt für Land- und Forstwirtschaft Berlin-Dahlem* **368**, 9–38.

Chambers, R. J., Sunderland, K. D., Wyatt, I. J. & Vickerman, G. P. (1983). The effects of predator exclusion and caging on cereal aphids in winter wheat. *Journal of Applied Ecology* **20**, 209–44.

Chiverton, P. A. (1988). Searching behaviour and cereal aphid consumption by *Bembidion lampros* and *Pterostichus cupreus*, in relation to temperature and prey density. *Entomologia Experimentalis et Applicata* **47**, 173–82.

Dennis, P. & Wratten, S. D. (1991). Field manipulation of populations of individual staphylinid species in cereals and their impact on aphid populations. *Ecological Entomology* **16**, 17–24.

Dingler, M. (1935). Die Tierwelt des Spargelfeldes. *Zeitschrift für Angewandte Entomologie* **21**, 291–327.

Dobson, R. (1961). Observations on natural mortality, parasites and predators of wheat bulb fly, *Leptohylemyia coarctata* (Fall.). *Bulletin of Entomological Research* **52**, 281–91.

Ekbom, B. S., Wiktelius, S. & Chiverton, P. A. (1992). Can polyphagous predators control the bird cherry-oat aphid (*Rhopalosiphum padi*) in spring cereals? *Entomologia Experimentalis et Applicata* **65**, 215–23.

Friedrichs, K. (1920). Untersuchungen über den Rapsglanzkäfer in Mecklenburg. *Zeitschrift für Angewandte Entomologie* **7**, 1–36.

Fritzsche, R. (1955). *Phaonia trimaculata* Bouché als Parasit des Großen Kohltriebrüßlers *Ceuthorhynchus napi* Gyll. und des Gefleckten Kohltriebrüßlers *Ceuthorhynchus quadridens* Panz. *Nachrichtenblatt für den Deutschen Pflanzenschutzdienst (Berlin)* **9**, 35–6.

Gabrys, B., Sobota, G., Gadomski, H. & Sarzynska, E. (1999). Czasowa i przestrzenna struktura populacji Carabidae w uprawie rzepaku ozimego. *Progress in Plant Protection* **39**, 429–31.

George, K. S. (1957). Preliminary investigations on the biology and ecology of the parasites and predators of *Brevicoryne brassicae* (L.). *Bulletin of Entomological Research* **48**, 619–29.

Goltermann, S. (1994). *Das Auftreten von Laufkäfern (Col., Carabidae) auf Winterrapsfeldern und deren Einfluß auf den Massenwechsel von* Meligethes aeneus *F. (Col., Nitidulidae)*. PhD thesis, University of Rostock, Germany.

Good, J. A. & Giller, P. S. (1991). The diet of predatory staphylinid beetles. A review of records. *Entomologist's Monthly Magazine* **127**, 77–89.

Gravesen, E. & Toft, S. (1987). Grass fields as reservoirs for polyphagous predators (Arthropoda) of aphids (Homoptera, Aphididae). *Journal of Applied Entomology* **104**, 461–73.

Griffiths, E. (1982). The carabid *Agonum dorsale* as a predator in cereals. *Annals of Applied Biology* **101**, 152–4.

Harenberg, A. (1997). *Auswirkungen extensiv geführter Anbausysteme in verschiedenen Fruchtfolgen (Raps-, Zuckerrübenfruchtfolge) und einer selbstbegrünenden Dauerbrache auf Spinnen (Arachnida: Araneae)*. PhD thesis, Technical University of Braunschweig, Germany.

Harris, K. M. (1973). Aphidiophagous Cecidomyiidae (Diptera): taxonomy, ecology and assessments of field populations. *Bulletin of Entomological Research* **63**, 305–25.

Hartke, A., Drummond, F. A. & Liebman, M. (1998). Seed feeding, seed caching, and burrowing behaviors of *Harpalus rufipes* DeGeer larvae (Coleoptera: Carabidae) in the Maine potato agroecosystem. *Biological Control* **13**, 91–100.

Hatley, C. L. & MacMahon, J. A. (1980). Spider community organization: seasonal variation and the role of vegetation architecture. *Environmental Entomology* **9**, 632–9.

Hengeveld, R. (1980). Qualitative and quantitative aspects of the food of ground beetles (Coleoptera: Carabidae): a review. *Netherlands Journal of Zoology* **30**, 555–63.

Hokkanen, H., Husberg, G.-B. & Söderblom, M. (1988). Natural enemy conservation for the integrated control of the rape blossom beetle *Meligethes aeneus* F. *Annales Agriculturae Fenniae* **27**, 281–93.

Janßen, W. (1963). *Untersuchungen zur Morphologie, Biologie und Ökologie von* Cantharis *L. und* Rhagonycha Eschsch. *(Cantharidae, Col.)*. Unpublished PhD thesis, University of Kiel, Germany.

Jourdheuil, P. (1960). Influence de quelques facteurs écologiques sur les fluctuations de population d'une biocénose parasitaire: étude relative à quelques Hyménoptères (Ophioninae, Diospilinae, Euphorinae) parasites de divers Coléoptères infeodes aux Crucifères. *Annales des Épiphyties* **11**, 445–658.

Kelm, M., Ferguson, A. W., Williams, I. H., Kenward, M. G. & Clark, S. J. (1994). Wzajemne zaleznosci pomiedzy chowaczen podobnikeim *Ceutorhynchus assimilis* Payk, pryszczarkiem kapustnikiem *Dasineura brassicae* Winn. jako skodnikami luszczyn rzepaku. *Materialy Sesji Instytutu Ochrony Roslin* **34**, 258–62.

Kirchner, H.-A. (1961). Zur Populationsdynamik des Kohlschotenrüsslers (*Ceutorhynchus assimilis* Payk.) und zur Möglichkeit einer Prognose seines Auftretens. *Nachrichtenblatt des Deutschen Pflanzenschutzdienstes (Berlin) NF* **15**, 41–6.

Klumpp, M. (1990). *Vergleichende Untersuchungen über die Fauna der Vegetationsschicht auf biologisch-dynamisch und 'konventionell' bewirtschafteten Ackerflächen – mit besonderer Berücksichtigung der* Platypalpus-*Arten (Diptera, Hybotidae) als Prädatoren*. PhD thesis, University of Gießen, Germany.

Kromp, B. (1999). Carabid beetles in sustainable agriculture: a review on pest control efficacy, cultivation impacts and enhancement. *Agriculture, Ecosystems* and *Environment* **74**, 187–228.

Krooß, S. (1996). *Staphyliniden in abgestuft extensiv bewirtschafteten Agrarökosystemen: Populationsdynamik, Gemeinschaftsstruktur und Bedeutung als Prädatoren.* PhD thesis, University of Göttingen, Germany.

Kurppa, S. (1989). Pests of cultivated plants in Finland during 1988. *Annales Agriculturae Fenniae* **28**, 97–102.

Langenstück, C., Heimbach, U. & Larink, O. (1998). Larven der Cantharidae (Insecta: Coleoptera) auf Ackerflächen in SO-Niedersachsen und Aspekte ihrer Biologie. *Braunschweiger Naturkundliche Schriften* **5**, 551–68.

Leuchs, F. (1956). *Über die Ursachen des Massenwechsels von* Meligethes aeneus *F. und einigen anderen Ölfruchtschädlingen.* PhD thesis, University of Bonn, Germany.

Lindroth, C. H. (1945). Die fennoskandischen Carabidae. Eine tiergeographische Studie. I. Spezieller Teil. *Meddelanden från Göteborgs Musei Zoologiska Avdelning* **109**, 1–709.

Luka, H., Pfiffner, L. & Wyss, E. (1998). *Amara ovata* und *A. similata* (Coleoptera, Carabidae), zwei phytophage Laufkäferarten in Rapsfeldern. *Mitteilungen der Schweizerischen Entomologischen Gesellschaft* **71**, 125–31.

Majerus, M. E. N. (1994). *Ladybirds.* Harper Collins: London.

Mamedov, A. A. (1995). On quantitative estimation of the effectiveness of natural enemies of *Oulema melanopus* L. (Coleoptera, Chrysomelidae) on winter wheat. *Entomological Review* **74**, 1–9.

McEwan, P., New, T. R. & Whittington, A. E. (2001). *Lacewings in the Crop Environment.* Cambridge University Press: Cambridge.

Nentwig, W. (1980). The selective prey of Linyphiid-like spiders and of their space webs. *Oecologia (Berlin)* **45**, 236–43.

Nilsson, C. (1989). The pollen beetle (*Meligethes aeneus* F.) in winter and spring rape at Alnarp 1976–1978. III. Mortality factors. *Växtskyddsnotiser* **52**, 145–50.

Nilsson, C. & Andreasson, B. (1987). Parasitoids and predators attacking pollen beetles (*Meligethes aeneus* F.) in spring and winter rape in southern Sweden. *IOBC/wprs Bulletin* **10**, 64–73.

Nuss, H. & Büchs, W. (2000). Zur Bedeutung epigäischer Raubarthropoden als Antagonisten von Blüten- und Schotenschädlingen des Winterrapses mit Verpuppung im Boden. *Mitteilungen der Deutschen Gesellschaft für Allgemeine und Angewandte Entomologie* **12**, 141–7.

Nyffeler, M. (1982). *Field Studies on the Ecological Role of the Spiders as Insect Predators in Agroecosystems (Abandoned Grassland, Meadows, and Cereal Fields).* Unpublished PhD thesis, Swiss Federal Institute of Technology (ETH), Zürich, Switzerland.

Perron, J. P., Leroux, E. J. & Lafrance, J. (1956). Notes on *Coenosia tigrina* (F.) (Diptera: Anthomyiidae), mainly on habits and rearing. *The Canadian Entomologist* **138**, 608–11.

Perry, J. N. (1995). Spatial analysis by distance indices. *Journal of Animal Ecology* **64**, 303–14.

Poehling, H. M., Dehne, H. W. & Sprick, P. (1985). Untersuchungen zur Bedeutung von Carabiden und Staphyliniden als Blattlausantagonisten in Winterweizen und deren Beeinträchtigung durch insektizide Wirkstoffe. *Mededelingen van de Faculteit Landbouwwetenchappen Rijksuniversiteit Gent* **50**, 519–30.

Prescher, S. & Büchs, W. (1998). Der Einfluß abgestufter Extensivierungsmaßnahmen im Raps- und Erbsenanbau auf nützliche Fliegen (Diptera, Brachycera). *Gesunde Pflanzen* **50**, 213–8.

Prescher, S. & Büchs, W. (1999). Fliegenzönosen (Diptera, Brachycera) aus abgestuft extensiv bewirtschafteten Raps- und Erbsenkulturen. *Verhandlungen der Gesellschaft für Ökologie* **29**, 265–9.

Scheller, H. V. (1984). The role of ground beetles (Carabidae) as predators on early populations of cereal aphids in spring barley. *Zeitschrift für Angewandte Entomologie* **97**, 451–63.

Scherney, F. (1958). Über die Wirkung verschiedener Insekticide auf Laufkäfer (Col., Carabidae). *Pflanzenschutz* **10**, 87–92.

Scherney, F. (1959). *Unsere Laufkäfer, ihre Biologie und wirtschaftliche Bedeutung*. A. Ziemsen: Wittenberg, Lutherstadt.

Sengonça, C., Kotikal, Y. K. & Schade, M. (1995). Attraktivität unterschiedlicher olfaktorischer Reize auf die Larven von *Chrysoperla carnea* (Stephens) (Neuroptera, Chrysopidae) im achtarmigen Olfaktometer. *Mitteilungen der Deutschen Gesellschaft für Allgemeine und Angewandte Entomologie* **10**, 605–8.

Shaw, M. W. (1970). Egg-laying by some dipterous pests of cultivated cruciferae in north-east Scotland. *Journal of Applied Ecology* **7**, 353–91.

Stark, A. (1988). Auftreten und Bedeutung räuberischer Dipteren in Getreidefeldern. *Tagungsberichte der Akademie der Landwirtschaftswissenschaften der DDR* **271**, 527–31.

Stark, A. (1994). Zum Beutespektrum und Jagdverhalten von Fliegen der Gattung *Platypalpus* (Empidoidea, Hybotidae). *Studia Dipterologica* **1**, 49–74.

Stark, A. & Wetzel, T. (1987). Fliegen der Gattung *Platypalpus* (Diptera, Empididae) – bisher wenig beachtete Prädatoren im Getreidebestand. *Zeitschrift für Angewandte Entomologie* **103**, 1–14.

Stubbs, A. & Chandler, P. (1978). *A Dipterist's Handbook. The Amateur Entomologist 15*. The Amateur Entomologist's Society: Hanworth.

Sunderland, K. D. & Vickerman, G. P. (1980). Aphid feeding by some polyphagous predators in relation to aphid density in cereal fields. *Journal of Applied Ecology* **17**, 389–96.

Sunderland, K., Fraser, A. M. & Dixon, A. F. G. (1986). Field and laboratory studies on money spiders (Linyphiidae) as predators of cereal aphids. *Journal of Applied Ecology* **23**, 433–47.

Symondson, W. O. C. & Liddell, J. E. (1993). The detection of predation by *Abax parallelepipedus* and *Pterostichus madidus* (Coleoptera: Carabidae) on Mollusca using a quantitative ELISA. *Bulletin of Entomological Research* **83**, 641–7.

Tischler, W. (1958). Synökologische Untersuchungen an der Fauna der Felder und Feldgehölze. *Zeitschrift für Morphologie und Ökologie Tiere* **47**, 54–114.

Varis, A.-L., Holopainen, J. K. & Koponen, M. (1984). Abundance and seasonal occurrence of adult Carabidae (Coleoptera) in cabbage, sugar beet and timothy fields in southern Finland. *Zeitschrift für Angewandte Entomologie* **98**, 62–73.

Warner, D. J., Allen-Williams, L. J., Ferguson, A. W. & Williams, I. H. (2000). Pest–predator spatial relationships in winter rape: implications for integrated crop management. *Pest Management Science* **56**, 977–82.

17

Pathogens of Oilseed Rape Pests

Heikki M. T. Hokkanen & Ingeborg Menzler-Hokkanen
University of Helsinki, Department of Applied Biology, Box 27, FIN-00014 Helsinki, Finland
Tariq M. Butt
University of Wales Swansea, School of Biological Sciences, Swansea SA2 8PP, UK

Introduction

Various pathogens are associated with pests of arable crops, and some of these antagonists are available commercially for use as biocontrol agents (BCAs) (see Copping, 2001). However, although some products (e.g. two including *Bacillus thuringiensis* ssp. *aizawai* encapsulated delta-endotoxins) are recommended for use on oilseed rape or other brassica crops, the insect targets of the former are lepidopterous pests: namely, cabbage white butterflies (*Pieris* spp.) and diamond-back moth (*Plutella xylostella*) in non-European countries.

In the context of this review, pathogens of oilseed rape pests include consideration of the following groups of organisms:

- entomopathogenic fungi
- entomopathogenic nematodes
- bacteria
- viruses
- protozoans.

Entomopathogenic Fungi

Several hundred species of fungi may (potentially) infect insects in oilseed rape fields, a few dozen of which might be of some practical importance. The imperfect fungi *Metarhizium anisopliae* and *Beauveria bassiana* have been studied most intensively; *Paecilomyces farinosus*, *P. fumosoroseus* and *Verticillium lecanii* may also be of interest (cf. Table 17.1). Most of these species are considered to be generalists because they will usually infect insects from different orders. In contrast, entomophthoralean fungi such as *Entomophthora muscae* and *Erynia* (= *Pandora*) *neoaphidis*, although occurring widely in natural ecosystems, are usually restricted to specific insect hosts. Within oilseed rape crops, these fungi infect pests such as cabbage root fly (*Delia radicum*) (Eilenberg, 1995), aphids (e.g. cabbage aphid, *Brevicoryne brassicae*) and some Lepidoptera (such as diamond-back moth). The problem in working with entomophthoralean fungi is that they are obligate pathogens and it has not been possible to grow them properly on artificial media.

Table 17.1 Hyphomycetous fungi known to be pathogenic to brassicaceous pests, as confirmed by various workers.

Pest	Pathogen					
	Bb	Em	Ma	Nr	Pf	Tc
Peach/potato aphid (*Myzus persicae*)	–	–	TB	–	–	–
Turnip aphid (*Lipaphis pseudobrassicae*)	TB	–	TB	–	–	–
Cabbage seed weevil (*Ceutorhynchus assimilis*)	TB	–	TB	–	–	–
Cabbage flea beetles (*Phyllotreta* spp.)	–	–	TB/HH	–	–	–
Cabbage stem flea beetle (*Psylliodes chrysocephala*)	TB	–	TB	–	–	–
Mustard beetle (*Phaedon cochleariae*)	TB	–	TB	–	–	–
Pollen beetle (*Meligethes aeneus*)	TB/HH	–	TB/HH	–	TB	–
Cabbage root fly (*Delia radicum*) adults	IK	JE	–	–	–	–
Cabbage root fly (*Delia radicum*) larvae	HH	–	HH	–	–	–
Turnip root fly (*Delia floralis*) adults	–	JE	–	–	–	–
Turnip root fly (*Delia floralis*) larvae	–	–	IK	–	–	IK
Cabbage moth (*Mamestra brassicae*)	IK/V	–	V	V	V	–
Diamond-back moth (*Plutella xylostella*)	V	–	–	–	V	–

Bb = *Beauveria bassiana*; Em = *Entomophthora muscae*; Ma = *Metarhizium anisopliae*; Nr = *Nomourea rileyi*; Pf = *Paecilomyces fumosoroseus*; Tc = *Tolypocladium cylindrosporum*.

TB = T. M. Butt; HH = H. M. T. Hokkanen; IK = I. Klingen; JE = J. Eilenberg; V = various other workers).

Their importance in natural ecosystems, however, is demonstrated by their being readily found (e.g. sourced from infected flies) within minutes by expert observers from almost any location (J. Eilenberg, *unpublished*). Most species of pathogens are essentially soil-dwelling organisms, and their future use in crop protection (for example, as spray-formulated biocontrol products) requires special attention.

Direct applications of entomopathogenic fungi (EPFs) to control any specific oilseed rape pest have not yet been made. However, encouraging field results using certain Finnish strains of *Metarhizium anisopliae* to control coleopterous pests on oilseed rape, e.g. cabbage seed weevil (*Ceutorhynchus assimilis*), cabbage stem flea beetle (*Psylliodes chrysocephala*) and pollen beetle (*Meligethes aeneus*) have been obtained (e.g. Butt *et al*., 1998; Carreck *et al*., 1999). Further, the potential impact of *Beauveria bassiana* treatments on the overwintering survival of pollen beetle adults in Finland has been assessed by Hokkanen (1993); at least at the northernmost edge of the pest's distribution, the additional 'stress' caused by *B. bassiana* on the beetles was considered likely to have a dramatic effect on survival. In the laboratory, *B. bassiana* readily killed the beetles; however, only indirect mortality via lower overwintering survival in the field was demonstrated. A further field study in Finland showed a significant decrease in the numbers of emerging new-generation cabbage flea beetles (*Phyllotreta* spp.) in turnip rape field after spray (–41%) and soil (–34%) treatment by *M. anisopliae* (I. Menzler-Hokkanen *et al*., *unpublished manuscript*). Interestingly, no significant decrease in the numbers of emerging pollen beetles were detected in either treatment (as with *B. bassiana*, see Hokkanen, 1993), even when 57% of pollen beetle larvae were infected in the spray-treated plots. Most likely this can be explained by the population dynamics of the pest, which even a large mortality acting at the larval stage is unlikely to affect (see Hokkanen, 2000).

There are no data to indicate the practical role of naturally occurring EPFs in current agro-ecosystems; perhaps the effect is often negligible because, despite the fact that these fungi can be isolated from almost any soil sample, their spore numbers are probably too small to cause infections (see Vänninen *et al.*, 1989; Vänninen, 1995). In addition, the eco-chemistry of the rhizosphere in *Brassica* crops is most peculiar, and is probably suppressive to most fungi (Kirkegaard & Sarwar, 1999; Klingen *et al.*, 2002), including EPFs. To develop a practical system, it would be necessary to search for specific adaptations to this environment.

Entomopathogenic Nematodes

Two genera of entomopathogenic nematodes (EPNs), *Heterorhabditis* and *Steinernema* (but also several species from other genera) may be of interest in the context of oilseed rape pests. Currently, species of both genera are used extensively as biopesticides to control crop pests, mainly in greenhouses but also on golf courses and some high-value outdoor crops (Copping, 2001). EPNs appear to be common in agricultural soils, but they often occur in too small numbers to have any practical impact on insect control. *Steinernema feltiae*, one of the most commonly used EPNs, is known to occur at an approximate density of 600–4,300 infective juveniles per m^2 in an experimental farm in Poland (Bednarek, 1998), and in Germany about 1,500 *S. feltiae* were detected in the upper 15 cm layer, with an average total of 3,000 individuals per m^2 (Sturhan, 1996). Using semi-quantitative methods, such as insect baiting, however, most samples from agricultural fields do not yield any EPNs; for example in a survey in Finland EPNs were isolated from only 10% of samples collected from oilseed rape fields (Vänninen *et al.*, 1989).

A few attempts have been made to control pests of oilseed crops by using EPNs. In the laboratory, these nematodes have no problem in killing larvae of most pests (Nielsen, 2000), whereas adults or pupae are not affected to any significant degree. The first field experiments include those of Christian Laumond (INRA, Antibes) in France in the late 1970s, in which blanket treatment of an oilseed rape field with a preparation of *Steinernema feltiae* subsequently reduced the numbers of emerging new-generation pests by at least 50% (at least in the case of *Meligethes* spp. and *Ceutorhynchus* spp., and probably also in the case of some other pests) (C. Laumond, *personal communication*). More detailed observations have also been made in France on the effects of *Steinernema* on fully fed larvae of cabbage stem flea beetle, cabbage stem weevil (*Ceutorhynchus pallidactylus*), rape stem weevil (*C. napi*), rape winter stem weevil (*C. picitarsis*) and pollen beetles. Brassica pod midge (*Dasineura brassicae*) larvae were not considered because EPNs are not effective against them. Attempts were also made to locate new strains of *Steinernema*, and a method of *in vitro* mass multiplication was developed (Laumond, 1978; Laumond & Bonifassi, 1987). Small plot tests (in the 1970s) and, later, field-scale tests (from 1981 to 1984) were done. The latter involved attempts at the biocontrol of oilseed rape pests, with several hectares receiving treatment (see Jourdheuil *et al.*, 1974a, b; Ballanger & Laumond, 1983). Although nematodes offered possibilities for regulating populations of insect pests in oilseed rape, the results of these tests did not give consistent results or high levels of biocontrol (Y. Ballanger, *pers. comm.*). The initial proposal was to treat at the end of the winter period, to obtain control of cabbage stem flea beetle larvae as these entered the soil to pupate, and then (in sequence) to control larvae of rape winter stem weevil, rape stem weevil, cabbage stem weevil, pollen beetle and, finally, cabbage seed weevil (Jourdheuil *et al.*, 1970). The assumption was that the nematode populations, with a range of hosts becoming available over time, would increase naturally and become more and more effective antagonists from one pest to the other. Unfortunately, the nematode populations did not appear to act in this way (Laumond, 1990). Later, however, Nielsen (2000) has pursued this avenue further, and demonstrated that infected larvae of cabbage seed weevil can produce a mean of 1,300 infective juveniles of *S. feltiae*, and that pollen beetle larvae similarly produce about 700 new infective juveniles per larva. With a field density of at least 2,000 pollen beetle larvae per m^2 in Finland (H. M. T. Hokkanen, *unpublished*), this alone would increase the density of infective juveniles in the field up to over one million per m^2.

A recent field study in Finland in 2001 (H. M. T. Hokkanen *et al.*, *unpublished*) has indicated an excellent potential for the control of the pollen beetle with EPNs (*S. feltiae*). Treatment of the field soil close to the pupation time of pollen beetle larvae (at the rate of 1 million infective juveniles per m^2) resulted in 92% control compared with untreated plots; also, the emergence of cabbage flea beetle adults was reduced by 56% by this treatment.

Practical difficulties with this approach are that the technique is still prohibitively expensive, and it will not prevent crop losses in the current year. However, novel delivery techniques (H. M. T. Hokkanen *et al.*, *unpublished*) and increasing market sizes may drastically impact the economics, and the prospects of long-term population suppression as part of an IPM strategy could justify nematode treatments even if immediate crop losses cannot be prevented.

In addition to the above-mentioned EPNs, an interesting nematode parasite (*Howardula phyllotretae*) has been found in the adults of cabbage flea beetles (Oldham, 1933); this organism is not pathogenic, but it obviously must have an impact on the total fitness of the beetle (Ekbom, 1991). Further examples of parasitism by nematodes are afforded by mermithids (*Hexamermis* spp.) and by *Parasitorhabditis* spp., which sometimes infest the prepupae and pupae of cabbage stem flea beetle (Balachowsky, 1963). In France several studies have been made of parasitic nematodes associated with *Baris* weevils (e.g., Jourdheuil, 1961; Laumond, 1970, 1978; Caudal, 1976), and a species of *Deladenus*, for example, has been recorded infesting adults of *Baris coerulescens*. Such nematodes are also associated with *Baris* spp. on non-brassicaceous hosts (C. Laumond, *unpublished report*).

With regards turnip sawfly (*Athalia rosae*), according to Hoffmann & Schmutterer (1983) the nematode *Mermis albicans* parasitizes the larvae, with high rates of parasitism having been observed. More recently, Sáringer *et al.* (1996) reported that spraying a suspension of the nematode *Steinernema carpocapae* onto oilseed rape foliage resulted in 40% mortality of larvae, whereas spraying the soil surface destroyed 67% of the developmental stages in the ground; laboratory infections of fully fed turnip sawfly larvae moving into the soil to pupate were more successful with *S. carpocapsae* than with *S. feltiae*.

Protozoans

The members of this group that are of potential interest in the biocontrol of crop pests are all intracellular parasites of insects. The vast majority of them do not cause direct mortality of the host; however, they can have direct impacts:

- increased mortality during the winter;
- shortened 'life-expectancy' in general;
- decreased fecundity.

Only certain Microsporida have been studied in any detail. The best information associated with an oilseed rape pest (namely, *Meligethes*) is that on *Nosema meligethi*, with a range of studies by three researchers: B. Ekbom (Sweden), H. M. T. Hokkanen (Finland) and J. J. Lipa (Poland) – see, for example, Lipa & Hokkanen (1991, 1992a, b); Hokkanen & Lipa (1995).

The key findings from these studies are as follows:

- *N. meligethi* is common in 'wild' adult populations of *Meligethes* but very rare in populations associated with oilseed rape crops.
- For some reason the current distribution of *N. meligethi* is limited to eastern Europe and Finland; the parasite is found very rarely in Sweden, Denmark and Austria but, despite intensive efforts and wide-ranging observations, not in other western European countries.
- *N. meligethi* infestations can be artificially augmented.
- *N. meligethi* appears to be specific to *Meligethes* spp.
- *N. meligethi* has a major impact on overwintering mortality of *Meligethes* adults.

Thus, potentially this disease plays an important role in the population dynamics of the pollen beetle under natural conditions, but for some reason appears incompatible with current oilseed rape cultivation practices. With regards oilseed rape pests other than pollen beetles, protozoans are also found at least in cabbage flea beetles (Ekbom, 1991).

Bacteria

The most significant insect pathogens to date in terms of their use as bio-pesticides are the hundreds of products based on *Bacillus thuringiensis* (see Copping, 2001). Unfortunately, the spectra of activity of the currently available *Bt* strains, as a rule, do not cover other than many Lepidoptera, certain Diptera and some Coleoptera. None of the coleopteran species of importance as pests in oilseed rape is susceptible to any of the *Bt* toxins currently available commercially.

H. M. T. Hokkanen (*unpublished*) recently screened about 100 unknown types of *Bt* strains, based on a vast collection of some 50,000 strains where the rest were of known *Bt* types, to assess their effect on pollen beetle adults. In that screening, one strain showed reasonably good results and is subject to further studies. It may be possible in the future to discover, or to engineer, more *Bt* strains that show activity against insect pests of oilseed rape crops. The first steps in this direction have already been taken. Lereclus *et al.* (1992) reported expansion of insecticidal host range of *Bacillus thuringiensis* by recombination to include the mustard beetle (*Phaedon cochleariae*). Three natural strains of *Bacillus thuringiensis tenebrionis* were also found in bioassays to cause 100% mortality in mustard beetles (Robert *et al.*, 1994), while another bacterium, *Bacillus firmus*, has been discovered in China to be pathogenic to the flea beetle *Phyllotreta vittata* (Huang *et al.*, 1992). Several recent patent applications and issued patents claim *Bt* activity, which includes important oilseed rape pests such as *Phyllotreta cruciferae* (Payne *et al.*, 2000) as well as other species of *Phyllotreta*, cabbage stem flea beetle, pollen beetles and ceutorhynchid weevils (Lambert *et al.*, 1994). However, detailed efficacy data to substantiate these claims are not available. It also appears that the main aim of the published patents is to incorporate these new *Bt*-toxin genes into transgenic plants, rather than to develop new biopesticides from them.

Viruses

Insect viruses, many of which have been commercialised as biopesticides, seldom are known to attack Coleoptera – the main group of pests on oilseed rape crops in Europe. Nor have viruses been found in Diptera, whereas they are common on Lepidoptera and Hymenoptera. Currently, therefore, there seems little prospect of developing viable virus-based biopesticides for use in biocontrol on European oilseed rape crops.

Classification

Information on the classification of pathogens is confused and not readily available. The following is that currently adopted by those involved in biocontrol studies on oilseed rape crops:

a) **Entomopathogenic fungi**
   Phylum Deuteromycotina, form-class Deuteromycetes (fungi imperfecti)
   - form-order Moniliales – genera *Beauveria*, *Metarhizium*, *Paecilomyces*, *Tolypocladium* and *Verticillium*

   Phylum Zygomycotina, class Zygomycetes
   - order Entomophthorales – genera *Entomophthora*, *Erynia* (= *Pandora*) and *Zoophthora*

b) **Entomopathogenic nematodes – phylum Nematoda**
   Class Rhabditida
   - family Allantonematidae – genus *Howardula*
   - family Heterorhabditidae – genus *Heterorhabditis*
   - family Neotylenchidae – genus *Deladenus*
   - family Rhabditidae – genus *Parasitorhabditis*
   - family Steinernematidae – genus *Steinernema*

   Class Mermithida
   - family Mermithidae – genera *Hexamermis* and *Mermis*

c) **Bacteria**
   Class Schizomycetes
   - order Eubacteriales – genus *Bacillus*

d) **Protozoans**
   Order Microsporida
   - family Nosematidae – genus *Nosema*

Isolation Techniques

Most insect pathogens have been obtained directly from diseased insects collected from the field. For some organisms such as obligate pathogens this is the only method (e.g. insect viruses and protozoa), whereas facultative pathogens can also be isolated directly from the environment (e.g. with bait insects or with selective growth substrates). Field manuals describing such techniques should be consulted for details (e.g. Lacey, 1997). Some of these aspects are discussed below, using EPFs as an example.

Isolation of entomopathogenic fungi

Most of the fungi being developed for the control of brassicaceous pests belong to the phylum Deuteromycotina, class Hyphomycetes (Table 17.1). These are the asexual or anamorph stages of mostly ascomyceteous fungi and, strictly, should be known as 'hyphomycetous anamorphs'. Some members of the Zygomycotina, order Entomophthorales, are known to play an important role in regulation of certain brassicaceous pests (Table 17.1). Details on initial handling and diagnosis of diseased insects have been recently reviewed (see Lacey, 1997). Most Hyphomycetes can be scraped directly off the cadaver, whereas insects infected with entomophthoralean fungi may be positioned to shower their conidia directly onto a nutrient surface. If sporulation or external hyphal growth has not yet taken place, diseased insects can be incubated in a humid chamber (such as a Petri dish lined with moist filter paper); this will encourage sporulation. Entomopathogenic fungi (EPFs) can be isolated from the soil by using insects as baits or by the use of selective media.

Insect baiting

Larvae of several insect species (e.g. *Galleria mellonella*, *Tribolium destructor* and *Tenebrio molitor*) are easy to rear and are highly susceptible to a wide range of EPFs. When incubated with the soil in humid containers EPFs will be able to infect the host relatively free of competitors (Zimmermann, 1986). Sporulating cadavers can be placed whole or dabbed on a selective medium for isolation of the pathogen (see below). Recently, other insects, in particular the expected target pest species, have also been used for the targeted isolation of pathogens.

Isolation on selective media

Most selective media contain either a fungicide and/or antibiotics that encourage growth of EPFs and discourage growth of other microorganisms. The most notable ones include oatmeal dodine agar, Veen's medium and a copper-based medium (Butt & Goettel, 2000). *Cordyceps militaris* and *Paecilomyces farinosus* are tolerant of high Cu levels (400 mg/litre), followed by *Metarhizium anisopliae* and *Beauveria bassiana*. Most other soil-borne fungi, including nematophagous species of *Verticillium*, are least tolerant. Mycozed cadavers can be placed directly on the selective media or the pathogens can be isolated by soil plating. Soil in cores collected from field sites may be either homogenized or may be stirred vigorously in water, and aliquots spread evenly over the selective medium. Alternatively, the soil may be sprinkled over the selective medium.

Once isolated, many fungi, especially from the Hyphomycetes, can be maintained *in vitro* on several media (e.g. potato dextrose agar, sabouraud dextrose agar). Conidia and mycelium should be stored in cryovials under nitrogen, or freeze-dried and stored in sterile glass ampules. Freshly harvested conidia can also be air dried and stored in a desiccator at 4°C or at room temperature. Several hyphomycetous fungi (e.g. *Verticillium lecanii*, *M. anisopliae*) can be stored as conidia bound to silica gel at –40°C. More details on isolation and storage of EPFs are given in Lacey (1997).

Identification of Insect Pathogens

While isolation of insect pathogens from a target ecosystem or insect is fairly straightforward, identification is often quite difficult, except for certain species (such as *Metarhizium anisopliae* – the 'green muscardine fungus' – which is easily identifiable owing to its green spores and peculiar sporulation structures). Usually, identification will need to be done or confirmed by experts of the specific pathogen group in question. On a world scale, however, expertise on pathogens is very limited, although (for example) for EPFs internationally recognized experts in Europe are available at several institutes. These include the Institut Pasteur, Paris, France (B. Papierok), the Biologische Bundesanstalt für Land und Forstwirtschaft, Darmstadt, Germany (G. Zimmermann), the Royal Veterinary and Agricultural University, Frederiksberg, Denmark (J. Eilenberg), and the University of Wales Swansea, UK (T. M. Butt).

This section will focus (by way of example) on the identification and characterisation of hyphomycetous fungi, because of their commercial potential. However, many of the approaches used to identify Hyphomycetes are also applicable to entomophthoralean fungi and other pathogens. The identification of the vast majority of Hyphomycetes is based on conidiogenesis. Most EPFs have diagnostic features making them readily identifiable. However, many of these fungi exist as species complexes that can be differentiated only on the basis of molecular, biochemical or pathological characteristics.

Morphological markers

Anamorphs show considerable structural diversity and can be classified or described according to three attributes: (1) the method of conidium production (conidiogenesis), (2) the locus of conidiogenesis, and (3) the sequence of conidiogenesis. Conidium production involves the transformation of part of a cell into a separable spore. Sometimes a hypha or a cell becomes separated into one or more segments by septa. The separate cells then thicken, swell and, finally, separate. This kind of conidiogenesis, where the septa appear before the conidium is initiated, is called *thallic*. When the new conidium is initiated, begins to swell or thicken, and *then* is cut off by a septum, the conidiogenesis is called *blastic*. When the wall of the conidium is continuous with the cell that produced it, it is called either *holothallic* (when thallic) or *holoblastic* (when blastic). When only the inner walls of the conidium-bearing cell are involved in conidiogenesis, the prefix 'entero-' is used, resulting in the terms *enterothallic* and *enteroblastic*. Although enterothallic types are rare, enteroblastic types are probably the most common of all and are represented by the ubiquitous conidiogenous cell.

Conidiogenesis usually can be traced to a particular point or locus on the conidiogenous cell. If this point remains stationary and gives rise to more conidia it is said to be *stable*. If subsequent conidia are produced at new points on the cell the conidiogenous locus is said to be *dynamic*. Dynamic loci may gradually elongate and be called *progressive*, or may move toward the base of the conidiogenesis cell and be *retrogressive*. Usually, it is necessary to examine several cells to 'reconstruct' the loci of successive spore productions. If more than one conidium is produced by the conidiogenous cell they can occur either *simultaneously* or *serially* (i.e. one after another). Descriptions of selected species of EPFs, based on conidiogenesis, are given in Box 17.1.

Morphological markers are restricted to those pathogens that are easily cultured on artificial media. Many morphological and antibiotic resistance variants have been identified in biocontrol fungi. These include colony variants in mycelial or conidial colour, growth rate, colony surface texture, topology and viscosity, and resistance to antibiotics. Morphological variants and antibiotic resistance may occur as spontaneous mutants or may be generated through mutagenesis. For example, acriflavine-resistant mutants occur

Box 17.1
Descriptions of selected species of entomopathogenic fungi (EPFs)

Beauveria bassiana

- *Field characters.* Forming dense white to cream covering on host surface, occasionally synnematous (i.e. forming erect fascicles).
- *Conidiophores.* Abundant, arising from the vegetative hyphae, 2 μm wide, bearing cluster of globose to flask-shaped phialides (3–6 × 3–5 μm) which may give rise to further phialides, with a well-developed sympodial, rachis (apical extension) up to 20 μm long by 1–2 μm wide; rachis denticulate, with one conidium per denticle (*c.* 1 μm long).
- *Conidia.* Blastic, hyaline, smooth, globose to broadly ellipsoidal, sometimes with an apiculate base, 2–3 × 2.0–2.5 μm, rarely reaching 3.5 μm in diameter.

Metarhizium anisopliae

- *Field characters.* Mycelium often wholly covering affected hosts; conidiophores in compact patches/bundles; conidia pale to dark green, often in compact bundles.
- *Conidiophores.* Abundant, arising from vegetative hyphae, branching irregularly, usually with 2–3 branches at each node, 4–14 × 1.5–3.0 μm. Branching repeatedly at broad angles and resembling a candelabrum (candleholder). Phialides cylindrical or clavate, abruptly tapering towards the apex and arranged in a dense hymenium.
- *Conidia.* Borne in chains at the apices of phialides, narrow cylindrical, tapering and truncate at both ends, usually 1.5–3.0 μm wide and 5–9 μm long.

Paecilomyces fumosoroseus

- *Field characters.* Conidiophores usually well developed, often synnematous. Fungus may appear as a fluffy, pink growth, covering infected host.
- *Conidiophores.* Phialides arranged singly or in whorls, with a swollen (flask-like, clavate to globose) base and a distinct neck; orientation of phialides gives cluster of spore chains a feathery to cottony appearance.
- *Conidia.* Unicellular, arranged in long chains, uncoloured, smooth and ovoid (4 μm long).

Tolypocladium cylindrisporum

- *Field characters.* Conidiophores irregularly branched
- *Conidiophores.* Phialides with flask-like to subglobose base, with the short, narrow necks often bent out of the axis of the base; occurring singly or in whorls on vegetative cells.
- *Conidia.* Aseptate, cylindrical, straight or slightly curved; released in slime drops.

Verticillium lecanii

- *Field characters.* White to grey mycelium radiating from cadaver.
- *Conidiophores.* Similar to vegetative hyphae 1–2 μm wide. Phialides borne singly, in pairs or whorls of 3–4. May occur in terminal cluster. These delicate structures may vary in size, but usually taper uniformly from the base to the truncate apex (i.e. they lack a swollen base or a distinct neck).
- *Conidia.* Produced singly and aggregating in heads at the tips of phialides; ellipsoidal to cylindrical with rounded ends, varying in size with the strain. Usually 2–3 × 1–2 μm but can be 2–10 μm long and 1–1.7 μm wide.

spontaneously in *Metarhizium anisopliae* at a mutation rate as high as 1×10^{-9} (Tinline & Noviello, 1971). The major problem associated with the use of morphological or antibiotic resistance mutants is that they are unstable and may revert back to a wild type. Colony morphology also varies considerably in many fungal species, and this morphology may not be stable upon subculturing since many fungi exhibit pleiomorphic deterioration of colony characteristics. Reasons for phenotypic instability in fungi are largely unknown, but the parasexual cycle is thought to be a major factor (Al-Aidroos, 1980; Messias & Azevedo, 1980; Riba & Ravelojoana, 1984). Diversity may also result from mutations, deletions, transposons, cytoplasmically transmitted genetic elements (e.g. cytoplasmic RNAs and invertrons), and DNA (deoxyribonucleic acid) and RNA (ribonucleic acid) polymerases that may act as mitochondrial retroposons (Bainbridge, 1987; Kistler & Miao, 1992).

Immunological markers

Immunological markers have been used primarily to distinguish species differences in fungi (e.g. Segers *et al.*, 1999). They and, in particular, enzyme-linked immunosorbent assays (ELISA) have been reported as suitable for the detection of many plant-pathogenic fungi (Schotts *et al.*, 1994). Serological tests have also been used to identify *Beauveria bassiana*, *Metarhizium anisopliae* and species of the Entomophthorales (Shimizu & Aizawa, 1988; Guy & Rath, 1990; Segers *et al.*, 1999). A potential problem in employing ELISA is the ability of the polyclonal antibody to react with genetically similar species. For example, Hajek *et al.* (1991) developed an ELISA with polyclonal antibodies to the plasma membrane of *Entomophaga maimaiga*, a pathogen of the larvae of gypsy moth (*Lymantria dispar*). However, a closely related fungus, *E. aulicae*, showed cross reactivity with the antisera.

Biochemical characterisation

Strains of EPFs differ in their physiology, and this is reflected in the enzymes and secondary metabolites they secrete. The secreted products can be used to discriminate between morphologically identical strains. The major drawback with enzymes and metabolite markers is that their production is often influenced by cultural conditions. Biochemical markers are, therefore, often used in conjunction with molecular markers for strain characterisation (Leal *et al.*, 1997; Segers *et al.*, 1999).

The most popular biochemical approach is the use of isozyme/allozyme electrophoresis. The term isozyme refers to isomers of functional types of enzyme, regardless of locus. Allozyme refers to isomers within a gene locus. Isomers within a locus are also called alleles. Starch gel electrophoresis is used to assay the allozymes of numerous gene loci. Sample enzyme extracts are placed in a pH-buffered starch gel and an electric field is applied. The enzymes migrate through the gel matrix, in response to the electric field, at a rate proportional to their net surface charge. After electrophoresis, the gels are cut into thin horizontal slices and each slice is placed into an enzyme-specific staining solution. The allozyme information read from the gels is recorded into data files for computer analysis. Multilocus allozyme information from a single 'tree' can provide a biochemical 'fingerprint' (i.e. a genotype) of that 'tree'. Allozyme data from populations of 'trees' can form a powerful database for analysing the genetic characteristics of a species.

Allozyme electrophoresis usually requires from 8 to 20 allozymes, depending on the degree of polymorphism and the number of alleles at each locus. St. Leger *et al.* (1992a) examined 120 isolates *of Metarhizium anisopliae* and found that they fell into 48 distinctive genotypic classes. A large amount of variability was found for *M. anisopliae* with less than 20% similarity among some classes (St. Leger *et al.*, 1992a). Similar results were found for *Beauveria bassiana* (St. Leger *et al.*, 1992b). Allozyme analysis has also been applied in an attempt to distinguish pathotypes *of Beauveria brongniartii*. Isolates showed high intraspecific variation, and little association was observed between genotypic class and virulence to the European cockchafer (*Melolontha melolontha*) larvae (Reineke & Zebitz, 1996). Allozyme markers provided a rapid method for identifying strains of *M. anispoliae* used in field trials against pests of oilseed rape (Inyang *et al.*, 2000).

Molecular markers

Base substitutions, such as insertions or deletions in the genome, can be detected with restriction fragment length polymorphisms (RFLP). A probe is required that will identify size variations in certain parts of the digested genomic DNA. Probes that have been used to detect inter- and intraspecific variation in fungi are ribosomal DNA (rDNA) (Pipe *et al.*, 1995), mitochondrial DNA (mtDNA) (Hegedus & Khachatourians, 1993a) and telomeric DNA (Couteaudier & Viaud, 1997).

Cloned DNA probes are portions of the genomic DNA identified through various subtractive or differentiation screening methods that hybridize to a particular strain or species of fungus. Kosir *et al.* (1991) used a cloned DNA probe to differentiate a strain of *Beauveria bassiana* virulent to migratory grasshoppers (*Melanopus sanguinipes*) (Acrididae), from a less virulent strain. Hegedus & Khachatourians (1993b) developed DNA probes that could differentiate *B. bassiana* from several other hyphomycetous EPFs.

Random amplification of polymorphic DNA (RAPD) utilizes the polymerase chain reaction (PCR) and single primers of between 15–20 nucleotides in length in order to detect variability at arbitrary regions of the genome. The technique is powerful, owing to its relative speed and the large number of loci that can be screened.

Problems associated with RAPDs include the appearance and disappearance of minor bands with different runs and variability between thermocyclers from different manufacturers. Higher template concentrations can cause reactions to become inconsistent or to fail, and lower ones cause faint and inconsistent bands. Careful standardization is necessary to obtain reproducible results, as small differences of procedure could lead to some variation (i.e. to the addition or removal of some bands). Differences that lead to this type of variation include the use of hot or cold starting procedures (Chou *et al.*, 1992), and whether PCR mixes were made whilst the tubes were at room temperature or whilst being held on ice. However, taking this into consideration, the appearance or absence of the major bands is a quick and easy method to identify a genet. Using RAPD analysis, wide genetic variation in isolates of *M. anisopliae* has been observed on a worldwide level (Leal *et al.*, 1994b; Fegan *et al.*, 1993; Tigano-Milani *et al.*, 1995). A RAPD study of Brazilian isolates of *M. anisopliae* by Fungaro *et al.* (1996) showed that there was large variability of soil isolates (47% similarity) compared with those isolates from a hemipteran insect (*Deois flavopicta* – Cercopidae) host (82% similarity).

Amplified restriction length polymorphisms (AFLPs) have been used to detect genetic variation in plant-pathogenic fungi (Majer *et al.*, 1996). AFLP is considered to be a reliable, reproducible technique that evaluates genomic variation at 10–100 times more sites than either RAPD or RFLP.

Targeted PCR refers to the amplification and analysis of particular regions of the genome, either by direct sequencing or by RFLP analysis of the amplified region. Neuveglise & Brygoo (1994) showed that 28S rDNA differed in size between two strains of *Beauveria brongniartii*. [rRNA genes include 5.8S, 18S (or 16S) and 28S (or 23S for different organisms) where 'S' means 'sedimentation coefficient' during the separation of these elements.] Curran *et al.* (1994) investigated the ITS (internal transcribed sequence) regions and 5.8S of *Metarhizium* rDNA. [Sequences between 18S and 5.8S rDNA = ITS1, and those between 5.8S and 28S = ITS2.] Primers for targeted PCR were developed from a specific cloned DNA probe, in order to differentiate *B. bassiana* from other EPFs (Hegedus & Khachatourians, 1995). These studies showed the evolutionary relationships of various strains; they may also be used in a broad taxonomic application for identifying a fungal species, but they have not been shown as useful in identifying the genet. One targeted PCR-based approach for identifying genets within insect and plant pathogens would be to fingerprint fungi, based on microsatellite variation. Identifying species-specific microsatellite markers is, potentially, time consuming, although their utility in identifying a fungal genet is effective.

If certain amplified regions are subjected to further analysis then the information may be used to identify the fungal genet. RFLP analysis of the rDNA in *M. anisopliae* showed that some strains could be differentiated according to their geographic origins (Pipe *et al.*, 1995). RFLP analysis of the Pr1 (extracellular protease encoding) gene amplified from various isolates of *M. anisopliae* also showed a correlation of RFLP profiles with geographic origin (Leal *et al.*, 1997). Because of the spatial distribution related to rDNA or Pr1 RFLP, these marker systems could be applied to detect an *M. anisopliae* isolate released into another geographic location.

Another approach to identify genets of *M. anisopliae* is to utilize a variety of biochemical and molecular techniques. Isozyme analysis, RAPD and protease production were used to investigate the relationships between 30 isolates of *M. anisopliae*. Similarly, an approach which utilized biochemical markers and a phylogenetic approach based on ITS nucleotide differences were used to differentiate various species and isolates within the genus *Verticillium*. However, identification of a fungal genet, based on several techniques, is not as efficient as a single discriminating technique.

In some cases, RFLP analysis is used to discriminate amongst fungal pathotypes. *Beauveria brongniartii* was analyzed by RFLP analysis of ITS regions amplified by PCR (Neuveglise *et al.*, 1994) and allowed separation of the strains into distinctive subgroupings; one of the subgroupings contained strains virulent to the white grub *Hopochelus marginali* (Scarabaeidae). Variability in the ITS regions was due to point mutations that occurred as frequently as 14.7% and 16.7% in the ITS1 and ITS2 regions, respectively. A method for rapidly screening a large number of isolates of *B. bassiana* and *Aspergillus niger* has recently been developed, based on direct PCR from colonies treated with Novozym 234 (a powder produced from the extracellular enzymes in *Trichoderma harzianum*) (van Zeijl *et al.*, 1998). This could allow the rapid screening of a large number of fungal isolates for identification after a biocontrol release.

Double-stranded RNA viruses, transposable elements and introns

The presence of double-stranded RNA (dsRNA) has been reported in a large number of fungi. In some cases dsRNAs are not associated with any modification of cell ultrastructure (Tooley *et al.*, 1989); some, however, may enclosed in lipid-rich vesicles (Dodds, 1980; Hanson *et al.*, 1985). In other cases, dsRNAs are associated with virus-like particles, and are indicative of mycovirus infection (Gillings *et al.*, 1993; Leal *et al.*, 1994a).

Two families of mycovirus have been recognised: the Totiviridae and the Partiviridae (Buck, 1988). The Totiviridae have isometric particles 40–45 nm in diameter and a single component dsRNA genome 4.5–6.0 kb in size. The Partiviridae have isometric particles 30–35 nm in diameter and a dsRNA genome of two components, with sizes ranging from 1.5 to 2.0 kb.

Occasionally, the presence of dsRNA is detrimental to pathogenicity or affects fungal phenotype (Anagnostakis, 1981; Fulbright, 1984). Usually dsRNAs appear to have no effect and, therefore, are potentially useful markers in population studies. However, this seems unlikely since their fidelity as markers is suspect and dsRNA transfer between fungal biocontrol strains or species is not fully understood. In natural populations of EPFs in Canada, dsRNAs were found in 38% and 17% of the *Metarhizium anisopliae* and *Beauveria bassiana* isolates, respectively (Melzer & Bidochka, 1998). In this study, dsRNA banding patterns of Canadian EPFs (*B. bassiana*, *M. anisopliae* and *Metarhizium flavoviride*) were compared with non-Canadian isolates, and it was found that dsRNA banding patterns could be distinctive for an isolate or could be shared among species from diverse geographic origins. Similarly dsRNAs may not be transmitted faithfully during subculturing of the fungus (Melzer & Bidochka, 1998).

Eukaryotic transposons can be divided into two major groups: class I and class II. Class I transposable elements are able to transpose via an RNA intermediate, a process which employs a reverse transcriptase. Class II elements, which transpose via DNA intermediates, demonstrate direct excision prior to their integration into target sequences (Kempken & Kück, 1998; Kempken, 1999). Maurer *et al.* (1997) isolated a class II transposable element, named *hupfer*, from *B. bassiana* by trapping it in the nitrate reductase structural gene. The transposable element had an open reading frame similar to the IS630- or *mariner-Tc*1-like transposases. A *Restless*-d1 transposon has also been reported in *Beauveria nivea* ATCC42437 (Kempken, 1999). Although transposable elements are common in fungi, the variability and stability of these elements is not well understood. Therefore, at present, their utility as genetic markers is unknown.

Group I introns, characterized by the ability to 'self-splice' from precursor transcripts, occur in a diverse range of eukaryotic and prokaryotic organisms, organelles and genes. They have been found in viruses, mitochondrial and chloroplast genomes, and ribosomal RNA genes (rDNA) of fungi, plants, protists and eubacteria. Very little is known about the occurrence of group I introns in EPFs. Neuveglise & Brygoo (1994) were the first to report the presence of these introns in an entomopathogenic fungus. They detected three introns within the 28S rDNA of *Beauveria brongniartii*. These introns were polymorphic in both size and restriction patterns, and were inserted at four different positions of the 28S rDNA (Neuveglise *et al.*, 1997). Five new group-I intron sequences were discovered within the LSU [large subunit (28S rDNA), in contrast with the SSU (small subunit) (5.8S rDNA)] of *M. anisopliae* (Mavridou *et al.*, 2000). More recently, C. S. Wang *et al.* (*pers. comm.*) identified four group-I introns inserted at four distinct sites of the 28S rDNA of *B. bassiana*, and found them to be useful markers to distinguish between strains. C. S. Wang *et al.* (*pers. comm.*) also examined the presence and distribution of these introns, using site-specific primers in a population of 125 *B. bassiana* strains, 120 of which were from China. They found that *c.* 7% of the strains were intron-less, whereas most contained one, two or three introns. Only one strain contained all four introns.

Future Biocontrol Options

Entomopathogenic fungi (EPFs)

While much of the commercial development, as well as research, concerning EPFs has focused on the imperfect fungi (Deuteromycetes) such as *Beauveria bassiana* and *Metarhizium anisopliae*, they occur widely in natural ecosystems and, potentially, the possibilities for their active management in the biocontrol of pests are considerable (Inglis *et al.*, 2001). Interestingly, agricultural soils appear usually to be almost deserts in terms of naturally occurring EPFs (see Hokkanen, 1993; Vänninen, 1995) and other antagonists such as EPNs or microsporidians (virtually an 'enemy-free space' for the pests), and yet in the nearby natural habitats these are common and easy to find. For example, in forest areas next to the fields it is possible to isolate on average 25,000 *M. anisopliae* propagules/g of soil (H. M. T. Hokkanen, *unpublished*). Therefore, research on the methods of re-establishing and managing these pathogen populations in the field ecosystem might result in a significant 'added natural mortality factor' on the pests.

Another possible way of actively utilizing EPFs within oilseed rape crops may involve the initiation of epizootics (sources of infection) earlier in the season, in which case their impact on pest populations would be expected to be greater. This could be done by using specific traps into which the pest insects are attracted, and in which the spores of the disease are transmitted to healthy hosts from diseased cadavers or heavily dosed individuals. Meadow *et al.* (2000) demonstrated that it is possible to transmit fatal doses of *B. bassiana* conidia from adult cabbage root flies (fly to fly) for a series of at least six flies. Also, T. M. Butt & L. Ibrahim (*unpublished observations*) obtained significant control of adult pollen beetles by exposing them to individuals heavily dosed with conidia of *M. anisopliae*. Pell *et al.* (2001) discuss the use of the entomopthoralean fungus *Zoophthora radicans* and auto-dissemination devices for the control of diamond-back moths. Effective control is also achieved by using feeding deterrents or repellents to drive pests to trap crops or to sites on the plant that are inoculated with fungal biopesticide (Amiri *et al.*, 1999). 'Push-pull' pest control strategies, which usually incorporate the above ideas, optimize the impact of fungal biopesticides and, concomitantly, reduce chemical pesticide inputs (Pickett *et al.*, 1995).

Improved formulation and delivery techniques for fungal biopesticides can help to increase their efficiency against oilseed rape pests. Although adequate moisture is very important for the activity and persistence of many biopesticides, rain can quickly become detrimental if it washes off the organisms from the foliage. Inyang *et al.* (2000) have shown, with different formulations of *M. anisopliae*, that simulated rain washes off conidia from oilseed rape leaves and reduces mortality of target pests. The number of conidia lost (compared with unexposed leaves) was reduced if the inoculum was formulated in viscous vegetable or mineral oils rather than in aqueous carriers. Detailed studies on host-pathogen interactions, especially the invasive and developmental processes of EPFs, will help, ultimately, in the design of better application strategies against oilseed rape pests (see, for example, Butt *et al.*, 1995; Inyang *et al.*, 1998, 1999a, 1999b; Ibrahim *et al.*, 1999; Amiri *et al.*, 1999; Amiri-Besheli *et al.*, 2000; Bandani *et al.*, 2000; Altre & Vandenberg, 2001).

Fungal biopesticides currently available in some countries include *B. bassiana*, *M. anisopliae* and *Paecilomyces fumosoroseus*. These can all be cultured easily on artificial media and can be formulated for use with conventional spraying equipment. Further, *Beauveria brongniartii*, for example, is used in Switzerland and Austria as a granulous preparation, which is incorporated into the soil for the control of the European cockchafer (not, of course, an oilseed rape pest). However, none of these agents is currently used against oilseed rape pests. At Rothamsted Research (UK), recent studies has shown good control results – comparable to chemical pesticides – by using *Metarhizium* vectored by honey bees for the control of pollen beetle (T. M. Butt & I. H. Williams, *unpublished*). The integration of this approach to the 'push-and-pull' strategy continues at Rothamsted Research as a research focus.

Entomopathogenic nematodes (EPNs)

EPNs show promise for the biocontrol of crop pests, and their mass production for commercial use is currently underway, with particular reference to the control of weevils, particularly species of *Otiorhynchus*, such as vine weevil (*O. sulcatus*) (see, for example, Ehlers & Hokkanen, 1996). Weevil pests (although *Ceutorhynchus* spp. rather than *Otiorhynchus* spp.) are very important in oilseed rape, and as they are usually very susceptible targets for infection by nematodes as well as by *Metarhizium*; thus, the key oilseed rape pest complex would appear to be a worthwhile avenue for investigation. At the University of Kiel, Germany, an active nematode research group is currently investigating the possibility of the reintroduction and management of nematodes in agroecosystems; oilseed rape is one target crop in this work (R.-U. Ehlers, *pers. comm.*). At the Helsinki University, Finland, research is underway to improve EPN delivery systems, in order to make nematode treatments less costly and more effective; this research focuses also on the oilseed rape ecosystem (H. M. T. Hokkanen, *unpublished*).

Bacteria and viruses

Because, in Europe, oilseed rape pests are mainly coleopterans, the selection of important insect pathogens is rather restricted. For example, as mentioned above, bacteria (the world's commercially most important insect pathogen group – including *Bacillus thuringiensis*) are not known to infect any of the key oilseed rape pests. Similarly, viruses (many of which have been commercialised as biopesticides) seldom infect beetle hosts. Currently, therefore, there seems little prospect of developing viable bacteria-based or virus-based biopesticides for use in biocontrol on European oilseed rape crops.

Protozoans

Apparently, few if any studies involving the specific, obligate protozoan parasite (*Nosema meligethi*) of pollen beetle have been done since those of J. J. Lipa & H. M. T. Hokkanen in the early 1990s (see references). This pathogen, however, could be an important long-term stabilizing factor in a biocontrol-based integrated pest management (IPM) system, and its study has now been incorporated as a small part into the new EU-research project MASTER (*Integrated pest management strategies incorporating bio-control for European oilseed rape pests*), with field studies starting in 2002.

Interactions with other Biocontrol Agents (BCAs)

The successful integration of BCAs into IPM systems requires specific knowledge of interactions between the key components. Such interactions might range from 'antagonistic' through 'neutral' to 'synergistic'. Also, currently, data on the possible impacts on non-target organisms is increasingly demanded.

Although abundant data exist on the impacts of EPFs on vertebrates, including safety to humans, there is far less information on the impact of such biopesticides on non-target invertebrates: predators, parasitoids, pollinators, detritivores, rare and endangered insects, etc.

There are no proper guidelines as to how to assess ecological safety and, historically, it has been traditional to ask whether there are known cases of epidemics by the candidate organism among the non-target species that might be at risk. New research to develop assessment methods is now underway, including the ERBIC project *Evaluating environmental risks of biocontrol introductions into Europe* (H. M. T. Hokkanen, *unpublished*), which aims to develop mathematical models that could help to estimate non-target impacts which, for EPFs, involve the following three important parameters:

- persistence;
- virulence to non-targets;
- dispersal to other habitats.

A case study in Finland used turnip rape as the model agro-ecosystem and pollen beetle as the model target pest, along with its parasitoids and predators and other potential non-targets in the system. BCAs suitable for use in this system were primarily EPFs (*Beauveria* and *Metarhizium*). This case study may be used to examine the three abovementioned parameters, and also synergy, as follows:

- **Persistence**
  The persistence of fungal propagules is influenced by a wide range of climatic (e.g. radiation, humidity), edaphic (e.g. soil texture, pH) and biotic (e.g. mycophagous organisms) factors (Inglis *et al.*, 2001). Studies over three years in Finland on four soil types in agricultural fields, demonstrated a rapid decline in inoculum density of augmented *B. bassiana*, such that it had disappeared by the following season, but a much slower decline by *M. anisopliae* that was still detectable after three years (Vänninen *et al.*, 2000).
- **Virulence to non-targets**
  Field studies were done over two years on the natural occurrence of EPFs on important non-target insects in turnip rape fields in Finland (H. M. T. Hokkanen *et al.*, *unpublished*). Non-targets collected from the study fields were as follows: ground beetles (Carabidae) – 32 species, 3,910 individuals; ladybirds (Coccinellidae) – 2 species, 504 individuals; adult rove beetles (Staphylinidae) – 7 species, 73 individuals; rove beetle larvae – 1 species, 73 individuals; springtails (Collembola) – 2 species, 274 individuals; spiders (Araneae) – 3 species, 35 individuals. In all, therefore, 47 species and 5,044 individuals were collected, of which only five (0.1%) were infected by EPFs; in all cases the pathogen was *Beauveria bassiana*. Earlier soil sampling had shown that single isolates of both *B. bassiana* and *Metarhizium anisopliae* were present, naturally, in the soil before treatment.

Similar assessments were made after *M. anisopliae* treatments (incorporated into soil and sprayed onto the crop), along with conventional chemical treatment and treated reference plots, on large, replicated field plots. There were no consistent differences in insect catches between the treatments; nor were infected non-targets ever found. These collections yielded ground beetles (17 species, 661 individuals), ladybirds: *Coccinella septempunctata* (68 individuals) and *Hippodamia tredecimpunctata* (23 individuals), adult rove beetles (7 species, 54 individuals), rove beetle larvae (1 species, 26 individuals) and springtails (1 species, 34 individuals), but not a single one was infected with the EPF.

Tests on two parasitoids of pollen beetle – *Diospilus capito* (Braconidae) and *Phradis morionellus* (Ichneumonidae) (see Chapter 4) – indicated that spray treatment applied against adults and larvae of the target pest also killed parasitoid larvae. Soil treatment killed neither the targets nor the parasitoids; however, the rate of latent infections was 80% in the target host and 80% in *D. capito* but only 20% in *P. morionellus* (Husberg & Hokkanen, 2000). In contrast, a thorough field test indicated no significant impact of *M. anisopliae* on the numbers of *D. capito* emerging from spray-treated or soil-incorporated plots, in comparison with the untreated control. However, numbers of *P. morionellus* were reduced by 23% and 31%, respectively, by these treatments (I. Menzler-Hokkanen *et al.*, *unpublished manuscript*). This still compares very favourably with the impact of chemical control (lambda-cyhalothrin) in the same experiment: on average, numbers of emerging pollen beetle parasitoids were reduced by 64%.

- **Dispersal to other habitats**
  EPFs are known to be dispersed by target organisms over long distances, but at very low concentrations. The publicized case of *Metarhizium anisopliae* being spread by Japanese beetle (*Popillia japonica*) in the Azores is an example (Lacey *et al.*, 1995). Adult pollen beetles can disperse up to 10 km in the autumn and/or in the spring; however, of 1,489 specimens examined in Finland following test application of an EPF against the earlier (parent) generation, none was infected by either *Beauveria bassiana* or *M. anisopliae* (H. M. T. Hokkanen, *unpublished*). The ability of pollinating insects within oilseed rape crops to spread EPFs is currently being used experimentally to aid infection within target pest populations (Butt *et al.*, 1998; Carreck *et al.*, 1999).
- Synergy
  There are very few published data on synergistic interactions; joint action between predators, parasitoids and microbial agents is sometimes acknowledged but seldom documented. Current literature, however, contains one paper on bacteria + EPFs (Glare, 1994), and two papers on EPNs + EPFs (*Beauveria*) (Barbercheck & Kaya, 1990, 1991).

Effects of Pesticides on Pathogens

For a number of years the IOBC/wprs Working Group on 'Pesticides and beneficial organisms' has had the mandate to assess the impact of pesticides on fungi (*Beauveria*) and nematodes (*Steinernema*) used in biocontrol programmes. The results of these standardized assessments have been published (e.g. by Hassan *et al.*, 1994); in addition, specific earlier studies have been done in Finland on the impact of the commonly used pesticides on several beneficial EPFs (Vänninen & Hokkanen, 1988).

As a rule, most pesticides appear to have only a minor impact on the functioning of EPFs in the field. A quite natural exception to this appears to be most of the broad-spectrum fungicides. The issue in a *Brassica* ecosystem, however, is quite complicated, because brassicaceous plants naturally produce glucosinolates in their rhizosphere; these are in turn broken down into isothiocyanates (well known as broad-spectrum fungicides and the active component in many commercial fungicides).

An important exception to the general harmlessness of most other pesticides to fungi appears to be the herbicide trifluralin, commonly used as a routine in oilseed rape as a pre-emergence herbicide. Trifluralin has been shown to inhibit strongly the growth and development of deuteromycetous EPFs (Vänninen & Hokkanen, 1988).

In contrast to the possible adverse effects of some pesticides on EPFs, members of at least two genera of EPNs (*Heterorhabditis* and *Steinernema*) appear to tolerate chemical pesticides very well. In fact, some of the studies in Finland show an improved survival of EPNs in an oilseed rape field that had received the full chemical treatment, compared with their survival in an untreated control (A. Vainio, *pers. comm.*). The explanation may lie in the suppression of nematode antagonists in the chemical treatment, indicating a good compatibility between chemical pesticides and EPNs.

There are several reports where the application of fungal biopesticides with sublethal doses of insecticides has substantially enhanced the efficacy of pathogens (Inglis *et al.*, 2001). Some pesticides (e.g. imidacloprid) may immobilize the pest, preventing it from grooming itself and thus removing fungal inoculum. Some pesticides (e.g. chitin synthesis inhibitors and growth regulators) may make the pest more susceptible to infection by weakening the cuticle or by delaying ecdysis.

Conclusions

Habitat management to encourage activity of resident (naturally occurring) predators (see earlier chapters in this monograph), EPFs, EPNs, etc. is clearly important as is the need to assess the impact on natural enemies of crop cultural controls (e.g. deep ploughing), soil tillage etc. in general.

Insect pathogens appear to be compatible with most other BCAs and can act synergistically in many cases. Further, they appear to pose little or no threat to most non-target organisms in an agricultural system, but whether this applies equally to forests and pastures is unknown. Most adverse impacts are likely to be transient and, therefore, should cause only limited concern (see Hokkanen & Hajek, 2002).

At least in the short term, insect pathogens are not expected to become available as commercial products for the control of key oilseed rape pests in Europe. However, pathogens (together with parasitoids and predators – and host plant resistance) may feature in future IPM programmes.

In recent years, genetic modification of crops (including the introduction of *Bt* genes into oilseed rape) has entered the arena of crop protection research (see, for example, Hokkanen & Wearing, 1994, 1995). However, as mentioned above, with regards oilseed rape pests such developments do not immediately indicate any likelihood of success within Europe. Nevertheless, genetic engineering of oilseed rape crops (along with parasitoids, predators and pathogens) might eventually provide an important component of future IPM strategies.

Acknowledgements

This work was part-funded by the European Commission Concerted Action: *Minimizing pesticide use and environmental impact by the development and promotion of bio-control for oilseed rape pests, FAIR 96-1314*, and by the following grants: EU-FAIR5-CT97-3489 (ERBIC: *Evaluating environmental risks of biocontrol introductions into Europe*), EU-QLK5-CT-2001-01447 (MASTER: *Integrated pest management strategies incorporating bio-control for European oilseed rape pests*), and EU-FAIR6-CT-98-4105 (BIPESCO).

References

Al-Aidroos, K. (1980). Demonstration of a parasexual cycle in the entomopathogenic fungus *Metarhizium anisopliae*. *Canadian Journal of Genetics and Cytology* **22**, 309–14.

Altre, J. A. & Vandenberg, J. D. (2001). Factors influencing the infectivity of isolates of *Paecilomyces fumosoroseus* against diamondback moth, *Plutella xylostella*. *Journal of Invertebrate Pathology* **78**, 31–6.

Amiri, B., Ibrahim, L. & Butt, T. M. (1999). Antifeedant properties of destruxins and their use with the entomogenous fungus *Metarhizium anisopliae* for improved control of crucifer pests. *Biocontrol Science and Technology* **9**, 487–98.

Amiri-Besheli, B., Khambay, B., Cameron, S., Deadman, M. L. & Butt, T. M. (2000). Inter- and intra-specific variation in destruxin production by the insect pathogenic *Metarhizium*, and its significance to pathogenesis. *Mycological Research* **104**, 447–52.

Anagnostakis, S. L. (1981). Stability of Double-stranded RNA components of *Endothia parasitica* through transfer and subculture. *Experimental Mycology* **5**, 236–42.

Bainbridge, B. W. (1987). *Genetics of Microbes*. 2nd edn. Blackie & Son: Glasgow.

Balachowsky, A. S. (ed.) (1963). *Entomologie Appliquée à l'Agriculture. Tome I. Coléoptères, Deuxième Volume*. Masson et Cie: Paris.

Ballanger, Y. & Laumond, C. (1983). Utilisation des nématodes entomopathogènes dans la lutte contre les insectes du colza. *6ème Congrès International sur le Colza, 17–19 mai, Paris* **2**, 1160–6.

Bandani, A. R., Khambay, B. P. S., Faull, J., Newton, R., Deadman, M. & Butt, T. M. (2000). Production of efrapeptins by *Tolypocladium* species and evaluation of their insecticidal and antimicrobial properties. *Mycological Research* **104**, 537–44.

Barbercheck, M. E. & Kaya, H. E. (1990). Interactions between *Beauveria bassiana* and the entomogenous nematodes, *Steinernema feltiae* and *Heterorhabditis heliothidis*. *Journal of Invertebrate Pathology* **55**, 225–34.

Barbercheck, M. E. & Kaya, H. K. (1991). Competitive interactions between entomopathogenic nematodes and *Beauveria bassiana* (Deuteromycotina: Hyphomycetes) in soilborne larvae of *Spodoptera exigua* (Lepidoptera: Noctuidae). *Environmental Entomology* **20**, 707–12.

Bednarek, A. (1998). The agricultural system, as a complex factor, affects the population of entomopathogenic nematodes (Rhabditida: Steinernematidae) in the soil. *IOBC/wprs Bulletin* **21**, 155–60.

Buck, K. W. (1988). From interferon induction to fungal viruses. *European Journal of Epidemiology* **4**, 395–9.

Butt, T. M., Carreck, N. L., Ibrahim, L. & Williams, I. H. (1998). Honey bee mediated infection of pollen beetle (*Meligethes* spp.) by the insect-pathogenic fungus, *Metarhizium anisopliae*. *Biocontrol Science and Technology* **8**, 533–8.

Butt. T. M. & Goettel, M. (2000). Bioassays of entomogenous fungi. In: A. Navon & K. R. S. Ascher (eds). *Bioassays of entomopathogenic microbes and nematodes*. CAB International: Wallingford.

Butt, T. M., Ibrahim, L., Clark, S. J. & Beckett, A. (1995). The germination behaviour of *Metarhizium anisopliae* on the surface of aphid and flea beetle cuticles. *Mycological Research* **99**, 945–50.

Carreck N., Butt, T., Ibrahim, L., Pell, J. & Williams, I. (1999). The use of honey bees to disseminate an insect pathogenic fungus for control of insect pests on oilseed rape. *Beekeeping News* **85**, 9–10.

Caudal, J. (1976). *Contribution à l'étude du parasitisme de* Baris cœrulescens *et* B. laticollis *(Coleoptera: Curculionidae) par* Deladenus *sp. (Nematode: Neotylenchidae)*. Rapport de DAA. Ecole Nationale Supérieure Agronomique de Rennes.

Chou, Q., Russel, M., Birch, D. E., Raymond, J. & Bloch, W. (1992). Prevention of pre-PCR mis-priming and primer dimerization improves low-copy-number amplifications. *Nucleic Acids Research* **20**, 1717–23.

Copping, L. G. (ed.) (2001). *The BioPesticide Manual. 2nd edn.* BCPC: Farnham.

Couteaudier, Y. & Viaud, M. (1997). New insights into population structure of *Beauveria bassiana* with regard to vegetative compatibility groups and telomeric restriction length polymorphisms. *FEMS Microbiology Ecology* **22**, 175–82.

Curran, J., Driver, F., Ballard, J. W. O. & Milner, R. J. (1994). Phylogeny of *Metarhizium*: Analysis of ribosomal DNA sequence data. *Mycological Research* **98**, 547–52.

Dodds, J. A. (1980). Association of type 1 viral-like dsRNA with club-shaped particles in hypovirulent strains of *Endothia parasitica*. *Virology* **107**, 1–12.

Ehlers, R.-U. & Hokkanen, H. M. T. (1996). Insect biocontrol with non-endemic entomopathogenic nematodes (*Steinernema* and *Heterorhabditis* spp.): conclusions and recommendations of a combined OECD and COST workshop on scientific and regulatory policy issues. *Biocontrol Science and Technology* **6**, 295–302.

Eilenberg, J. (1995). Insect pathogenic fungi on cabbage root flies (*Delia radicum*) in Denmark. *Nordic Association of Agricultural Scientists (NJF) Rapport* **107**, 139–43.

Ekbom, B. (1991). The effect of abiotic and biotic factors on flea beetle populations in spring rape in Sweden. *IOBC/wprs Bulletin* **6**, 21–7.

Fegan, M., Manner, J. M., Maclean, D. J., Irwin, J. A. G., Samuels, K. D. Z., Holdom, D. G. & Li, D. P. (1993). Random amplified polymorphic DNA markers reveal a high degree of genetic diversity in the entomopathogenic fungus *Metarhizium anisopliae*. *Journal of General Microbiology* **139**, 2075–81.

Fulbright, D. W. (1984). Effect of eliminating dsRNA in hipovirulent *Endothia parasitica*. *Phytopathology* **74**, 722–4.

Fungaro, M. H. P., Vieira, M. L. C., Pizzirani-Kleiner, A. A. & Azevedo, J. L. de (1996). Diversity among soil and insect isolates of *Metarhizium anisopliae* var. *anisopliae* detected by RAPD. *Letters in Applied Microbiology* **22**, 389–92.

Gillings, M. R., Tesoriero, L. A. & Gunn, L. V. (1993). Detection of double-stranded RNA and virus-like particles in Australian isolates of *Pythium irregulare*. *Plant Pathology* **42**, 6–15.

Glare, T. R. (1994). Stage-dependent synergism using *Metarhizium anisopliae* and *Serratia entomophila* against *Costelytra zealandica*. *Biocontrol Science and Technology* **4**, 321–9.

Guy, P. L. & Rath, A. C. (1990). Enzyme-linked immunosorbent assays (ELISA) to detect spore surface antigens of *Metarhizium anisopliae*. *Journal of Invertebrate Pathology* **55**, 435–6.

Hajek, A. E., Butt, T. M., Strelow, L. I. & Gray, S. M. (1991). Detection of *Entomophaga maimaiga* (Zygomycetes Entomophthorales) using enzyme-linked immunosorbent assay. *Journal of Invertebrate Pathology* **58**, 1–9.

Hanson, D. R., Alfen, N. K. van, Gillies, K. & Powell, W. A. (1985). Naked dsRNAs associated with hypovirulence of *Endothia parasitica* is packaged in fungal vesicles. *Journal of General Virology* **66**, 2605–14.

Hassan, S. A., Bigler, F., Bogenschütz, H., Boller, E., Brun, J., Calis, J. N. M., Coremans-Pelseneer, J., Duso, C., Grove, A., Heimbach, U., Helyer, N., Hokkanen, H., Samsoe-Petersen, L., Sauphanor, B., Stäubli, A., Sterk, G., Vainio, A., Veire, M. van de, Viggiani, G. & Vogt, H. (1994). Results of the sixth joint pesticide testing programme of the IOBC/WPRS working group 'Pesticides and beneficial organisms'. *Entomophaga* **39**, 107–19.

Hegedus, D. D. & Khachatourians, G. G. (1993a). Identification of molecular variants in mitochondrial DNAs of members of the genera *Beauvaria*, *Verticullium*, *Paecilomyces*, *Tolypocladium*, and *Metarhizium*. *Applied and Environmental Microbiology* **59**, 4283–8.

Hegedus, D. D. & Khachatourians, G. G. (1993b). Construction of cloned DNA probes for the specific detection of the entomopathogenic fungus *Beauverai bassiana* in grasshoppers. *Journal of Invertebrate Pathology* **62**, 233–40.

Hoffmann, G. M. & Schmutterer, H. (1983). *Parasitäre Krankheiten und Schädlinge an Landwirtschaftlichen Kulturpflanzen*. Verlag Eugen Ulmer: Stuttgart.

Hokkanen, H. M. T. (1993). Overwintering survival and spring emergence in *Meligethes aeneus*: effects of body weight, crowding, and soil treatment with *Beauveria bassiana*. *Entomologia Experimentalis et Applicata* **67**, 241–6.

Hokkanen, H. M. T. (2000). The making of a pest: recruitment of *Meligethes aeneus* onto oilseed brassicas. *Entomologia Experimentalis et Applicata* **95**, 141–9.

Hokkanen, H. M. T. & Hajek, A. E. (eds) (2002). Assessment of environmental safety of biological insecticides. Kluwer Academic Publishers: Dordrecht.

Hokkanen, H. M. T. & Lipa, J. J. (1995). Occurrence and dynamics of *Nosema meligethi* (Microsporida) in populations of *Meligethes aeneus* (Coleopera, Nitidulidae) in Finland. *Entomologica Fennica* **6**, 11–8.

Hokkanen, H. M. T. & Wearing, C. H. (1994). The safe and rational deployment of *Bacillus thuringiensis* genes in crop plants: conclusions and recommendations of OECD workshop on ecological implications of transgenic crops containing Bt toxin genes. *Biocontrol Science and Technology* **4**, 399–403.

Hokkanen, H. M. T. & Wearing, C. H. (1995). Assessing the risk of pest resistance evolution to *Bacillus thuringiensis* engineered into crop plants: a case study of oilseed rape. *Field Crops Research* **45**, 171–9.

Huang, Y. X., Huang, R. G. & Li, K. H. (1992). A toxicity test of *Bacillus firmus* against *Phyllotreta vittata* (Col: Chrysomelidae). *Chinese Journal of Biological Control* **8**, 182.

Husberg, G.-B. & Hokkanen, H. M. T. (2000). Effects of *Metathizium anisopliae* treatment on the pollen beetle *Meligethes aeneus* and its parasitoids *Phradis morionellus* and *Diospilus capito*. *BioControl* **46**, 261–73.

Ibrahim, L., Butt, T. M., Beckett, A. & Clark, S. J. (1999). Germination of oil-formulated conidia of the insect-pathogen, *Metarhizium anisopliae*. *Mycological Research* **103**, 901–7.

Inglis, D. G., Goettel, M. S., Butt T. M. & Strasser, H. (2001). *Use of hyphomycetous fungi for managing insect pests*. In: T. M. Butt, C. W. Jackson & N. Magan. (eds). *Fungi as biocontrol agents: Progress, Problems and Potential*. CAB International: Wallingford.

Inyang, E., Butt, T. M., Beckett, A. & Archer, S. (1999a). The effect of crucifer epicuticular waxes and leaf extracts on the germination and virulence of *Metarhizium anisopliae* conidia. *Mycological Research* **103**, 419–26.

Inyang, E., Butt, T. M., Doughty, K. J., Todd, A. D. & Archer, S. (1999b). The effects of isothiocyanates on the growth of the entomopathogenic fungus *Metarhizium anisopliae* and its infection of the mustard beetle. *Mycological Research* **103**, 974–80.

Inyang, E., Butt, T. M., Ibrahim, L., Clark, S. J., Pye, B. J., Beckett, A. & Archer, S. (1998). The effect of plant growth and topography on the acquisition of conidia of the insect pathogen *Metarhizium anisopliae* by larvae of *Phaedon cochleariae*. *Mycological Research* **102**, 1365–74.

Inyang, E., Butt, T. M., McCartney, A., Oyejola, B., Ibrahim, L., Pye, B. J. & Archer, S. (2000). Effect of formulation, application and rain on the persistence of the entomogenous fungus *Metarhizium anisopliae* on oilseed rape. *Mycological Research* **104**, 653–61.

Jourdheuil, P. (1961). Influence de quelques facteurs écologiques sur les fluctuations de population d'une biocénose parasitaire. *Annales des Épiphyties, Série* **C**: 1–223.

Jourdheuil, P., Laumond, C., Bonifassi, E. & Millot, P. (1970). Recherches préliminaires sur l'utilisation des nématodes en vue de la lutte contre les insectes des crucifères. *Journées Internationales sur le Colza, 26–30 mai, Paris*, pp. 364–71.

Jourdheuil, P., Laumond, C., Bonifassi, E. & Millot, P. (1974a). Essai de lutte biologique avec les nématodes néoaplectanidae contre les insectes des crucifères. *4ème Congrès International sur le Colza, 4–8 juin, Giessen*, pp. 579–86.

Jourdheuil, P., Laumond, C., Bonifassi, E. & Millot, P. (1974b). Essai de lutte biologique avec les nématodes néoaplectanidae contre les insectes des crucifères. *Informations Techniques CETIOM* **35**, 12–16.

Kempken, F. (1999). Fungal transposons: from mobile elements towards molecular tools. *Applied Microbiology and Biotechnology* **52**, 756–60.

Kempken, F. & Kück, U. (1998). Transposons in filamentous fungi – facts and perspectives. *BioEssays* **20**, 652–9.

Kirkegaard, J. A. & Sarwar, M. (1999). Glucosinolate profiles of Australian canola (*Brassica napus annua* L.) and Indian mustard (*Brassica juncea* L.) cultivars: implications for biofumigation. *Australian Journal of Agricultural Research* **50**, 315–24.

Kistler, H. C. & Miao, V. P. W. (1992). New modes of genetic change in filamentous fungi. *Annual Review of Phytopathology* **30**, 131–52.

Klingen, I., Hajek, A., Meadow, R. & Renwick, J. A. A. (2002). Effect of brassicaceous plants on the survival and infectivity of insect pathogenic fungi. *BioControl* **47** (in press).

Kosir, J. M., MacPherson, J. M. & Khachatourians, G. G. (1991). Genomic analysis of a virulent and less virulent strain of the entomopathogenic fungus *Beauveria bassiana*, using restriction fragment length polymorphisms. *Canadian Journal of Microbiology* **37**, 534–41.

Lacey, L. A. (ed.) (1997). *Manual of Techniques in Insect Pathology*. Academic Press: San Diego.

Lacey, L. A., Amaral, J. J., Coupland, J., Klein, M. G. & Simoes, A. M. (1995). Flight activity of *Popillia japonica* (Coleoptera: Scarabaeidae) after treatment with *Metarhizium anisopliae*. *Biological Control* **5**, 167–72.

Lambert, B. J., Jansens, S. K. & Peferoen, M. (1994). Bacillus thuringiensis *strains toxic to* Diabrotica *species*. U.S. Patent number 05369027 issued to Plant Genetic Systems, N.V., Belgium, 1994-11-29.

Laumond, C. (1970). Hétérogonie et adpatations morphologiques chez une Sphaerulidae (Nematoda), parasite de *Baris coerulescens*. *Compte-rendu de l'Académie des Sciences, Paris* **271**, 1575–7.

Laumond, C. (1978). Les nématodes entomoparasites, perspectives d'utilisation en lutte biologique. *Bulletin Technique d'Information, Ministère de l'Agriculture*, pp. 332–3, 453–7.

Laumond, C. (1990). Utilisation des nematodes entomoparasites Steinernematidae et Heterorhabditidae: données actuelles. *Rencontres Caraïbes en Lutte biologique* **58**, 215–21.

Laumond, C. & Bonifassi, E. (1987). Production de masse de nématodes entomopathogènes *Neoaplectana* (*Steinernema*) pour leur utilisation en lutte biologique. *Conférence Internationale sur les Ravageurs en Agriculture*, pp. 3, 6, 519–2.

Leal, S. C. M., Bertioli, D. J., Ball, B. & Butt, T. M. (1994a). Presence of double stranded RNAs and virus-like particles in the entomopathogenic fungus *Metarhizium anisopliae*. *Biocontrol Science and Technology* **4**, 89–94.

Leal, S. C., Bertioli, D., Butt, T. M. & Peberdy, J. (1994b). Characterization of isolates of the entomogenous fungus *Metarhizium anisopliae* by RAPD-PCR. *Mycological Research* **98**, 1077–81.

Leal, S. C. M., Bertioli, D. J., Butt, T. M., Carder, J. H., Burrows, P. R. & Peberdy, J. F. (1997). Amplification and restriction endonuclease digestion of the Pr1 gene for the detection and characterization of *Metarhizium* strains. *Mycological Research* **101**, 257–65.

Lereclus, D., Vallade, M., Chaufaux, J., Arantes, O. & Rambaud, S. (1992). Expansion of insecticidal host range of *Bacillus thuringiensis* by *in vivo* genetic recombination. *Bio-Technology* **10**, 418–21.

Lipa, J. J. & Hokkanen, H. M. T. (1991). A haplosporidian *Haplosporidium meligethi* sp. n., and a microsporidian *Nosema meligethi* I. & R., two protozoan parasites from *Meligethes aeneus* F. (Coleoptera, Nitidulidae). *Acta Protozoologica* **30**, 217–22.

Lipa, J. J. & Hokkanen, H. M. T. (1992a). *Nosema meligethi* I. & R. (Microsporida) in populations *of Meligethes* spp. in Europe. *Biocontrol Science and Technology* **2**, 119–25.

Lipa, J. J. & Hokkanen, H. M. T. (1992b). Safety of *Nosema meligethi* I. and R. (Microsporida) to *Apis mellifera* L. and *Coccinella septempunctata* L. *Journal of Invertebrate Pathology* **60**, 310–11.

Majer, D., Mithen, R., Lewis, B. G. & Pieter, V. (1996). The use of AFLP fingerprinting for the detection of genetic variation in fungi. *Mycological Research* **100**, 1107–11.

Mavridou, A., Cannone, J. & Typas, M. A. (2000). Identification of group-I introns at three different positions within the 28S rDNA gene of the entomopathogenic fungus *Metarhizium anisopliae* var. *anisopliae*. *Fungal Genetics and Biology* **31**, 79–90.

Maurer, P., Rejasse, A., Capy, P., Langin, T. & Riba, G. (1997). Isolation of the transposable element hupfer from the entomopathogenic fungus *Beauveria bassiana* by insertion mutagenesis of the nitrate reductase structural gene. *Molecular and General Genetics* **256**, 195–202.

Meadow, R., Vandenberg, J. D. & Shelton, A. M. (2000). Exchange of inoculum of *Beauveria bassiana* (Bals.) Vuill. (Hyphomycetes) between adult flies of the cabbage maggot *Delia radicum* L. (Diptera: Anthomyiidae). *Biocontrol Science and Technology* **10**, 479–85.

Melzer, M. J. & Bidochka, M. J. (1998). Diversity of double-stranded RNA viruses within populations of entomopathogenic fungi and potential implications for fungal growth and virulence. *Mycologia* **90**, 586–94.

Messias, C. L. & Azevedo, J. L. (1980). Parasexuality in the deuteromycete *Metarhizium anisopliae*. *Transactions of the British Mycological Society* **75**, 473–7.

Neuveglise, C. & Brygoo, Y. (1994). Identification of group-I introns in the 28s rDNA of the entomopathogenic fungus *Beauveria brongniartii*. *Current Genetics* **27**, 38–45.

Neuveglise, C., Brygoo, Y. & Riba, G. (1997). 28S rDNA group-I introns: a powerful tool for identifying strains of *Beauveria brongniartii*. *Molecolar Ecology* **6**, 373–81.

Nielsen, O. (2000). *Interactions of entomopathogenic nematodes and insects from cruciferous crops*. PhD thesis, The Royal Veterinary and Agricultural University, Copenhagen, Denmark.

Oldham, J. N. (1933). On *Howardula phyllotretae* n. sp., a nematode parasite of flea beetles (Chrysomelidae, Coleoptera) with some observations on its incidence. *Journal of Helminthology* **11**, 119–36.

Payne, J. M., Michaels, T. E., Bradfisch, G. A., Muller-Cohn, J. & Fu, J. (2000). Bacillus thuringiensis *isolates, toxins, and genes selectively active against certain coleopteran pests*. U.S. Patent number 6071511 issued to Mycogen Corporation, U.S.A., 2000-06-06.

Pell, J. K., Eilenberg, J., Hajek, A. E. & Steinkraus, D. C. (2001). *Biology, ecology and pest management potential of Entomophthorales*. In: T. M. Butt, C. W. Jackson & N. Magan. (eds). *Fungi as biocontrol agents: progress, problems and potential*. CAB International: Wallingford.

Pickett, J. A., Butt, T. M., Doughty, K. J., Wallsgrove, R. M. & Williams, I. H. (1995). Minimising pesticide input in oilseed rape by exploiting natural regulatory processes. *Proceedings of the 9th International Rapeseed Congress, Cambridge, 4–7 July 1995*, **2**, 565–71.

Pipe, N. D., Chandler, D., Bainbridge, B. W. & Heale, J. B. (1995). Restriction fragment length polymorphisms in the ribosomal RNA gene complex of isolates of the entomopathogenic fungus *Metarhizium anisopliae*. *Mycological Research* **99**, 485–91.

Reineke, A. & Zebitz, C. P. W. (1996). Protein and isozyme patterns among isolates of *Beauveria brongniartii* with different virulence to European cockchafer larvae (*Melolontha melolontha* L.). *Journal of Applied Entomology* **120**, 307–15.

Riba, G. & Ravelojoana, A. M. (1984). The parasexual cycle in the entomopathogenic fungus *Paecilomyces fumoso-roseus* (Wize) Brown and Smith. *Canadian Journal of Microbiology* **30**, 922–6.

Robert, P., Chaufaux, J. & Marchal, M. (1994). Sensitivity of larval *Oxythyrea funesta* (Coleoptera: Scarabaeidae, Cetoniinae) to three strains of *Bacillus thuringiensis* (subsp. *tenebrionis*). *Journal of Invertebrate Pathology* **63**, 99–100.

St. Leger, R. J., Allee, L. L., May, B., Frank, D. C., Staples, R. C. & Roberts, D. W. (1992a). World-wide distribution of genetic variation among isolates of *Beauveria spp. Mycological Research* **96**, 1007–15.

St. Leger, R. J., May, B., Allee, L. L., Frank, D. C., Staples, R. C. & Roberts, D. W. (1992b). Genetic differences in allozymes and in formation of infection structures among isolates of the entomopathogenic fungus *Metarhizium anisopliae*. *Journal of Invertebrate Pathology* **60**, 89–101.

Sáringer, G., Fodor, A., Nadasy, M., Lucskai, A. & Georgis, R. (1996). Possibilities of biological control using entomopathogenic nematodes against *Leptinotarsa decemlineata* L. and *Athalia rosae* L. larvae. *Mededelingen – Faculteit Landbouwkundige en Toegepaste Biologische Wetenschappen, Universiteit Gent* **61**, 961–6.

Schotts, A., Dewey, F. M. & Oliver, R. P. (1994). *Modern Assays for Plant Pathogenic fungi: Identification, Detection and Quantification.* CAB International: Wallingford.

Segers, R., Butt, T. M., Carder, J., Keen, J. N., Kerry, B. R. & Peberdy, J. (1999). The subtilisins of fungal pathogens of insects, nematodes and plants: distribution and variation. *Mycological Research* **103**, 395–402.

Shimizu, S. & Aizawa, K. (1988). Serological classification of *Beauveria bassiana*. *Journal of Invertebrate Pathology* **52**, 348–53.

Sturhan, D. (1996). Seasonal occurrence, horizontal and vertical dispersal of entomopathogenic nematodes in a field. *Mitteilungen aus der Biologischen Bundesanstalt für Land- und Forstwirtschaft, Berlin-Dahlem* **317**, 35–45.

Tigano-Milani, M. S., Gomes, A. C. M. M. & Sobral, B. W. S. (1995). Genetic variability among Brazilian isolates of the entomopathogenic fungus *Metarhizium anisopliae*. *Journal of Invertebrate Pathology* **65**, 206–10.

Tinline. R. D. & Noviello, C. (1971). Heterokaryosis in the entomogenous fungus, *Metarrhizium anisopliae*. *Mycologia* **63**, 701–12.

Tooley, P. W., Hewings, A. D. & Falkenstein, K. F. (1989). Detection of double-stranded RNA in *Phytophtora infestans*. *Phytopathology* **79**, 470–4.

Vänninen, I. (1995). Distribution and occurrence of four entomopathogenic fungi in Finland: effect of geographical location, habitat type and soil type. *Mycological Research* **100**, 93–101.

Vänninen, I. & Hokkanen, H. (1988). Effect of pesticides on four species of entomopathogenic fungi *in vitro*. *Annales Agriculturae Fenniae* **27**, 345–53.

Vänninen, I., Husberg, G.-B. & Hokkanen, H. M. T. (1989). Occurrence of entomopathogenic fungi and entomoparasitic nematodes in cultivated soils in Finland. *Acta Entomologica Fennica* **53**, 65–71.

Vänninen, I., Tyni-Juslin, J. & Hokkanen, H. (2000). Persistence of augmented *Metarhizium anisopliae* and *Beauveria bassiana* in Finnish agricultural soils. *BioControl* **45**, 201–22.

Zeijl, C. M. J. van, Kamp, E. H. M. van de, Punt, P. J., Selten, G. C. M., Hauer, B., Gorcom, R. F. M. van & Hondel, C. A. M. J. J. van den (1998). An improved colony-PCR method for filamentous fungi for amplification of PCR-fragments of several kilobases. *Journal of Biotechnology* **59**, 221–4.

Zimmermann, G. (1986). The '*Galleria* bait method' for detection of entomopathogenic fungi in soil. *Journal of Applied Entomology* **102**, 213–5.

Glossary

Abdomen The third of the three main divisions of the body of an insect; the second of the two major divisions of the body of a spider (*see also* **Opisthosoma**).
Abdominal Pertaining to the abdomen.
Adipose tissue Fatty tissue, as in connective tissue in which the cells contain large globules of fat.
Aestivation A period of diapause (q.v.) in summer.
Agro-ecosystem An ecosystem (q.v.) based on an agricultural environment.
Alate A winged individual – as applied to a winged aphid.
Alula The outermost of three flap-like outgrowths of the wing of certain flies (*see* **Squama**).
Anamorph The asexual stage of a fungus.
Anellus (*pl.* **anelli**) A small, sometimes hardly visible, ring segment of most chalcidoid antennae, located behind the pedicellus and forming the first one or more flagellomeres.
Antenna (*pl.* **antennae**) One of a pair of sensory 'feelers' on the head of an insect.
Antennal Pertaining to the antenna (q.v.).
Anthesis The stage of flowering.
Aphicide An insecticide active specifically against aphids.
Aphidophagous Aphid-feeding.
Aptera (*pl.* **apterae**) A wingless insect – applied usually to a wingless aphid.
Apterous Wingless.
Areolet The small cell in the wing of Ichneumonidae situated between the two recurrent veins.
Arista A bristle-like outgrowth, as on the antennae of certain insects.
Autecology The ecology of individual species, as opposed to the ecology of communities.
Autolysis Destruction of the cells of a body by the action of its own serum.
Biocontrol (= **biological control**) Control of organisms by biological means.
Biodiversity The range of interacting organisms that form communities in an environment.
Bivoltine With two generations annually.
Brachypterous Having shortened wings.
Brassicaceous Pertaining to a member of the plant family Brassicaceae (= Cruciferae).
Carapace The combined head and thorax of a spider.
Carina A ridge or a keel.
Carpel The female reproductive organ of a flowering plant.

Cauda The tail-like structure of an aphid.
Cell Of an insect wing, an area of the membrane partly (= open cell) or entirely (= closed cell) surrounded by veins.
Cephalothorax The fused head and thorax of a spider.
Chelicera (*pl.* **chelicerae**) One of a pair of jaws of a spider.
Chloroplast A specialized strucure (plastid) in the cytoplasm of a plant cell, containing the green pigment chlorophyll necessary for photosynthesis.
Chorion The protective, waterproof outer layer of an insect egg.
Chromosome One of the pair of strands in the nucleus of a cell that bears genes in a linear order.
Clypeal Pertaining to the clypeus (q.v.).
Clypeus The lowest part of the face of an insect, immediately above the labrum.
Colline A landscape with more or less smooth hills (not too steep slopes and not too high).
Condyle A lateral, ridge-like swelling on the base of the chelicera of certain spiders.
Conidiogenesis The production of conidia.
Conidiophore A specialized conidia-producing fungal hypha.
Conidium (*pl.* **conidia**) The asexual spore of certain fungi, cut off externally at the apex of specialized hyphae called conidiophores.
Cornutus An elongate projection on the gastral petiolus.
Costa The anterior-most vein of an insect wing, running along the costal margin.
Costal Pertaining to the costa (q.v.).
Costula A transverse ridge on the propodeum.
Coxa (*pl.* **coxae**) The first segment of the leg of an insect.
Cultivar A variety of cultivated plant, produced by breeding or selection.
Cuticle The outer, three-layered, non-cellular 'skin' of an insect.
Cytoplasm The living material within a cell, excluding the nucleus and plastids.
Cytoplasmic Pertaining to cytoplasm.
Detritivore An organism that feeds on detritus.
Deuterotoky (= **amphitoky**) Parthenogenesis in which the female produces offspring of both sexes.
Diapause A period of rest (*see* **Aestivation** *and* **Hibernation**).
Diel Daily.
Diploid With two sets of chromosomes.
Ectoparasitic Of a parasite or parasitoid that feeds externally.
Ectoparasitoid A parasitoid that feeds externally.
Ecosystem A community of interacting organisms, plus the environment in which they live and with which they also interact.
Elytral Pertaining to the elytra.
Elytral interval A longitudinal ridge on an elytron, as between two striae.
Elytral stria (*pl.* **elytral striae**) A longitudinal groove in an elytron.
Elytron (*pl.* **elytra**) The hardened, horny fore wing of a beetle (and certain other insects), a pair of which often form a protective cover over the abdomen.
Emarginate With a distinct indentation or notch.
Endoparasitic Of a parasite or parasitoid that feeds internally.
Endoparasitoid A parasitoid that feeds internally.
Entomopathogenic A pathogen (e.g. fungus or nematode) feeding on insects.

Epigaeic Ground-dwelling.
Epigeal Pertaining to epigaeic.
Eubacteria Rod-shaped or spherical microorganisms of the order Eubacteriales.
Eukaryotic organism An organism other than a bacterium or a blue-green alga (cf. Prokaryotic organism).
Femur (*pl.* femora) The third, and frequently the largest, segment of the leg, located between the trochanter and the tibia.
Filiform Spindle-shaped.
Flagellomere The individual component ('segments') of the flagellum of an antenna.
Flagellum The third (distal) part of the antenna of certain insects, beginning beyond the pedicellus.
Foliar Pertaining to the leaves.
Fossorial Adapted for digging.
Frass The solid excreta produced by an insect, especially a caterpillar.
Frons The upper part of the face of an insect which, in flies, occupies most of the front of the head other than the eyes.
Gamete One of the male or female sex cells that unite at fertilization to form a zygote.
Gaster In Hymenoptera Parasitica, the part of the abdomen behind the petiolus.
Gastral Pertaining to the gaster.
Gene One of the units of heredity that occupies a fixed position (locus) on a chromosome.
Genet One or more individuals produced by asexual reproduction from a single zygote.
Genetic Pertaining to genes.
Geniculate Abruptly bent, elbowed.
Genome An entire set of hereditary factors contained in a haploid number of chromosomes of an individual.
Glabrous Without hair, hairless.
Graminicide A herbicide active specifically against members of the family Poaceae (cereals and grasses).
Haemocyte An insect blood cell.
Haltere The club-shaped or drumstick-like balancing organ on the metathorax of a fly, formed from the modified hind wing.
Haplodiploidy The system in which haploid (unfertilized) eggs give rise to males and diploid (fertilized) eggs to females.
Haploid With one set of chromosomes.
Hemimetabolous insect An insect with incomplete metamorphosis, i.e. no pupal stage in the lifecycle (cf. holometabolous insect).
Herbicide A pesticide active against weeds.
Hibernation A period of diapause (q.v.) in winter.
Holometabolous insect An insect with complete metamorphosis, i.e. with a pupal stage in the lifecycle (cf. Hemimetabolous insect).
Homonym A specific or generic name that has been used for two or more different organisms. The homonym first published is designated as 'senior' and that published later is designated as 'junior'.
Humulus (*pl.* humuli) The sometimes clearly pronounced 'shoulder' on the prothorax of a fly – also termed 'humeral callus'.
Hydrogeology The study of the geological aspects of the Earth's water.

Hyperparasitism Parasitism of a parasite or parasitoid by another.
Hyperparasitoid Parasitism of a host by one or more individuals arising from two or more females of the same species.
Hypha (*pl.* **hyphae**) The filament of a fungal body, composed of one or more cylindrical cells, and increasing in length by growing at the tip and giving rise to new hyphae by lateral branching.
Hyphal Pertaining to a hypha.
Hypostomal carina A ridge on the back of the head, delimiting the oral cavity.
Idiobiont A parasitoid that, by death or paralysis, prevents its host from developing any further (*see* **Koinobiont**).
Inoculum The infective material that will produce disease.
Insecticide A pesticide active against insects.
Instar One of usually several larval or nymphal stages in the development of an insect.
Internode The section of a plant stem between two successive nodes.
Interspecific Within one and the same species.
Intraspecific Between different species.
Koinobiont A parasitoid that allows its host to continue to develop (*see* **Idiobiont**).
Labial palp One of a pair of palps arising from the labium.
Labium The 'lower lip' of the mouth-parts of an insect.
Labrum The 'upper lip' of the mouth-parts of an insect, often concealed by the clypeus.
Lamella (*pl.* **lamellae**) A thin leaf-like flap or plate.
Larva (*pl.* **larvae**) The immature growth stage of a holometabolous insect, which is very different in appearance from the adult into which it metamorphoses during a pupal stage (cf. nymph).
Littoral Pertaining to the sea shore.
Locule A compartment of an anther or an ovary of a plant.
Locus (*pl.* **loci**) The fixed position of a gene on a chromosome.
Macropterous Fully winged.
Maxilla (*pl.* **maxillae**) One of two parts of the mouthparts of an insect, located just behind the mandibles.
Maxillary palp A palp arising from the maxilla.
Mentum A ventral plate derived from the labium.
Mespisternum The anterior part of the mesopleuron.
Mesonotum The dorsal surface of the second thorax segment of an insect.
Mesopleuron (*pl.* **mesopleura**) The lateral and ventral part of the mesothorax.
Mesoscutum The mesonotum without the scutellum.
Mesothorax The second part of the thorax.
Metanotum The dorsal part of the metathorax.
Metatarsus (*pl.* **metatarsi**) The penultimate segment of the leg of a spider.
Micropyle A small pore in the chorion of an insect egg, through which sperm may enter.
Mitochiondria Minute rod-shaped or granular bodies in the cytoplasm of cells, that function in the metabolism of fat, glycogen and proteins.
Mitochondrial Pertaining to mitochondria.
Molluscicide A pesticide active against molluscs, such as slugs.
Moniliform Bead-like.
Multivoltine With several generations annually.
Mutagenesis The act of changing the genetic constitution of a cell by altering the structure of the hereditary material deoxyribonucleic acid (DNA).

Mycelial Pertining to mycelium (q.v.).
Mycelium A collective term for the mass of hyphae forming the vegetative part of the body of a fungus.
Nematicide A pesticide active specifically against nematodes.
Nematophagous Feeding on nemtodes.
Neonate Newly emerged, as from an egg.
Notaulus (*pl.* **notauli**) One of a pair of longitudinal grooves on the mesonotum of certain Hymenoptera Parasitica, dividing the mesonutum into a central and two lateral sections.
Nymph The immature growth stage of a hemimetabolous insect, usually similar in general appearance to the adult and often sharing similar feeding habits (cf. larva).
Occiput The posterior part of the back of the head.
Ocellus (*pl.* **ocelli**) Light-sensitive, simple, usually oval or round eyes; normally, in an adult insect, three of them form a triangle between the compound eyes at the top of the head.
Oligostenohygric Restricted to a very narrow range of (soil) humidity.
Oligostenothermic Restricted to a very narrow range of temperature.
Omni-directional As in an insect trap, able to collect target organisms arriving from all directions.
Opistosoma An alternative term for the abdomen of a spider.
Organelle In a protist (q.v.), a specialists part of a cell (such as a flagellum), analogous to an organ in higher organisms.
Ovariole One of several egg-producing tubules forming the ovary.
Oviposition Egg laying.
Ovipositor The tube-like or needle-like egg-laying structure of an insect; in the Hymenoptera Parasitica, sometimes extending beyond the apex of the gaster.
Palp One of a pair of segmented, sometimes leg-like, sensory structures arising from the maxillae or the labium of an insect, or from the head of a spider.
Parasite An organism that lives within or upon another (the host), obtaining food from it but giving nothing in return.
Parasitoid A parasite in which, usually, the larval stage is parasitic and the adult free-living – a parasitoid (unlike a parasite) usually kills its host.
Parthenogenetic Reproducing asexually.
Pathogenicity The infective capacity of a pathogen.
Pathotype A population of a pathogen that can attack only some of the range of hosts capable of being attacked by the whole species.
Pedicellus The second segment of an antenna, located between the scape and the flagellum.
Petiolus The narrow 'waist' or stalk found in certain Hymenoptera Parasitica, joining the gaster with the propodeum.
Phenology Timing of life stages in relation to climate.
Photo-eclectors A non-selective trap allowing insects emerging from the soil to be collected by attracting them to daylight.
Phototactic The reaction to light, being either positive (i.e. active) or non-phototactic (i.e. inactive).
Phylogenetic Pertaining to phylogeny.
Phylogeny The relationships between organisms, as determined by their evolutionary history.
Pinaculum (*pl.* **pinacula**) A small chitinized plate on the body of an insect larva.

Plastid A specialized structure in the cytoplasm of a plant cell, such as a chloroplast (q.v.).
Plurivoltine Continuously brooded.
Polyphagous Feeding on a wide range of material or hosts.
Pooter A hand-held device for collecting small invertebrates by suction.
Postscutellum A small, usually poorly developed section of the mesonotum, located immediately behind the scutellum – well developed in certain flies.
Prepupa (*pl.* **prepupae**) The stage of development of a holometabolous insect between the final-instar larva and the pupa.
Prokaryotic organism A bacterium or a blue-green alga (cf. Eukaryotic organism).
Proleg A false (pseudo) leg, as on the abdomen of a caterpillar.
Pronotum The often enlarged shield-like dorsal surface of the prothorax (q.v.).
Propodeum The first tergum of the abdomen, fused with the metanotum, located in front of the petiolus and thus appearing to be part of the thorax.
Prothoracic Pertaining to the prothorax.
Prothorax The first (of three) main sections of the thorax of an insect.
Protist A member of the third kingdom of primarily single-celled organisms (Protista), parallel to the plant (Plantae) and animal (Animalia) kingdoms, mainly plant-like algae and animal-like protozoans.
Pterostigma A usually darkly pigmented area present on the fore wing margin towards the apex of the costal vein.
Ptilinal suture The remains of the ptilinum (q.v.) on the face of certain flies.
Ptilinum A temporary sac everted from a fissure between the antennae of certain flies that helps the adult to break out of its puparium.
Pubescent Covered in short, soft hairs.
Pupa (*pl.* **pupae**) The pre-adult stage of a holometabolous insect in which metamorphosis occurs.
Pupal Pertaining to the pupa (q.v.).
Puparium (*pl.* **puparia**) In certain flies, a hardened, barrel-shaped structure formed from the cast-off skin of the final-larval instar, within which the pupa is formed.
Pupation The act of becoming a pupa.
Pyrethroid A synthetic insecticide structurally similar to a natural pyrethrin.
Riparian Pertaining to river banks.
Rostrum The snout of a beetle or weevil.
Saprophagous Feeding on dead or decaying tissues of plants or animals.
Scape The basal segment of the antenna.
Scutellum The middle part of the mesonotum or metanotum.
Semiochemical A chemical used in interspecific or intraspecific communication.
Seta (*pl.* **setae**) A small bristle.
Setiferous Seta-bearing.
Siliqua The elongated fruit (pod) of a brassicaceous plant, formed from an ovary of two united carpels and divided into two elongate chambers by a central septum. Dehiscence is by separation of the carpels from the base upwards, leaving the septum entire.
Siphunculus (*pl.* **siphunculi**) One of a pair of pores or tube-like structures on the abdomen of an aphid, through which alarm pheromones and other defensive compounds are discharged.

Spermatheca An organ in a female that receives and stores sperm following mating.
Spinnerets The silk-producing structures on the abdomen of a spider, arranged in three pairs: anterior, median and posterior.
Sporulation The act of spore production, as in a fungus.
Squama (*pl.* **squamae**) A membranous flap at the base of the wing of certain flies (*see* **Alula**).
Stenophagous Feeding on a restricted range of hosts.
Sternaulus A curved furrow dividing the lower part of the mesopleuron.
Striae Grooves running across or along the body cuticle (such as those on the elytra of beetles).
Style A slender bristle present at the apex of some antennae.
Superparasitism Parasitism of a host by more than one species of parasite.
Supra-orbital The area immediately above the eye.
Suture A groove on the body surface which usually divides one sclerite (plate) from another; also, the division between the paired elytra of a beetle.
Synovigenic Emerging from hibernation with undeveloped eggs.
Tarsal Pertaining to the tarsus.
Tarsal formula In certain beetles, the specified number of tarsal segments (tarsomeres) on the fore leg, mid-leg and hind leg.
Tarsomere One of several 'segments' of the tarsus.
Tarsus (*pl.* **tarsi**) The distal part ('foot') of the leg of an insect, subdivided into several segments (tarsomeres); also, the final segment of the leg of a spider.
Tegula (*pl.* **tegulae**) A small lobe covering the base of the fore wing.
Teratocyte A deformed cell.
Tergite A dorsal sclerite on the abdomen of an insect.
Thermophilic Heat-loving.
Thorax The second of the three main divisions of the body of an insect, from which arise the legs and the wings.
Tibia (*pl.* **tibiae**) One of the main sections of the leg of an insect, located between the femur and the tarsus.
Trochanter The second segment of the leg of an insect, between the coxa and the femur.
Univoltine Single-brooded, i.e. with just one generation annually.
Vertex The top of the head of an insect, posterior to the frons.
Zygote The cell that results from the fusion of a male and a female gamete.

Author Index

General Index

NOTE – Synonyms for specific (trivial) names in the body of tables are excluded, unless cited elsewhere or in common usage.